Interpreting Earth History

Eighth Edition

Interpreting Earth History

A Manual in Historical Geology

Eighth Edition

Scott Ritter
Brigham Young University

Morris Petersen
Brigham Young University

Long Grove, Illinois

For information about this book, contact:
Waveland Press, Inc.
4180 IL Route 83, Suite 101
Long Grove, IL 60047-9580
(847) 634-0081
info@waveland.com
www.waveland.com

Photo Credits:

Exercise 1: Alexander Petrenko
Exercise 2: revital/Shutterstock.com
Exercise 3: Kenneth Keifer/Shutterstock.com
Exercise 4: Lee Prince/Shutterstock.com
Exercise 5: Raduga11/Shutterstock.com
Exercise 6: Scott M. Ritter
Exercise 7: Bertl 123/Shutterstock.com
Exercise 8: Kenny Tong/Shutterstock.com
Exercise 9: Michal Ninger/Shutterstock.com
Exercise 10: Vladimir Sazonov/Shutterstock.com
Exercise 11: Florin Stana/Shutterstock.com
Exercise 12: Wollertz/Shutterstock.com
Exercise 13: Joy Stein/Shutterstock.com
Exercise 14: LesPalenik/Shutterstock.com
Exercise 15: Matthijs Wetterauw/Shutterstock.com
Exercise 16: Tom Grundy/Shutterstock.com
Exercise 17: Sumikophoto/Shutterstock.com
Exercise 18: Pictureguy/Shutterstock.com
Exercise 19: Patrick Poendl/Shutterstock.com
Exercise 20: Pichugin Dmitry/Shutterstock.com

10-digit ISBN 1-4786-1145-6
13-digit ISBN 978-1-4786-1145-5

Printed in the United States of America

14 13 12 11 10 9 8

Contents

Preface

Interpreting Earth History was written to provide deeper learning activities for historical geology students at the college and university level. Material is organized in much the same sequence as chapters in most popular historical geology textbooks and it is expected that students will use the explanatory text to augment, not replace, textbook content. The purpose of the manual is to provide students the opportunity to engage with geological data from a variety of sources (maps, fossils, rocks, etc.) and at a variety of scales to discern and explain geological patterns.

Of special concern to instructors is the number of exercises, time, and resources required for each lab, and sequence of topics. Each lab is written as a stand-alone activity so that it can be assigned in concert with the sequence of topics adopted by individual instructors. Some exercises can be done outside of the lab as homework assignments. Others require access to rock and fossil specimens provided by the instructor and are best done in a laboratory setting. Most courses will not have time to include all of the exercises contained in this manual. The intent is to provide a wide selection of exercises from which instructors may choose depending upon their teaching style, availability of materials, and other course needs.

The eighth edition of *Interpreting Earth History* includes many of the exercises incorporated in previous editions, but is now in full color. Color images enhance the student's ability to see and recognize geological patterns. It also makes it easier to see compositional (anatomical) and textural attributes of rocks and fossils. Selected chapters have been expanded to provide additional deeper learning. Two exercises (14 and 17) are new to this edition. Exercise 14 provides students an overview of the Precambrian history of the Canadian Shield as well as insights into the development of the stable platform. Similarly, exercise 17 provides a framework for understanding the stratigraphic, structural, and depositional history of North America during the Phanerozoic Eon.

The modifications and improvements to this edition of *Interpreting Earth History* reflect critiques by students and instructors who have found this manual to be a valuable companion to the study of historical geology. We are appreciative to all who have adopted this manual in their courses and who continue to provide constructive feedback.

Scott Ritter
Morris Petersen

Relative Dating and Unconformities

Establishing Sequences of Events

Exercise 1

Learning Objectives

After completing this exercise, you will be able to:

1. understand the differences between relative and absolute (radiometric) dating;
2. define the principles of relative dating, which include original horizontality, superposition, cross-cutting relationships, inclusions, and faunal succession;
3. establish the order of geological events that conspired to form the given relationships shown on block diagrams and images depicting geological features, as well as list the principle(s) that enabled you to establish the correct order of events;
4. recognize the four types of unconformities on block diagrams and images of actual field areas; and
5. explain the nature and relative duration of processes that create each type of unconformity.

Introduction

The discovery of "deep time" is one of geology's greatest contributions to human understanding. The conceptual foundations laid by eighteenth- and nineteenth-century geologists working in relatively small geographic areas paved the way for development of the modern high-resolution geological timescale (figure 1.1), which spans 4.6 billion years of Earth history and applies to geological features anywhere on Earth. The succession of eons, eras, and periods was constructed during the early part of the nineteenth century using the principles of relative dating that are the focus of this exercise. The absolute timescale (numerical scale) was added after the discovery of radioactivity and the development of techniques that were able to reliably measure small amounts of radiogenic isotopes in geological materials. The numerical scale, the subject of exercise 2, was developed during the latter half of the twentieth century.

Principles of Relative Dating

In this exercise, we are concerned only with a relative sequence of geological events; that is, event A preceded event B or geological feature B is younger than feature A, but older than feature C. To establish the correct order of events, geologists use five simple, but powerful, concepts. First, sedimentary rock layers are horizontal when first deposited. Any marked variation from the horizon-

EON		ERA	Duration in millions of years	Millions of years ago
PHANEROZOIC		*Cenozoic*	65.5	65.5
		Mesozoic	185.5	251
		Paleozoic	291	542
"PRECAMBRIAN"	PROTEROZOIC	*Neo-proterozoic*	458	1000
		Meso-proterozoic	600	1500
		Paleo-proterozoic	900	2500
	ARCHEAN	*Neoarchean*	300	2800
		Mesoarchean	400	3200
		Paleoarchean	400	3600
		Eoarchean	400	4000
	HADEAN			

ERA	PERIOD		EPOCH	Duration in millions of years	Millions of years ago
CENOZOIC	*Quat.*		*Pleistocene*	2.59	2.59
	Tertiary	*Neogene*	*Pliocene*	2.74	5.33
			Miocene	17.7	23
		Paleogene	*Oligocene*	10.9	33.9
			Eocene	21.9	55.8
			Paleocene	9.7	65.5
MESOZOIC	*Cretaceous*			80	145.5
	Jurassic			54.1	199.6
	Triassic			51.4	251
PALEOZOIC	*Permian*			48	299
	Carboniferous	*Pennsylvanian*		19.1	318.1
		Mississippian		41.1	359.2
	Devonian			56.8	416
	Silurian			27.7	443.7
	Ordovician			44.6	488.3
	Cambrian			55.7	542
"PRECAMBRIAN"				4058	4600

FIGURE 1.1 Modern geological timescale showing relative order and ages/durations of eons, eras, periods, and Cenozoic epochs. (Based upon Ogg, Ogg, and Gradstein, 2008.)

tal indicates subsequent movement of the Earth's crust. This relationship is called the **principle of original horizontality**.

Second, those rocks that are highest in a normal, undisturbed stratigraphic succession are youngest, or, conversely, those that are lowest in the undisturbed succession were deposited first and are oldest. This major principle is known as the **principle of superposition**. For example, rocks exposed along the rim of the Grand Canyon are younger than the rocks exposed at the level of the Colorado River in the bottom of the canyon. In areas that have undergone intense folding and faulting, layers may have been overturned. In these cases, the position of a layer in a stratigraphic succession is not indicative of its relative age.

Third, geologic structures or rock bodies that cross-cut other structures or bodies are younger than the features that are cut—the **principle of cross-cutting relationships**. Geologically speaking, faults or igneous dikes that offset or cross-cut series of strata are younger than the strata that are disrupted by faulting or intrusion. If an igneous dike is offset across a fault trace, this relationship indicates that the fault became active subsequent to the dike's emplacement. Consider the timing of events in figure 1.2A. The purple bed, layer 2, was deposited as part of a single horizontal stratum. As a result of faulting, the right fault block moved down relative to the block on the left, thereby offsetting the formerly continuous layer. Since layer 2 is offset along the trace of the fault, movement of the fault occurred after deposition of layer 2. How much time passed between deposition of layer 2 and its subsequent offset by faulting is impossible to tell from figure 1.2A. The faulting could have occurred 1,000 years or 1,000,000 years after deposition of layer 2. Essentially the same relationships are shown in figure 1.2B, but here deposition was renewed after faulting. Layers 4 and 5 have not been cut by the fault and hence are younger than the most recent fault movement. Relationships in figure 1.2B permit us to conclude that deposition of layer 1 preceded deposition of layer 2 (superposition) and that layer 3 was deposited subsequent to layer 2 (superposition). However, prior to deposition of layer 4, the fault became active, thereby offsetting layers 1 through 3 (cross-cutting relationships). Layer 4 represents erosional material derived from layer 3 on the left (upthrown) side of the fault, but deposited on the down-dropped side of the fault. Since the trace of the fault does not cut across layering in layer 5, this layer must be younger than the most recent movement on the fault (cross-cutting relationships). The principles of superposition and cross-cutting relationships permit us to easily discern the proper succession of geological events portrayed by the patterns in figure 1.2.

The **principle of inclusions** is a fourth way to determine relative ages. Simply put, a rock body represented by fragments (inclusions) embedded within another rock must be older than the rock that encloses the fragments. In figure 1.3A, fragments of metamorphic rock (dark) are embedded within

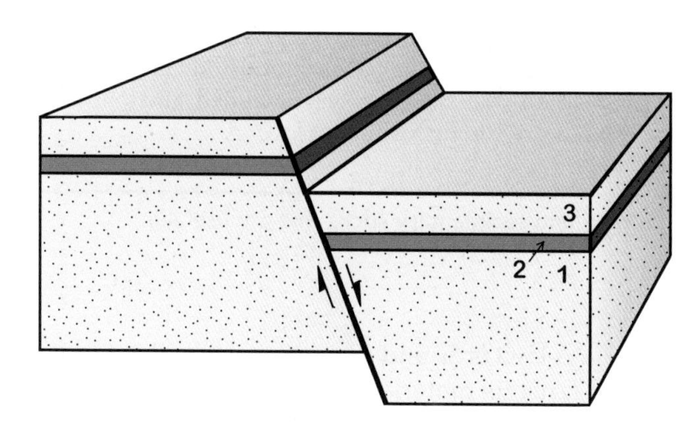

A. Relationships subsequent to faulting of units 1, 2, and 3.

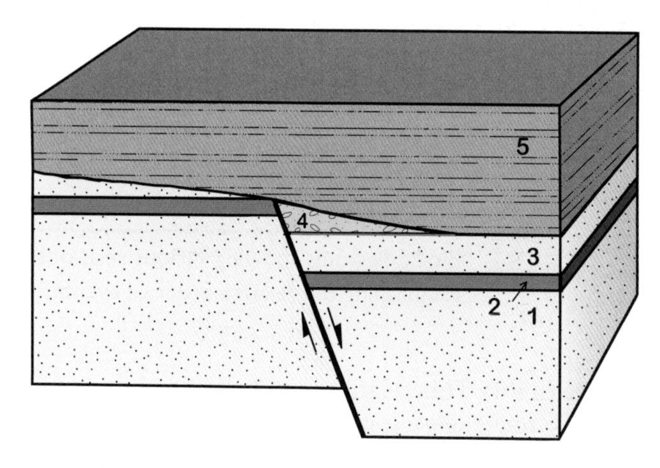

B. Relationships subsequent to erosion of the upthrown block and burial of both blocks by renewed sedimentation. Cross-cutting relations indicate that the fault has not moved subsequent to deposition of sedimentary units 4 and 5.

FIGURE 1.2 Block diagrams showing relationships of normal faulting.

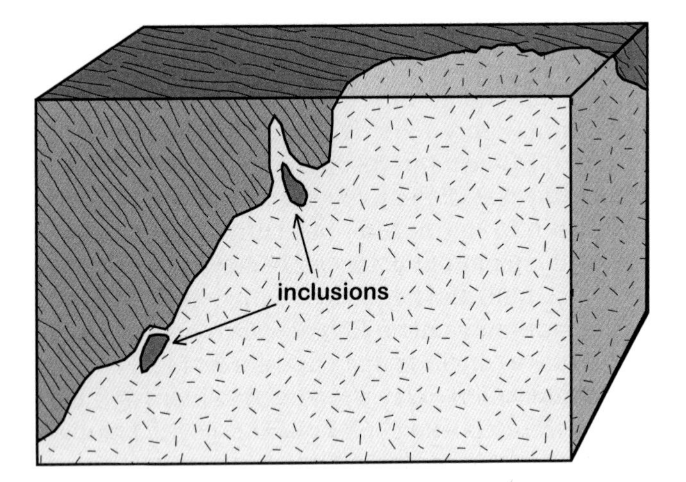

A Inclusions of foliated metamorphic (dark) rock "floating" in the mass of granite (light) indicate that the metamorphic rock is older.

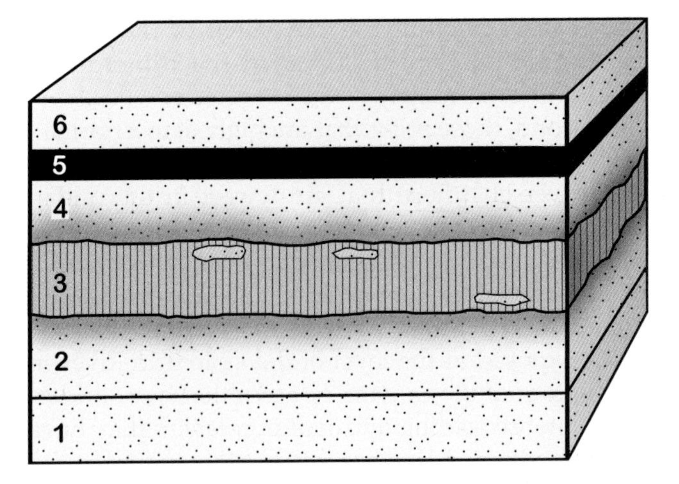

B. An igneous sill has baked both the underlying and overlying strata. Inclusions of sandstone from layers 2 and 4 indicate that igneous layer 3 post-dates both of the adjacent sandstone layers.

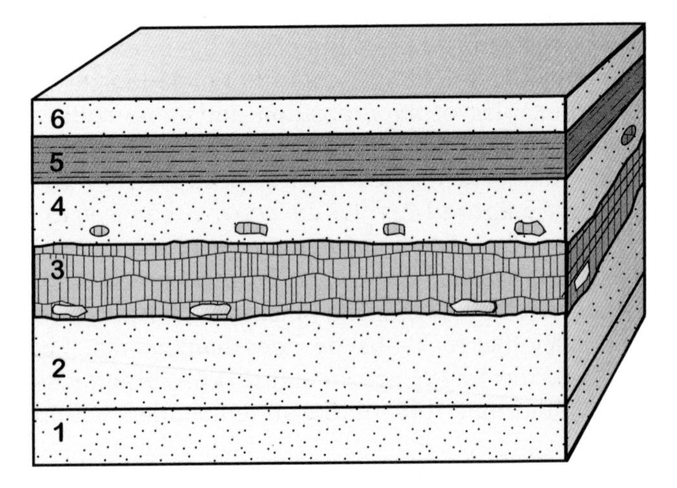

C The law of inclusions indicates that igneous layer 3 was formed prior to deposition of sandstone layer 4.

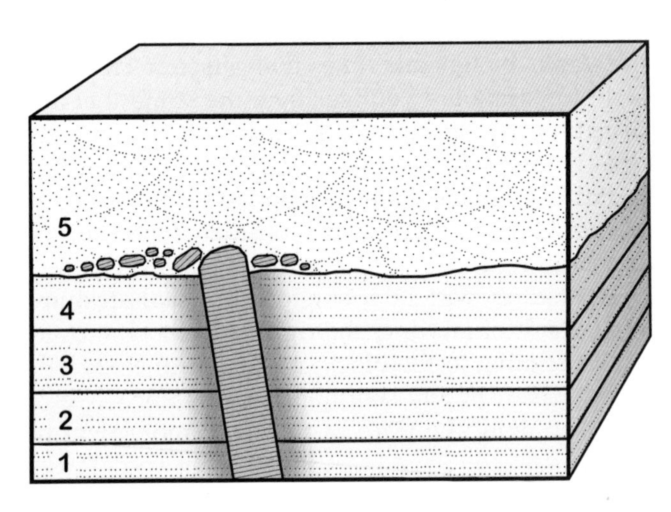

D. A dike has intruded beds 1 through 4, but is overlain (cross-cut) by layer 5. Inclusions of the dike rock were incorporated into the base of sandstone layer 5, also indicating that the dike was intruded and partially eroded prior to deposition of layer 5.

FIGURE 1.3 Block diagrams showing various relationships of igneous and sedimentary rocks that are useful in establishing the relative order of events.

granite (light). This relationship indicates that metamorphic rocks were torn from the wall of a magma chamber and enclosed within the magma as it was emplaced. In figure 1.3B, a layer of dark igneous rock (layer 3) is located between two layers of sandstone. This relationship may have occurred in one of two ways. Either the igneous layer formed as a surface flow subsequent to deposition of layer 2, but before deposition of layer 4, or the igneous layer was intruded as an igneous sill after deposition of layers 2 and 4. A lava flow and a horizontal sill (sheet of intruded igneous material) appear similar in outcrop and on geological maps, but have quite different age relationships. The enclosure of sandstone fragments (inclusions) of layer 4 within igneous rocks of layer 3 indicates that layer 3 is an intrusive body emplaced after layer 4 was deposited. Rocks in contact with the intrusion may be baked. Baking of the top of layer 2 and the base of bed 4 (indicated by shading) provides further evidence that the igneous sheet is a sill rather than a buried basalt flow. Compare figure 1.3B with relationships shown in figure 1.3C. The inclusion of volcanic cobbles and boulders from layer 3 in the lower part of sandstone layer 4 indicates that in this case the dark igneous layer was a surface basalt flow that was extruded and crystallized before deposition of layer 4.

The fifth and final principle of relative dating is known as the **law of faunal succession**. In 1805, the British canal surveyor William "Strata" Smith noted that fossils occurred with such specificity within strata of southwestern England that he could use fossils to recognize and correlate sedimentary strata throughout all of England. Once understood, this orderly succession of fossils was used to divide geological time into the eons, eras, and periods that we know today. Time boundaries between geological periods are based upon the first appearance of fossils in strata. For example, the base of the Devonian System is defined as the first appearance of a graptolite species known as *Monograptus uniformis*. Each era, period, and epoch hosted unique species of plants and animals. Marine rocks of Paleozoic age can be recognized by the presence of trilobites. No trilobites have ever been found in Mesozoic or Cenozoic strata, neither by William Smith nor by the thousands of geologists and paleontologists that have followed him. Instead, Mesozoic rocks are characterized by the remains of organisms, such as dinosaurs, that lived during the Mesozoic Era.

The law of faunal succession is particularly useful for making long-range correlations. For example, it would be impossible to correlate sedimentary or volcanic rock layers exposed in the Grand Canyon in Arizona to age-equivalent strata in southern Russia using superposition, original horizontality, cross-cutting relations, or inclusions because these principles show the relative age relationship between rock bodies that occur in geographically contiguous areas. No sedimentary layer, lava flow, fault, or fold can be traced globally. However, if portions of Arizona and southern Russia were covered by shallow oceans during the Permian Period, and these geographically distinct basins were both populated by individuals of one or more widely dispersed species that existed only during the Permian Period, fossil remains of this species (faunal succession) could be used to establish time equivalence between sedimentary layers deposited in the two basins. It is just such paleontological relationships that permit us to recognize rocks of a particular age (Cambrian, Ordovician, etc.) on a global scale.

Unconformities in the Rock Record

The sedimentary rock record does not encode an unbroken history of deposition in any one place. A drop in sea level may cause sedimentation to cease for a period of time, or uplift and erosion may remove large volumes of rock from a given region. Surfaces between superjacent bodies of rock that reflect missing pages or chapters of Earth history are called **unconformities**. The angular unconformity at Siccar Point in southeastern Scotland (figure 1.4) is perhaps the most famous since it was discovered and described by James Hutton (the originator of uniformitarian geology) in the late 1700s.

Since Hutton's time, unconformities have been recognized and studied around the world. In some rock successions, the amount of time reflected by the unconformities is greater than the time represented by the actual rocks. The four principal types of unconformities are **angular unconformities, nonconformities, disconformities,** and **paraconformities**. Perhaps the easiest to recognize is the angular unconformity. This occurs when there is a degree of angular discordance between the layered rocks located above and below the plane of the unconformity. In figure 1.5A, horizontal rocks of Early Tertiary age straddle nearly vertical rocks of Jurassic age. Strata below the unconformity were tilted and eroded prior to deposition of the horizontal beds. Since we know that the rocks below and above the unconformity are Jurassic and Early Tertiary in age, respectively, we can determine that uplift and erosion of the Jurassic strata took place during the Cretaceous Period. A minimum of 80 million years of time (duration of the Cretaceous Period) is represented by this unconformity—far more time than it took to deposit the Jurassic and Early Tertiary rocks shown in figure 1.5A.

A second type of unconformity is called a **nonconformity**. In this case, layered sedimentary rocks overlie an erosion surface developed on metamorphic and igneous rocks. Because the crystalline rocks that underlie nonconformities form deep in the crust where magmatism and regional metamorphism occur, the nonconformity reflects a period of tectonic mountain building followed by a prolonged period of regional erosion.

To understand the complexity and meaning of nonconformities, let's consider the surface between the Precambrian Vishnu Schist and Cambrian Tapeats Sandstone exposed in the Grand Canyon (figure 1.5B). The Vishnu Schist (dark rocks in the lower

FIGURE 1.4 Historically significant angular unconformity exposed as Siccar Point in southeastern Scotland. The angular discordance between Silurian (below) and Devonian (above) strata shown here corroborated James Hutton's inference that the Earth was old and that its formative processes were cyclical.

part of figure 1.5B) began as a succession of marine shale and siltstone deposited in a Precambrian sea that occupied the Grand Canyon area over 1.8 billion years ago. The area was subjected to mountain building from 1.8 to 1.7 billion years ago (radiometric ages), at which time the fine-grained sediments were altered to schist and intruded by veins of granitic magma. During the ensuing 1.27 billion years, the tectonic highlands were taken down to their metamorphic-igneous roots by weathering and erosion, resulting in production of a relatively flat surface underlain by deeply weathered schist and granite. Approximately 530 million years ago, Cambrian seas spread across this surface, reworking unconsolidated materials into a basal conglomerate (basal Tapeats Sandstone) that was covered by subsequent layers of sand (Tapeats Sandstone), clay (Bright Angel Shale), and limestone (Muav Limestone). Radiometric dating of key beds indicates that the nonconformity between the Vishnu Schist and basal Tapeats Sandstone represents approximately 1.27 billion years of "missing" time. Compare the duration of this nonconformity with that of the angular unconformity shown in figure 1.5A.

Disconformities comprise a third type of unconformity. These are more difficult to recognize than the preceding two types of unconformities because the sedimentary strata above and below the disconformity are parallel to one another. By definition, a disconformity is a surface of buried erosional relief between parallel layers of sedimentary rock. This means that the surface underlying the disconformity was carved by shallow to deep stream channels prior to deposition of the overlying strata. Figure 1.5C shows a disconformity developed within the Paleogene Colton Formation of central Utah. It is not possible to tell how much time is represented by this disconformity, but it certainly reflects less time than either of the two unconformities described above.

Finally, an unconformity between sets of parallel sedimentary strata that shows no evidence of erosional relief is defined as a **paraconformity**. The suspected paraconformity surface may be overlain by a pebble conglomerate or by a concentration of insoluble minerals such as phosphates and sulfides. In some cases the paraconformity is physically indistinguishable from a simple bedding plane. The most certain evidence of a paraconformable relationship is juxtaposition of fossils of distinctly different ages above and below the unconformable surface. Figure 1.5D shows paraconformable strata exposed in a road cut in southwestern Missouri. The recess in the cliff (white arrow) indicates the position of the paraconformity between the Early Ordovician Cotter Dolomite and Early Mississippian strata (Bachelor Formation and Compton Limestone). Ages of these formations are determined by fossil content. Hence this seemingly simple bedding plane represents a hiatus that encompasses part of the Ordovician and the entirety of the Silurian and Devonian Periods. Without the aid of fossils, the significance of this surface could be easily overlooked.

The relationships shown in figure 1.5D suggest that the shallow oceans that covered southwestern Missouri during deposition of the Cotter Dolomite withdrew from the area, probably owing to regional uplift. By Early Mississippian time, the area subsided below sea level once again and sedimentation resumed. The parallel arrangement of strata above and below the paraconformity indicates that strata below the paraconformity in this area were not

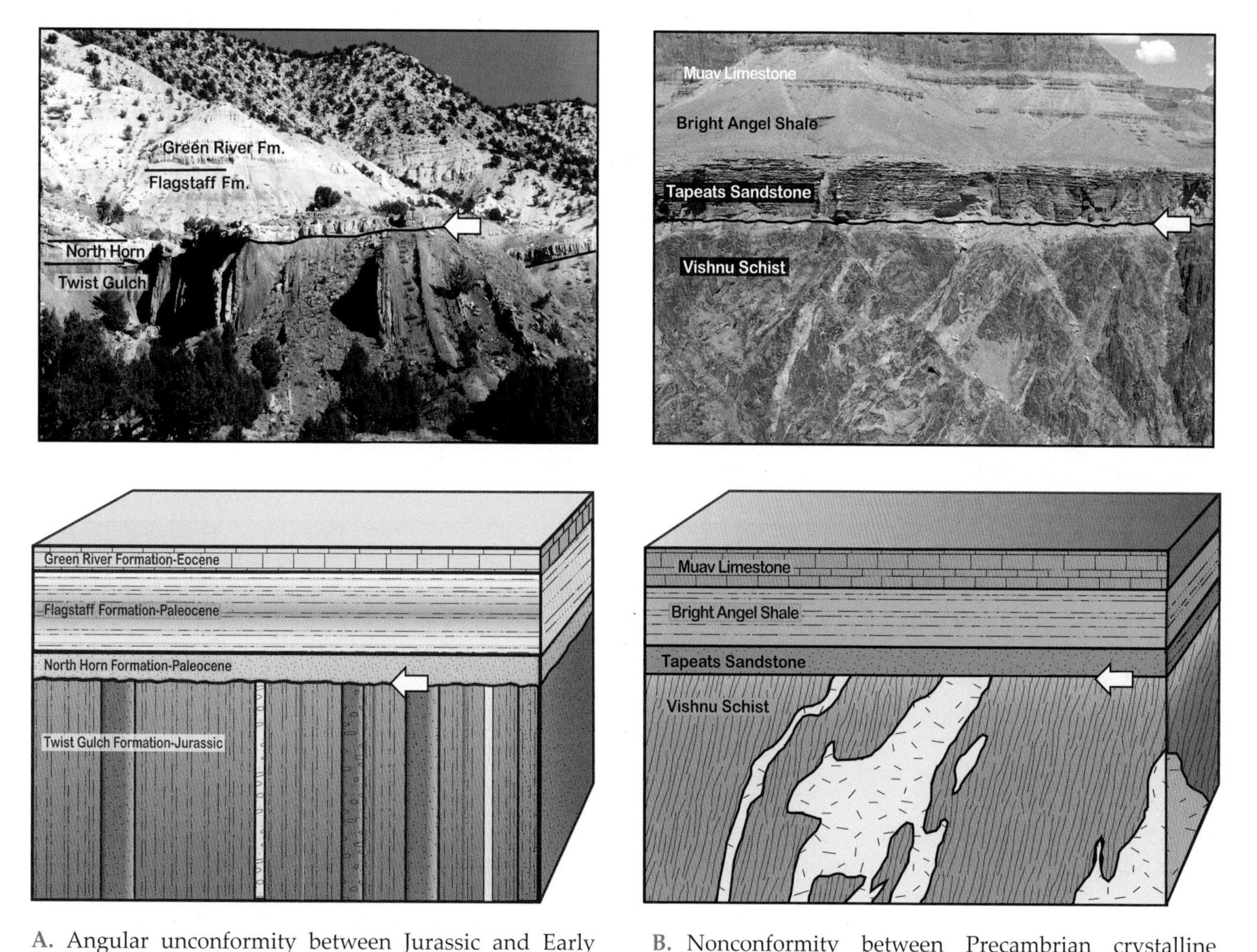

A. Angular unconformity between Jurassic and Early Tertiary strata exposed in Salina Canyon, central Utah.

B. Nonconformity between Precambrian crystalline rocks (foliated schist and granite veins) and horizontally bedded deposits of the Cambrian Tapeats Sandstone located in the lower part of the Grand Canyon of northern Arizona.

FIGURE 1.5 Types of unconformities. *(continued)*

tilted or folded during the period of non-deposition (Ordovician through Early Mississippian time). The duration of the paraconformity shown in figure 1.5D may be unusually long for this type of unconformity. Paraconformities typically represent hiatuses of much shorter duration, perhaps on the order of thousands to tens of thousands of years.

Deformation (folding, faulting), metamorphism, igneous activity, and regional thinning of strata in conjunction with unconformities are evidence for major periods of mountain building that have affected the continental borders of North America during the geological past. The nature of the sediments related to erosional surfaces and to fault scarps, or other features of relief, may also aid in defining the relative time of formation of particular features. For example, the clastic wedges of the Devonian Catskill delta and the major Cretaceous belts of coarse conglomerates, coal-bearing sandstone, and shale in western North America effectively date the time of major uplift of the Acadian Mountains in the east and Sevier Highlands of the west, respectively. Associated igneous and metamorphic rock bodies permit radiometric dating of these orogenic (mountain building) events.

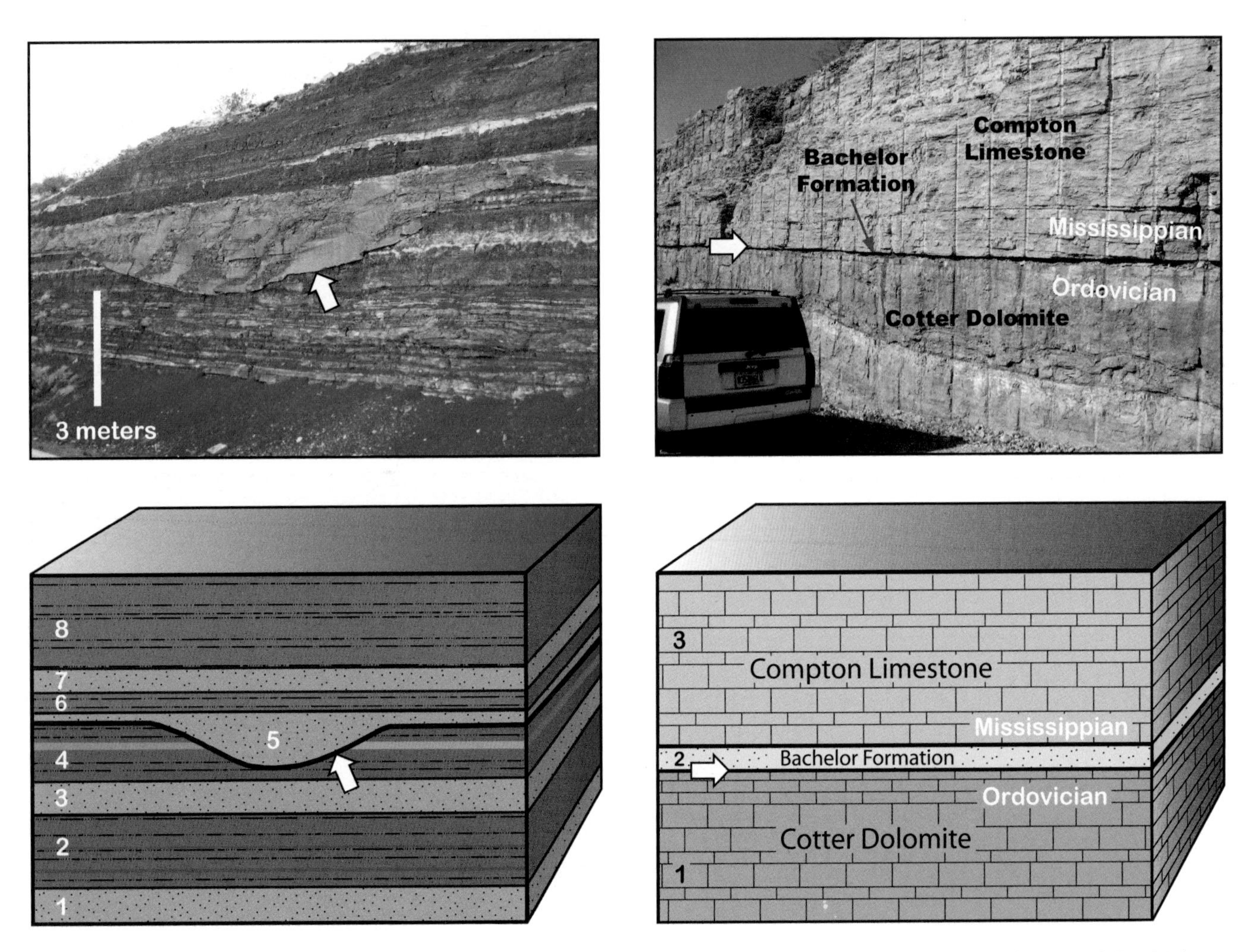

C. Disconformity (white arrow) in the Paleogene Colton Formation, central Utah. The sand lens in the upper part of the outcrop photo is over 2 m thick and fills relief scoured into underlying siltstone and shale.

D. Paraconformity between parallel beds of the Ordovician Cotter Dolomite and the Mississippian Bachelor–Compton Formations.

FIGURE 1.5 Types of unconformities.

PROCEDURE

PART A

Using the dating techniques discussed above, determine the sequence of geologic events represented in each of the block diagrams in figure 1.6.

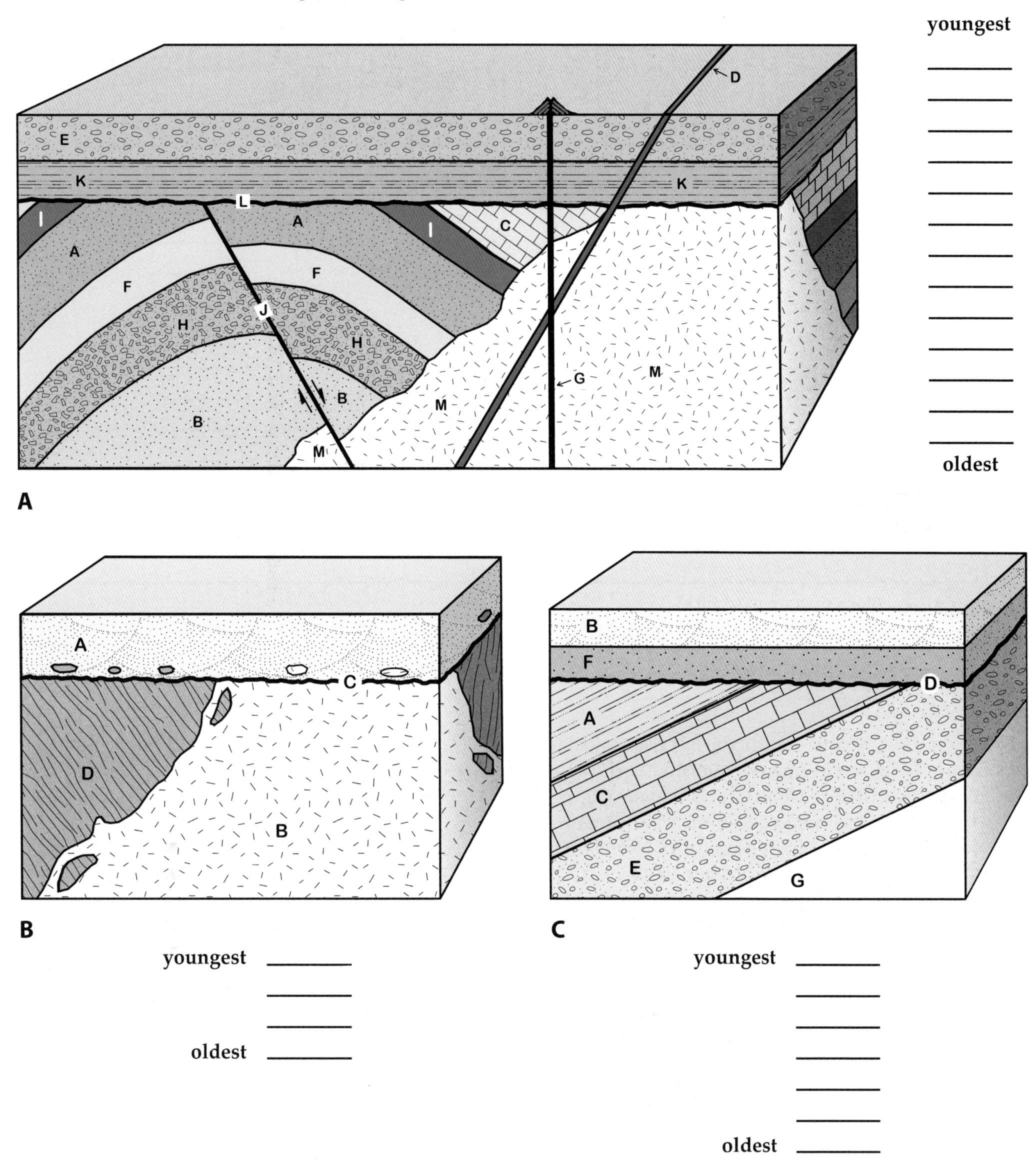

FIGURE 1.6 Block diagram exercise. *(continued)*

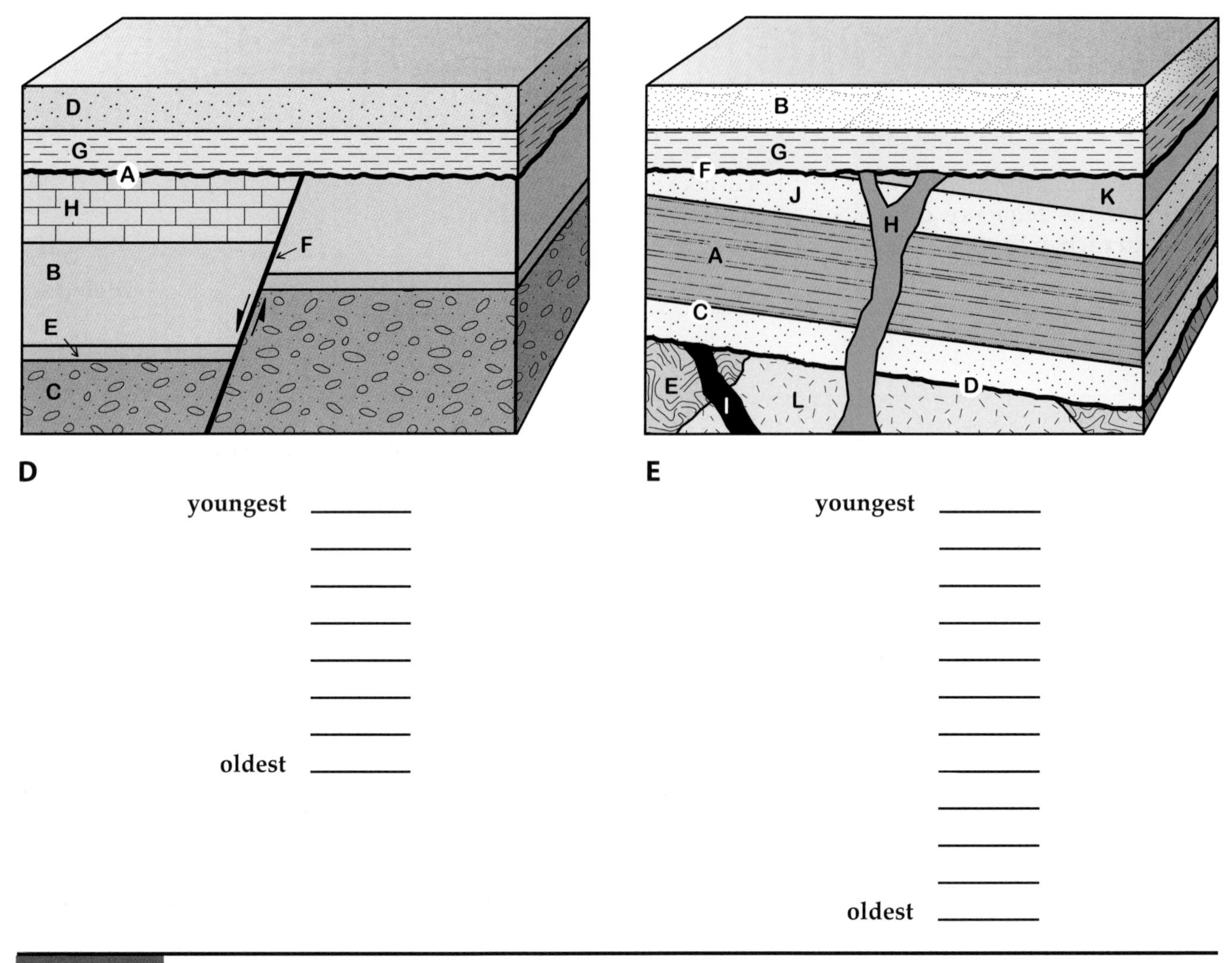

FIGURE 1.6 Block diagram exercise.

PART B

The Grand Canyon is one of the most spectacular laboratories of historical geology in the world. Use the cross section of the eastern Grand Canyon (figure 1.7) to answer the following questions. For questions 1 through 5, indicate which principle(s) of relative dating guided you in arriving at your answers.

1. What is the oldest body of rock in the Grand Canyon?

   Principle(s)

2. What is the oldest sedimentary layer in the Grand Canyon?

   Principle(s)

3. What is the youngest Precambrian formation in the canyon?

   Principle(s)

4. What is the oldest Paleozoic formation exposed in the canyon?

Principle(s)

5. What is the name and lithology of the youngest formation?

Principle(s)

6. What formations comprise the Cambrian System in this area?

7. Visually trace the unconformity at the base of the Tapeats Sandstone from left (west) to right (east) across the diagram. How does the nature of the unconformity change from west to east?

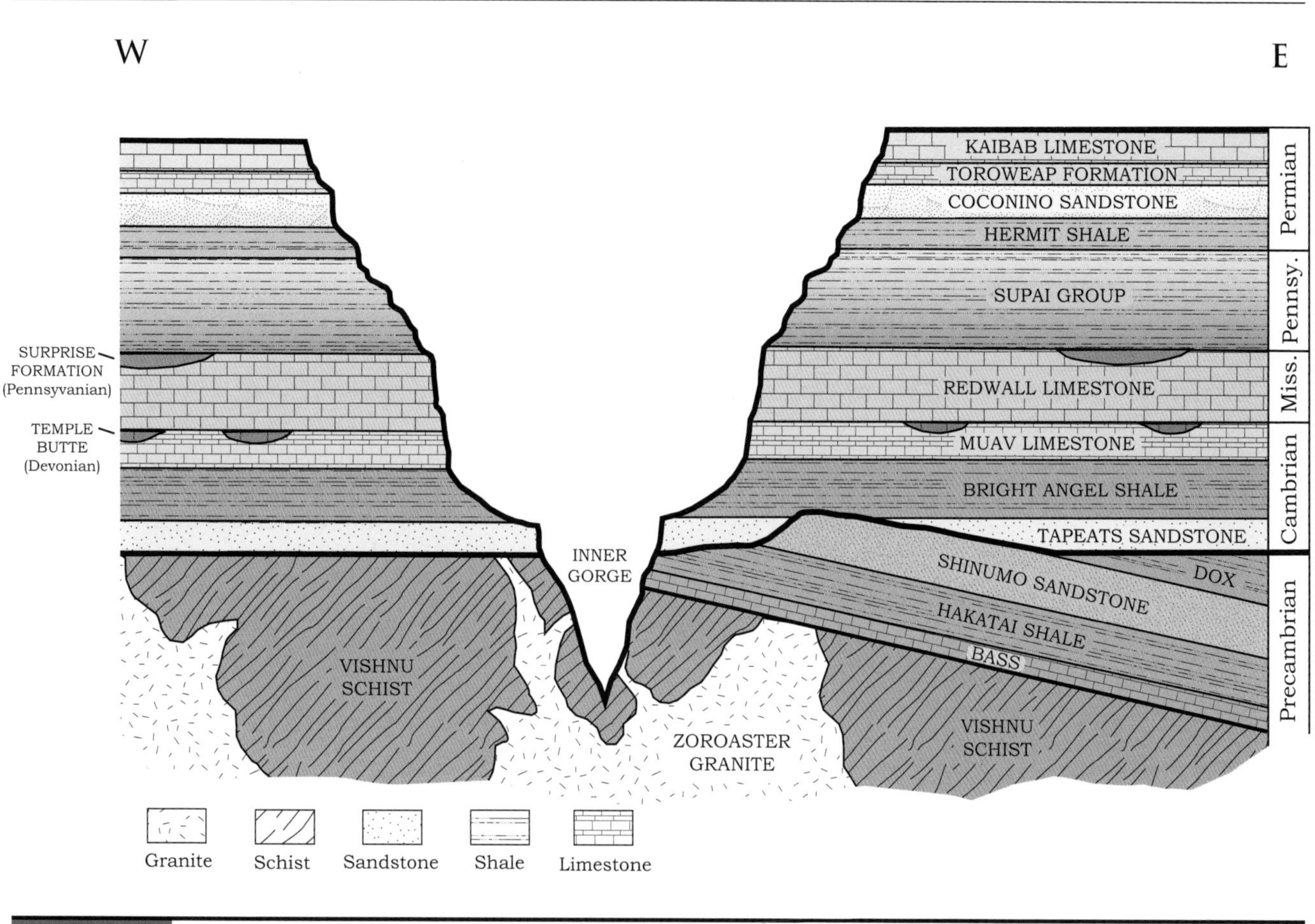

FIGURE 1.7 Geological relationships in the eastern portion of the Grand Canyon, Arizona.

8. What Paleozoic Systems are not represented in the Grand Canyon?

9. Briefly discuss the nature of the Devonian System in this region in terms of its regional extent, lithology, and history.

10. What was occurring in this area during the Mississippian Period?

11. What is the nature of the boundary between the Redwall Limestone and the Supai Group?

12. What was happening geologically in the Grand Canyon during the Ordovician Period?

13. Briefly outline the geological history of the Grand Canyon from Precambrian time until now based upon the patterns revealed in figure 1.7.

14. Which unconformity depicted in figure 1.7 represents the greatest hiatus (most complex and prolonged sequence of geological processes) in the Grand Canyon? Explain your answer in terms of the processes involved with the production of these types of unconformities.

PART C

One of the best-kept secrets of the US Park Service is Grand Canyon National Monument. Located approximately 170 km downstream from Grand Canyon National Park, the monument area displays a dramatic close-up view of the Grand Canyon, several episodes of faulting, and a spectacular display of recent as well as ancient volcanic activity, including a lava cascade that spilled into the Grand Canyon adjacent to Vulcan's Throne and is preserved now as a frozen lava fall.

At this location, it is possible to apply the rules of relative dating to reconstruct the local geologic history with great clarity. Superposition and cross-cutting relations are clearly illustrated by the geologic relationships that can be observed in the rocks of the monument area.

Figure 1.8 illustrates, in (A) photo, (B) sketch, and (C) cross section, the rocks in the immediate vicinity of Vulcan's Throne at Grand Canyon National Monument. Study the photograph and the two line drawings of the area and answer the following questions.

1. Using superposition and cross-cutting relations, establish the proper chronologic sequence for the following list of topographic/geologic features seen in this area:

_____ Erosion of Grand Canyon

_____ Erosion of ancient Toroweap Valley

_____ Basalt cascades

_____ Deposition of Supai Formation

_____ Deposition of Muav Limestone

_____ Deposition of Redwall Limestone

_____ Deposition of Temple Butte Formation

_____ Most recent eruption of Vulcan's Throne

_____ Early displacement of Toroweap Fault

_____ Late displacement of Toroweap Fault

_____ Deposition of Bright Angel Shale

_____ Basalt filling ancient Toroweap Valley

_____ Lava dams (remnants of which are preserved against the lower walls of canyon, labeled Intracanyon Flows on the cross section)

2. The elevation of the Colorado River below Vulcan's Throne is 1,675 ft. How deep was the lake behind the highest lava dam adjacent to Vulcan's Throne? Assume the lava dam was level with its remnant on the canyon wall below Vulcan's Throne.

3. What is the total displacement of the Toroweap Fault in this area?

4. What is the thickness of the lava fill in ancient Toroweap Canyon?

5. If the basalt at the top of ancient Toroweap Valley yields a radiometric date (K/Ar) of 15,000 years, and the lowermost flows in the same valley yield an age of 1.2 million years, calculate the rate of basalt filling in the valley (ft/my).

FIGURE 1.8A Photograph of Vulcan's Throne and immediate area looking northeast. The difference in elevation from the Colorado River to the top of Vulcan's Throne is 3,422 ft. (Courtesy of W. K. Hamblin.)

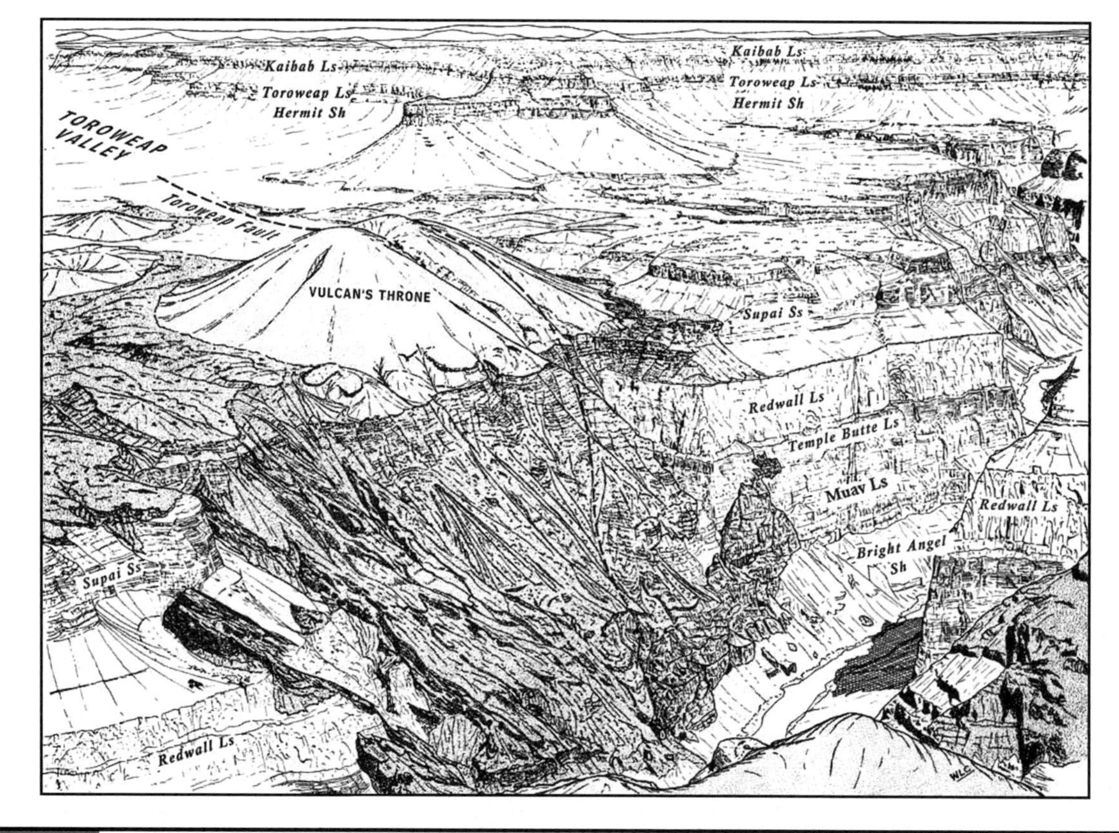

FIGURE 1.8B Sketch of the area included in figure 1.8A. (Courtesy of W. K. Hamblin.)

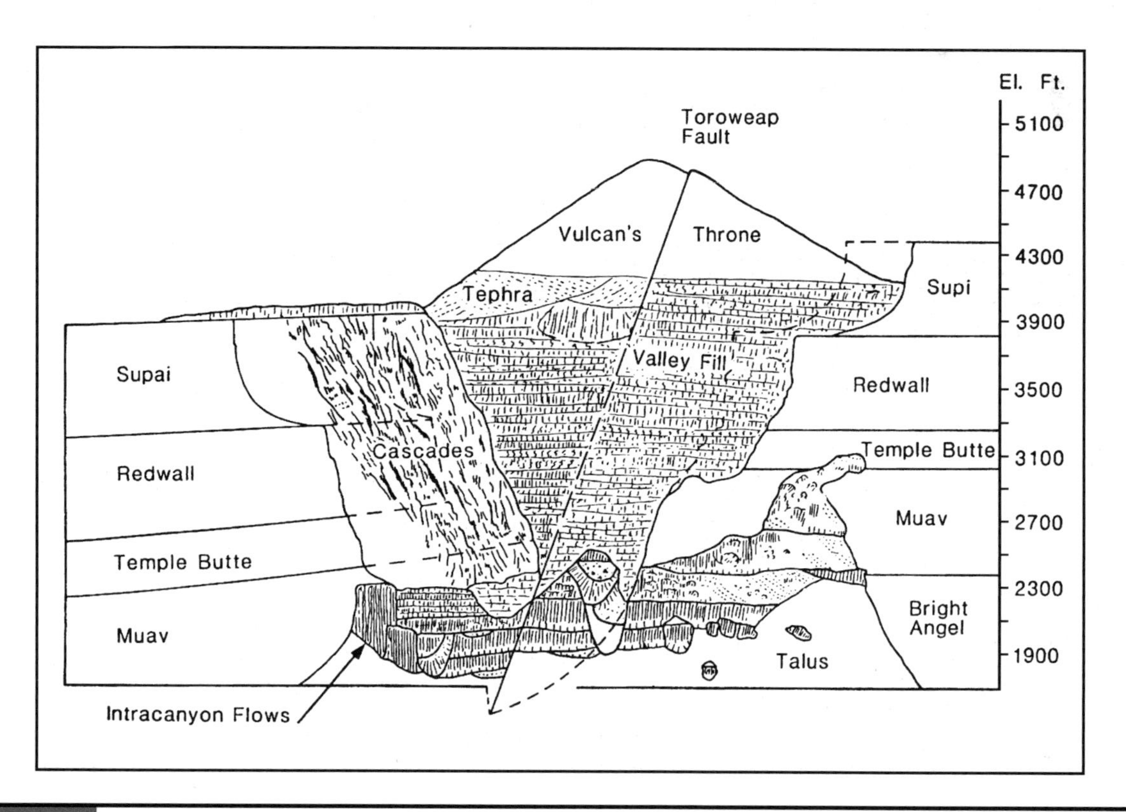

FIGURE 1.8C Line drawing of the wall of Grand Canyon below Vulcan's Throne. The basalt cascade shown in these illustrations is derived from a volcano located between 4 and 5 miles north-northwest of Vulcan's Throne. This view is looking to the north. (Courtesy of W. K. Hamblin.)

Exercise 2

Radiometric Ages

Establishing the Absolute Ages of Geological Events

Learning Objectives

After completing this exercise, you will:

1. be able to explain why radioactive decay provides a means for calculating the ages of geological materials;
2. be able to define the concept of half-life as it pertains to radioactive materials;
3. have familiarized yourself with the parent-daughter isotope pairs most commonly used to calculate ages of geological materials;
4. be able to indicate the types of source materials that contain radiogenic uranium, potassium, rubidium, and carbon;
5. be able to calculate the absolute ages of materials given the ratio of parent-daughter isotopes and the appropriate decay constant; and
6. know how radiometric dating of chiefly igneous bodies can be used to determine the ages and durations of eons, eras, periods, and epochs (which were chiefly determined on the basis of sedimentary rocks).

The age of Earth is approximately 4.6 billion years. This age, which was based upon radiometric dating, is dramatically older than ages calculated by previous means. For example, in 1899, John Joly attempted to estimate the age of Earth by measuring the amount of salt the rivers of the world carry to the sea annually and comparing that to the total salt in the ocean. Since the oceans' salt is derived from weathering of rocks on Earth's surface, Joly estimated 90 million years had elapsed since the first freshwater oceans condensed on Earth's surface. Since Earth is much older than this, the oceans should be much saltier than they are. Clearly there are mechanisms by which salt is removed from the oceans.

Several scientists have approached the age of Earth by studying the thickness of sedimentary rocks on Earth's surface and comparing them with modern rates of accumulation for these rock types. The resulting ages range from 17 million years to as much as 1.5 billion years. With a modern understanding of plate tectonics we recognize the problems with this approach. Deposition of sediment is interrupted and sediment is recycled.

Lord Kelvin approached this problem by studying the cooling of Earth. He measured the increase in temperature of the crust by depth in deep mines. We call the increase in temperature with depth in Earth's interior the geothermal gradient. He calculated how long it would take for a molten Earth to cool so that the geothermal gradient in Earth's crust matched his measurements. He estimated that this would take 24 million years. Kelvin did not understand that the natural decay of radioactive isotopes produces heat. He also did not know that convection in the mantle heats the lithosphere from below, raising the geothermal gradient he measured. Thus, his estimate was also far too young.

In 1896, a French physicist, Henry Becquerel, discovered radioactivity, a process whereby isotopes of certain elements, uranium for example, spontaneously break down to form isotopes of new and lighter elements such as lead. Using the principle of radioactivity to study the decay of uranium to lead, in 1907 an American chemist, Bertram Boltwood, calculated ages for rocks in various parts of the geologic column. In spite of his rough calculations, Boltwood's calculated dates were remarkably close to presently accepted ages for the same rocks.

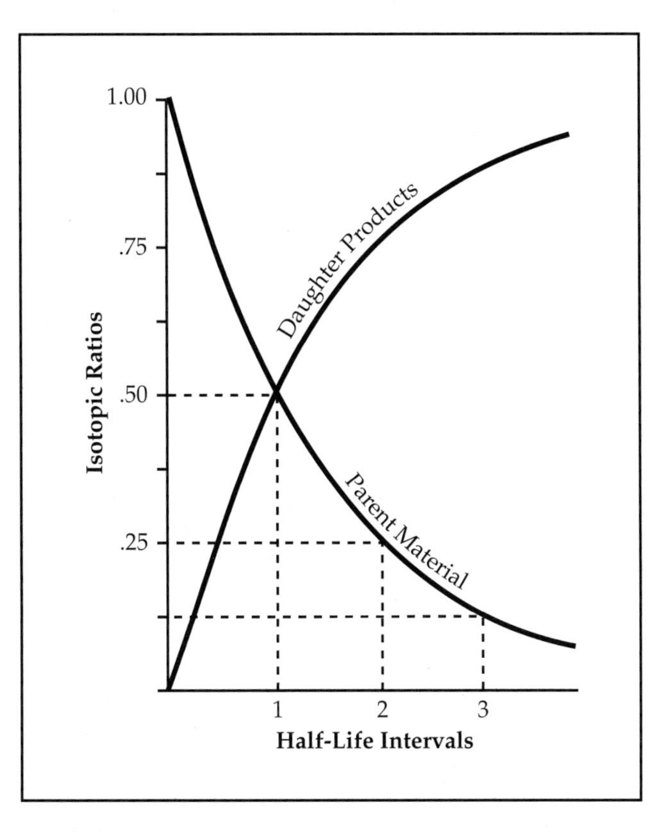

FIGURE 2.1 Graph showing decay of parent material with simultaneous buildup of daughter products over half-life intervals of time.

Radiometric ages are calculated in number of years and are referred to as absolute rather than relative ages. The basis for radiometric ages is natural radioactivity—the spontaneous decay of certain isotopes, called **parent isotopes**, to produce unique end products called **daughter products**, which accumulate as the parent material decays, as shown in figure 2.1. The decay rate, as well as the type of radioactive particle emitted, is unique for each radioactive isotope. The time period required for half of the atoms to decay is called the **half-life**.

Modes of Decay

Most radioactive isotopes decay by one of the following types of radiation:

- **Alpha radiation:** Alpha particles are composed of two protons and two neutrons from the parent isotope's nucleus. The loss of an alpha particle, therefore, reduces the parent isotope by four mass units (atomic weight) and reduces the atomic number by two. For example, parent uranium-238 decays to daughter thorium-234. The decrease in mass number is four. Thorium has an atomic number of 90, whereas uranium has 92. Thus, the atomic number, or number of protons, decreases by two. All of this results from the emission of one alpha particle. Thorium-234, in turn, is also radioactive, and a series of decays eventually lead to the formation of lead-206, which is stable.

- **Beta radiation:** Beta decay is a high-energy electron released from the nucleus, converting a neutron into a proton. The parent material changes only by the addition of a single proton in its nucleus, or an atomic number increase of 1. Mass number remains the same because the mass of the emitted electron is insignificant compared to the masses of the neutrons and protons. Beta radiation can also result by capture of an electron by the nucleus. The result of this change is the same except the atomic number decreases by 1. Rubidium-87 decaying to strontium-87 is an example of losing an electron and increasing the atomic number by one. Potassium-40 decaying to argon-40 illustrates electron capture and a decrease in the atomic number by one.

- **Gamma radiation:** Energy release is similar to X-rays.

Limitations to Radiometric Age Determination

Radiometric ages have limitations, depending upon a number of factors. Some methods are more precise, some less, depending upon a variety of conditions. Some issues associated with radiometric ages include:

1. For many, but not all, methods of measuring the absolute age of a rock or mineral, the sample must have remained closed. This means that neither parent nor daughter isotope has been gained or lost from the sample by reheating, chemical weathering, or some other process.
2. The equipment used, usually a mass spectrometer, must have sufficient accuracy and precision.
3. The result must be properly interpreted and the technique(s) must be appropriate for a specific geologic problem. For example, one technique applied to a metamorphosed pluton might reveal the age of metamorphism. Another technique applied to the same rock might estimate the age of the protolith. These ages could be very different, so care must be taken to chose the correct method and then interpret the results carefully.

The Radioactive Decay Law

The principle of radiometric age determination can be expressed by the following simple relationship:

$$\frac{A}{A_o} = \exp(-\lambda t)$$

where

A = the activity per unit weight of some radioactive parent isotope at some time t in a sample of interest

A_o = the activity per unit weight of the parent isotope when the sample formed

λ = the decay constant

t = the age of the sample

The age of the sample is thus:

$$t = -\frac{1}{\lambda}\ln\left(\frac{A}{A_o}\right)$$

It is very valuable to understand what the decay constant (λ) represents. We can do this by assuming that one half-life of the parent isotope has passed:

$$\frac{A}{A_o} = \frac{1}{2} = \exp\left(-\lambda t_{1/2}\right)$$

Thus the decay constant (λ), after rearrangement, is defined as:

$$\lambda = \frac{\ln(2)}{t_{1/2}}$$

The decay constant (λ) is really just another way to express the half-life. This relationship is true for all radiometric techniques.

Table 2.1 lists the radioactive parent isotopes most commonly used for radiometric age determination. Also listed are the decay products (daughter isotopes), calculated half-lives, and sources of radioactive geological materials. Using table 2.1, we can apply radiometric dating to calculate the age of a

Parent Isotope	Daughter Isotope	Duration of Half-Life (*T*)	Source Materials
Uranium-238	Lead-206	4.5 billion years	Zircon, uraninite
Uranium-235	Lead-207	713 million years	Zircon, uraninite
Potassium-40	Argon-40	1.3 billion years	Biotite, muscovite, whole volcanic rock
Rubidium-87	Strontium-87	48.6 billion years	Mica, feldspar, hornblende
Thorium-232	Lead-208	14.0 billion years	Igneous rocks
Carbon-14	Nitrogen-40	5,730 years	Wood, bone, coral

TABLE 2.1 Chart showing various isotopes used in radiometric dating, their daughter isotopes, duration of one half-life, and source materials for the parent isotope.

material. For example, let's say that a zircon crystal originally contained 100 million uranium-238 atoms and zero lead-206 atoms. It now contains 84.1 million atoms of U-238 and 15.9 atoms of Pb-206. Since 84% of the parent isotope remains, using figure 2.2 we see that one-quarter of a half-life has elapsed. We can now calculate the age of the material as follows:

$$\begin{pmatrix}\text{half-lives}\\ \text{elapsed}\end{pmatrix} \times \begin{pmatrix}\text{duration of}\\ \text{half-life}\end{pmatrix} = \text{age in years}$$

or

$$0.25 \times 4.5 \text{ billion years} = 1.125 \text{ billion years}$$

Percentage of Parent Isotopes	Percentage of Daughter Isotopes	Number of Half-Lives Elapsed	Age in Years (see figure 2.2 for values of ***T***)
100	0.0	0.0	0.000 x ***T***
98.9	1.1	1/64	0.015 x ***T***
97.9	2.1	1/32	0.031 x ***T***
95.8	4.2	1/16	0.062 x ***T***
91.7	8.3	1/8	0.125 x ***T***
84.1	15.9	1/4	0.250 x ***T***
70.7	29.3	1/2	0.500 x ***T***
50.0	50.0	1	1.000 x ***T***
35.4	64.6	1.5	1.500 x ***T***
25.0	75.0	2	2.000 x ***T***
12.5	87.5	3	3.000 x ***T***
6.25	92.75	4	4.000 x ***T***
3.13	96.87	5	5.000 x ***T***

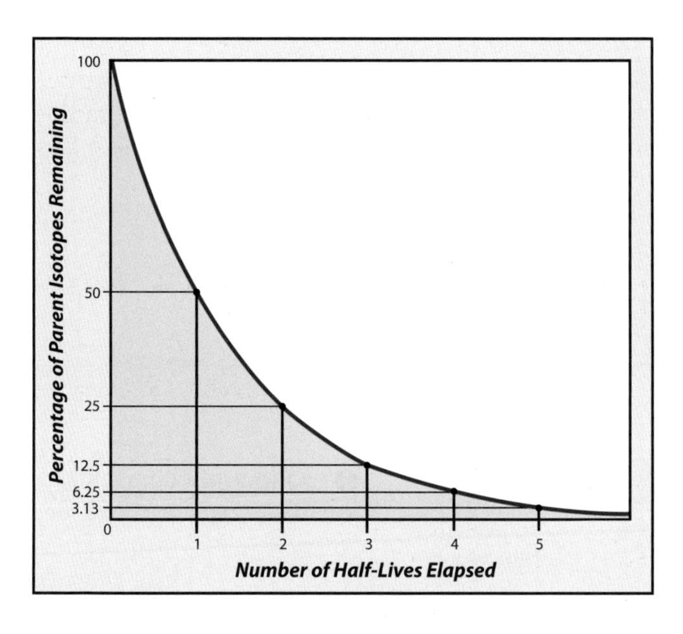

FIGURE 2.2 Graph showing decay curve for radiogenic isotopes.

PROCEDURE

The eons, eras, periods, and epochs of the geological timescale (figure 1.1) were constructed by applying the principles of relative dating to fossil-bearing sedimentary rock sequences in England and elsewhere. Sedimentary rocks are not conducive to radiometric dating because they are comprised of sedimentary particles derived from older source materials. Radiometric dating of sedimentary rock particles indicates the age of the igneous or metamorphic rock from which the particle was derived, not the age of sedimentary deposit to which it now belongs. Depositional ages are calculated by extrapolation from the ages of associated igneous and metamorphic rock bodies.

Determining the Absolute Age of the Carboniferous–Permian System Boundary

Although radiometric dating of each eon, era, and period boundary is based upon somewhat different isotopes, source materials, and stratigraphic relationships, the methodology that you use to date the beginning of the Permian Period is similar to that used in defining the ages of other geologic periods, eras, and eons. In this part of the exercise, you will:

- Determine the placement of the Carboniferous–Permian boundary in the deep-water Usolka River section (figures 2.3 and 2.4) using stratigraphic occurrences of key conodont species.
- Determine the radiometric age of the Carboniferous–Permian boundary by integrating radiometric dates derived from ash beds (bentonites) with the conodont data derived from marine limestones.

The Carboniferous–Permian boundary is defined by the first occurrence of the conodont species *Streptognathodus isolatus* in the sedimentary succession at the Aidaralash Creek type locality in northern Kazakhstan, located a few hundred kilometers south of the Usolka River locality. Strata of beds 1 through 39 (Late Carboniferous–Early Permian) at the type locality consist of deep-water silt and clay, with occasional sand and very coarse sand lenses and coarse ammonoid-, fusulinid-, and conodont-rich limestone beds. Horizons containing pebble and small cobble-size limestone concretions also yield conodonts. Ash beds and other rock types containing radiometrically dateable igneous materials are rare. However, time-equivalent strata exposed along the Usolka River in southern Russia contain conodont-bearing strata

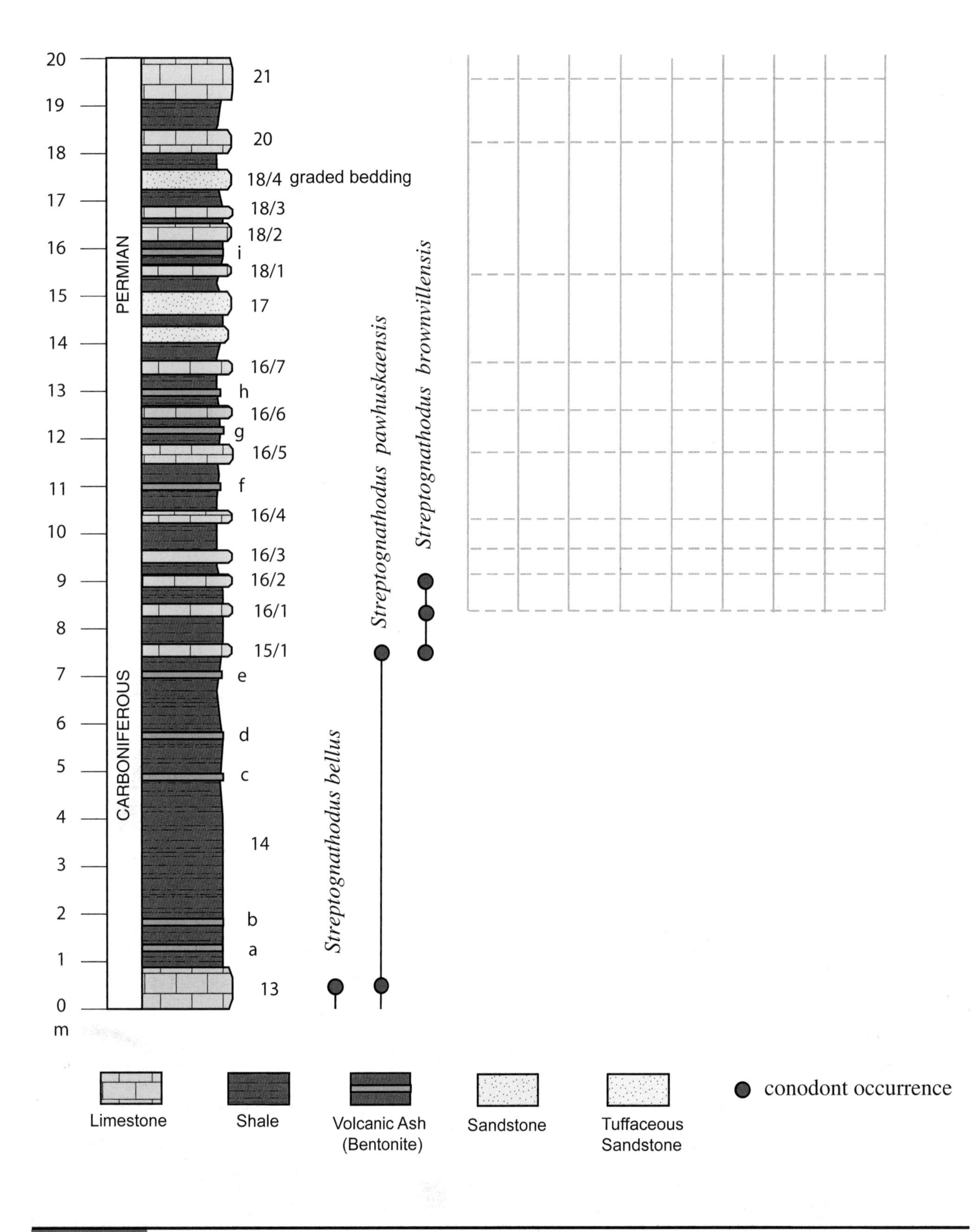

FIGURE 2.3 Stratigraphy of Carboniferous–Permian boundary strata at the Usolka River section, Russia.

interbedded with volcanic ash layers called bentonites and ash-rich sandstone beds. Laminated green and black shales are the most common rock type at Usolka. The rain of siliciclastic silt and clay was interrupted from time to time by deposition of distal carbonate turbidites (thin limestone beds), sandy turbidites (graded quartz sandstone), and slurries of ash and quartz sand (tuffaceous sandstone). Volcanic ash falls derived from tectonic uplift of the Ural Mountains accumulated as bentonites in this quiet-water setting. Although the deep-water setting was hostile to most forms of life, conodont elements are very abundant in the fine-grained limestones and in selected bentonite beds. These were transported into the basin by turbidity currents or swept from the water column by toxic ash falls.

Radiometric Ages on the Web

The **International Commission on Stratigraphy (ICS)** (www.stratigraphy.org) is the largest scientific body within the International Union of Geological Sciences (IUGS). One of its major objectives is the establishment of a standard, globally applicable stratigraphic scale, which it seeks to achieve through the coordinated contributions of a network of Subcommissions and Working Groups. The following information concerning the Carboniferous–Permian boundary was modified from the information provided in the section entitled "Chart/Time Scale" on their official website. This location contains information regarding the current definitions of all system and stage boundaries.

Conodont Data

1. The list below shows the occurrences of key conodont species in the Usolka River section. Using this data, plot the ranges of all conodont species on figure 2.3. Ranges of *S. bellus*, *S. pawhuskaensis*, and *S. brownvillensis* have already been done by way of example.

Sample 13	*S. bellus, S. pawhuskaensis*
Sample 15/1	*S. pawhuskaensis, S. brownvillensis*
Sample 16/1	*S. brownvillensis, S. simplex*
Sample 16/2	*S. brownvillensis*
Sample 16/3	*S. simplex, S. wabaunsensis*
Sample 16/4	*S. wabaunsensis*
Sample 16/5	*S. wabaunsensis, S. isolatus*
Sample 16/6	*S. wabaunsensis, S. isolatus, S. cristellaris, S. longissimus*
Sample 16/7	*S. cristellaris*
Sample 18/1	*S. longissimus, S. constrictus*
Sample 20	*S. longissimus, S. barskovi*
Sample 21	*S. constrictus, S. barskovi*

FIGURE 2.4 Outcrop view of the Usolka River section, Russia.

Radiogenic Isotope Data

The Carboniferous–Permian boundary beds at Usolka contain nine bentonite beds (labeled "a" through "i"), two tuffaceous sandstone beds, and a graded quartz sandstone bed (figure 2.3). Each of these beds yields zircon crystals (figure 2.5) that contain small amounts of uranium-238 (parent) and lead-206 (daughter). Table 2.2 shows the ratio of Pb-206/U-238 in zircon crystals from ash beds e, f, g, and i and from a graded sandstone bed labeled 18/4.

2. Using the isotopic data listed in table 2.2, calculate the average Pb-206/U-238 values for each sample using data from bed 18/4 as an example. Record your answers in the row labeled "sample average" for each column in table 2.2. Use the average Pb-206/U-238 values to approximate the age of each ash bed using figure 2.6.

3. Do the radiometric ages make sense stratigraphically? Are superpositionally younger strata also radiometrically younger? If not, which date(s) seem(s) to be wrong and why?

FIGURE 2.5 Scanning electron microscope image of a zircon crystal.

4. Use the relationships shown in figure 2.3 along with the conodont and isotopic data provided to calculate the age of the Carboniferous–Permian boundary. Suggest an absolute age for the boundary between the Carboniferous and Permian Systems.

_____________ million years.

Compare your estimate with the date shown on figure 1.1.

	Bentonite "e"	Bentonite "f"	Bentonite "g"	Bentonite "i"	Bed 18/4
	0.04762	0.04753	0.04729	0.04717	0.1712
	0.04758	0.04747	0.04726	0.04720	0.1731
	0.04767	0.04763	0.04735	0.04723	0.1698
	0.04753	0.04738	0.04732	0.04712	
	0.04745	0.04745	0.04724	0.04718	
	0.04772	0.04754	0.04728	0.04719	
		0.04751	0.04737	0.04714	
		0.04741			
		0.04759			
Sample average					0.1714

TABLE 2.2 Lead-206/uranium-238 ratios of individual zircon crystals from Usolka River samples.

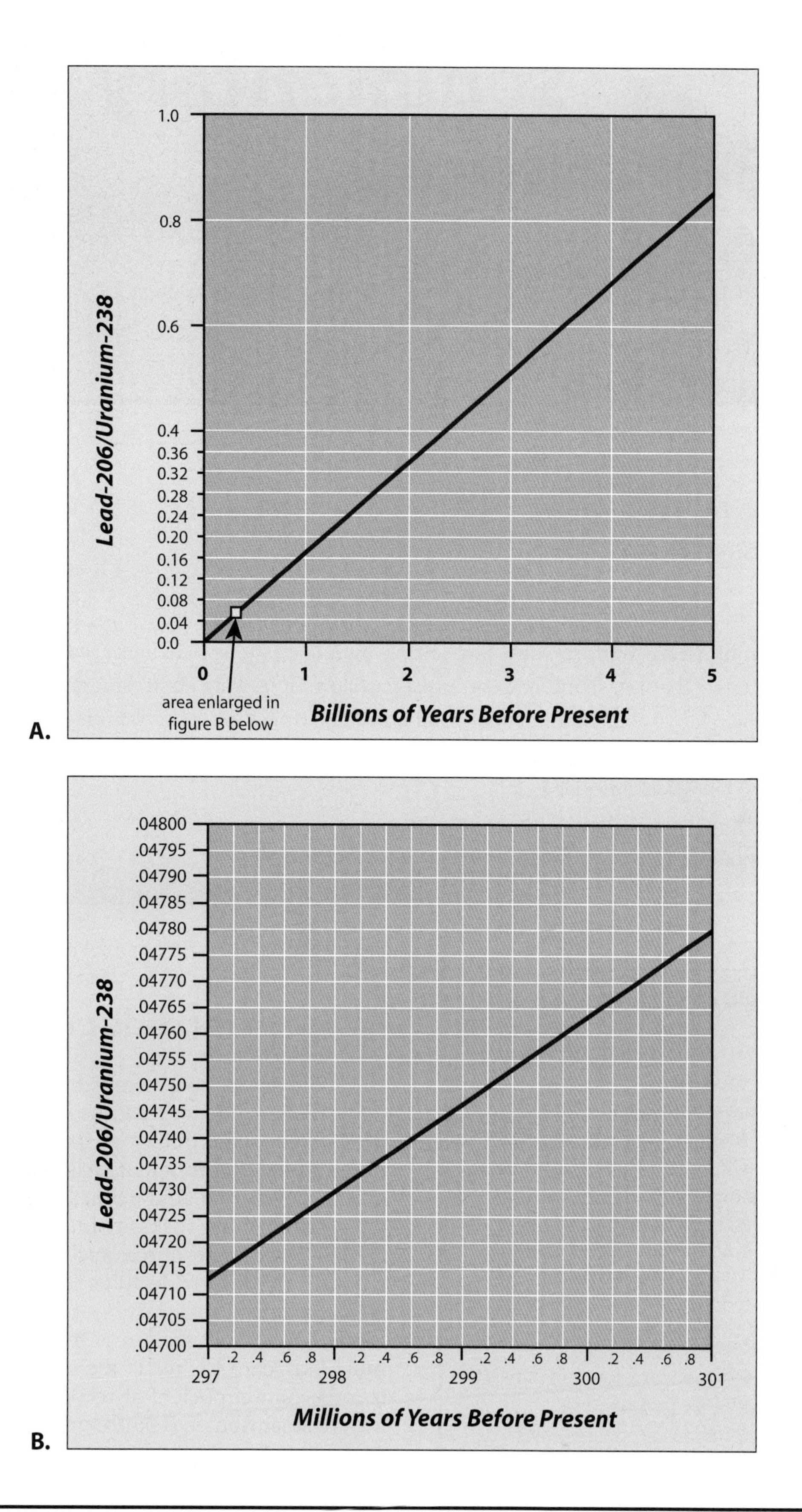

Figure 2.6 Graphs showing lead-206/uranium-238 ratio relationship to age of sample.

Exercise 3

Analysis of Sedimentary Rocks

Learning Objectives

After completing this exercise, you will be able to:

1. compare and contrast the origin, texture, and composition of clastic and chemical sedimentary rocks;
2. see and characterize the size, shape, and sorting of particles comprising clastic sedimentary rocks;
3. approximate the relative abundance of quartz, feldspar, and lithics in a hand sample of sandstone;
4. classify clastic sedimentary rocks according to the Dott classification scheme;
5. characterize the texture of limestone;
6. determine the grain composition of limestone; and
7. classify limestone samples using the Dunham classification scheme.

Materials you will need for this exercise:

1. Study set of clastic rocks with a range of compositions and textures
2. Study set of limestones with a range of compositions and textures
3. Study set of sedimentary structures that includes cross-bedding, asymmetrical and symmetrical ripple marks, graded bedding, mudcracks, and biogenic features (trace fossils)
4. Grain-size card
5. 10X hand lens
6. Dilute hydrochloric acid (HCl)
7. A reference set of common rock-forming minerals and rocks for comparative purposes
8. A reference set of common carbonate grains including ooids, intraclasts, and a variety of fossils

Introduction

Sedimentary rocks are extremely useful to students of Earth history because they store information about such things as paleogeography, past depositional environments, climate change, paleoecology, evolution, and extinction. From a practical view, sedimentary rocks are important because they contain essential resources such as oil, gas, coal, and metals. In order to "decipher" the history encoded in sedimentary rock layers or understand how to locate and extract resources from them, geologists must first learn to "read" the rocks. Your challenge is to become proficient at recognizing and describing those attributes (texture and composition) that reveal a rock's depositional and post-depositional history. You will begin acquiring this proficiency in this exercise by carefully describing, classifying, and interpreting a suite of sedimentary rocks. In the process, you will become acquainted with the descriptive terminology and classification schemes used by practicing sedimentary geologists.

Clastic and Chemical Sedimentary Rocks

Sediment results from the mechanical and chemical decomposition of preexisting igneous, metamorphic, or sedimentary rock. Sediment is produced by a suite of processes collectively called weathering. Mechanical weathering (frost wedging, exfoliation) breaks bedrock into pieces that can be carried downslope by gravity, water, wind, and ice. Chemical weathering of feldspar-rich granite and gneiss produces clay (clastic) particles that are carried downstream in suspension as well as dissociated ions that are carried in solution (chemical sediment). **Clastic** (or **detrital**) **sedimentary rock** forms when mechanically derived material is compacted and cemented in a basin of deposition. **Chemical sedimentary rocks** such as rock salt and limestone form when chemically derived sediment carried in solution is precipitated through evaporation or biogenic activity. These two major families of sedimentary rocks will be discussed separately, beginning with clastic rocks.

Describing Clastic Sedimentary Rocks

Clastic sedimentary rocks are composed of fragments (**clasts**) of minerals, rocks, or fossils that have been transported from the site of sediment production (**provenance**) to the site of deposition by wind, water, and/or ice. The nature of the source area, the transport history, and the environment of deposition are indicated by the texture, composition, and layering of the resulting sedimentary rock. Hence, the process of deciphering a rock's history begins with a thorough description of these attributes.

Texture refers to size, sorting, and shape of a rock's constituent clasts or particles. In nature, mechanical weathering produces particles that range in size from clay and silt at one extreme to house-size boulders at the other. By convention, geologists divide clastic sediment into three size categories (mud, sand, and gravel) using the Wentworth scale. Note that these household terms have specific meanings to geologists: "mud" refers to particles that are less than 1/16 mm in size, "sand" to particles ranging from 1/16 to 2 mm, and "gravel" to material coarser than 2 mm. Each of these can be further subdivided as shown on figure 3.1. In practice, grain size is readily evaluated by comparing particles in a hand sample to a grain-size

Particle Size (mm)	Particle Name	Sediment Name		Rock Name & Symbol
256	BOULDER	GRAVEL		BRECCIA OR CONGLOMERATE
128	COBBLE			
64				
32				
16				
8				
4	PEBBLE			
2				
1.0	COARSE	SAND		SANDSTONE
0.5				
0.25	MEDIUM			
0.125	FINE			
0.0625 (1/16 MM)	SILT	MUD	SILT	SILTSTONE
0.0039 (1/256 MM)	CLAY		CLAY	SHALE OR CLAYSTONE

FIGURE 3.1 Wentworth grain-size scale for clastic sediment showing size ranges (in mm) along with corresponding particle, sediment, and rock names.

comparator such as the readily available "grain-size card" (figure 3.2). This is done by holding the grain-size card next to the sample, comparing the grain-size scale on the card with the grains in the rock (using a 10X hand lens), and recording the results in a notebook. We cannot overemphasize the importance of using a hand lens to inspect the texture and composition of sedimentary rocks!

Sorting is a measure of the variation in grain size within a sediment sample or sedimentary rock (figure 3.3). Poorly sorted sediment displays a wide range of grain sizes, while well-sorted sediment shows little variation in the diameter of constituent grains. Sorting provides insights into the medium of sediment transport. Sediment that has been moved by wind and water is often rather well sorted. Sediment deposited rapidly by debris flows or floods is more poorly sorted. Since ice has no means of sorting the material that it carries, glacial deposits are among the most poorly sorted sediments in the rock record.

Shape, ranging from well rounded to very angular, refers to the smoothness or sharpness of a particle's corners and edges. This is a qualitative description that can be categorized using a standard shape comparator, such as that shown in figure 3.4. Rounding is caused by impact with other clasts, which knocks off sharp edges. The degree of roundness is controlled by the composition of the grain, the grain size, and the distance that the sediment has been transported. When first weathered from an igneous or metamorphic source rock, the majority of grains are angular. Because quartz lacks cleavage, angular quartz grains become increasingly rounded as they are transported downstream or spend long periods washing back and forth on a beach. Feldspar crystals, however, retain angular corners and edges even during long periods of transport because they possess cleavage planes that meet at sharp angles. Studies have shown that larger clasts become rounded more quickly during transport than do smaller grains of similar composition.

Composition is an important attribute of sedimentary rocks in terms of classification and interpretation. Clast composition is readily ascertained in coarse-grained rocks where clasts can be easily investigated with the naked eye or a hand lens. Determina-

FIGURE 3.2 Grain-size card.

Very Coarse Sand	(2.00 – 1.00 mm)	
Coarse Sand	(1.00 – 0.50 mm)	
Medium Sand	(0.50 – 0.25 mm)	
Fine Sand	(0.25 – 0.125 mm)	
Very Fine Sand	(0.125 – 0.062 mm)	

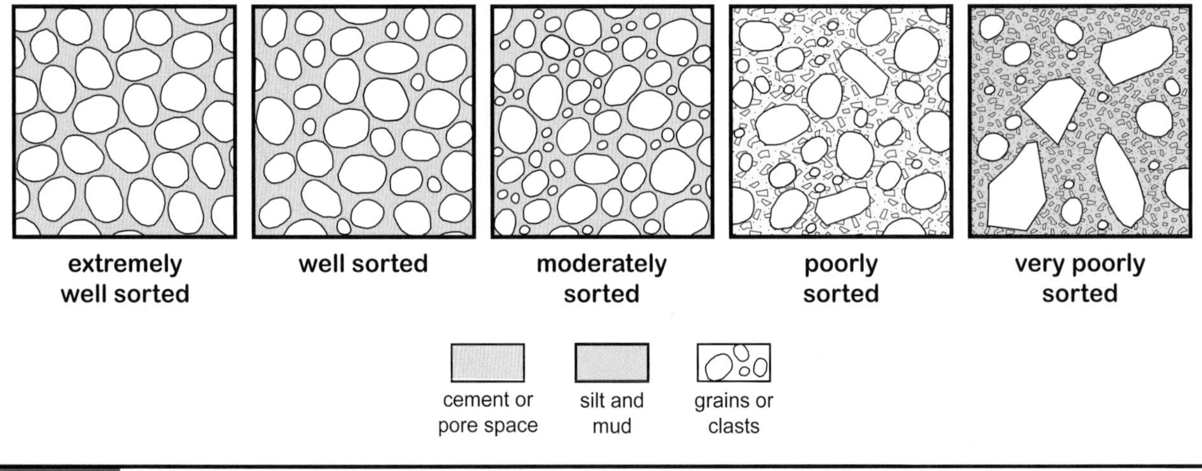

FIGURE 3.3 Images for visually estimating sediment sorting.

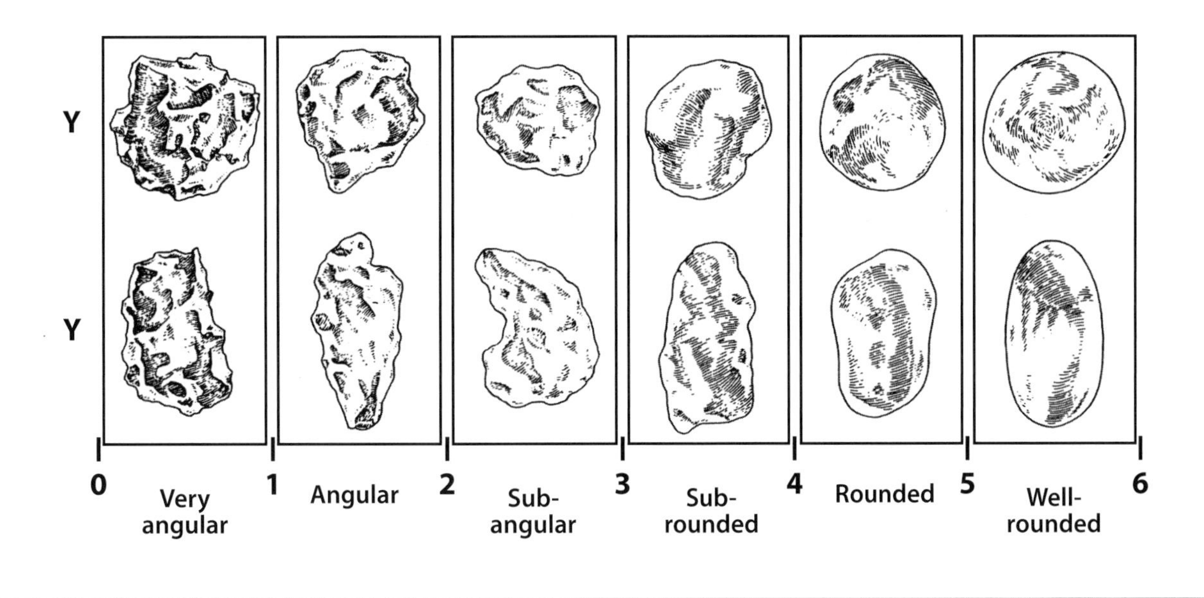

FIGURE 3.4 Images for visually estimating degree of particle rounding. (Based on Powers, 1953.)

tion of grain composition becomes increasingly difficult with decreasing grain size, requiring thin-section analysis for sandstones or x-ray diffraction for the finest-grained rocks. Visual evaluation of grain composition requires some familiarity with the attributes of common rock-forming minerals (quartz, feldspar, mica, pyroxene, and amphibole) and recognition of basic rock types. The mineral and rock composition of sandstone grains reveals information about the nature of the sediment source area—whether is it predominantly igneous, metamorphic, or sedimentary in composition. Since the majority of rock-forming minerals belong to the silicate family of minerals, clastic rocks are commonly referred to as **siliciclastic** sedimentary rocks or simply **siliciclastics**. More will be said about common sedimentary rock constituents later.

The **color** of a sedimentary rock can provide clues regarding the environment in which it was deposited. The most telling colors are red (indicating the presence of oxidized iron) and black (indicating the presence of carbon). Iron is present in most sedimentary rocks, whether in constituent mineral grains, the matrix, or the cement. When oxidized, even a small amount of iron can turn the rock red. Since oxygen is abundant in the atmosphere, most sedimentary rock deposited in non-marine settings (such as river flood plains and alluvial fans) are red, pink, tan, and yellow. By contrast, iron-bearing sedimentary rocks deposited in shallow marine settings, especially shales and siltstones, are typically green or gray.

Black shales and limestones form in areas where organic matter is abundant and oxygen levels are low. In these the sedimentary carbon is not oxidized. These occur in bodies of water, both oceans and lakes, where circulation is poor, and organic productivity is high. The presence of sulfur in these anoxic settings may favor the precipitation of iron pyrite (FeS_2). Black, pyrite-bearing shales represent deposition in deep marine and stagnant shallow lagoonal settings. Gray shales are typically indicative of deposition in shallow to moderately deep marine environments.

When characterizing the color of a sedimentary rock, you should state the main color, preceded by a modifier such as *light* gray, *medium* gray, or *dark* gray. If the specimen is red with hues of brown it should be described as brownish red, light brownish red, medium brownish red, etc. If the rock is primarily gray with hues of green the color should be recorded as greenish gray, light greenish gray, medium greenish gray, etc. Weathering may alter the surface of a rock, so when possible describe the color on both a fresh and on a weathered surface. The color that you enter on the lab sheet is qualitative. Different people may come up with slightly different names for the same color. For purposes of this exercise, describe the colors that *you* see as simply and accurately as possible.

Types of Clastic Sedimentary Rocks

Clastic sedimentary rocks are classified on the basis of grain size and composition (figure 3.1). Coarse-grained rocks comprised of gravel-size particles are classified as either **conglomerate** (rounded clasts) or **breccia** (angular clasts). Sandstone is formed by compaction and cementation of sand-size particles and are variably called **arenites** if over

95% of grains are sand-size, or **wackes** if the mud-size matrix comprises between 5 and 50% of the rock. Sedimentary rocks with over 50% mud-sized particles (silt and clay) are called **mudrocks**. If the majority of the particles falls within the silt range the rock is called a **siltstone**. **Claystone** is comprised predominantly of clay-size particles. If the clay-rich rock is platy or fissile, it is classified as **shale**.

Classification of Sandstones

The popular sandstone classification scheme of Dott (1964) is based upon two defining parameters: the percentage of mud-sized matrix and the composition of the sand framework grains (figure 3.5). Three principal types of sandstones are defined based upon the relative abundance of quartz, feldspar, and lithics (rock fragments) in the sand-size fraction (figures 3.6, 3.7, and 3.8). Sandstones in which the framework sand grains are comprised almost exclusively of quartz (90 to 100%) are classified as **quartz arenites** or **quartz wackes**. Sandstones with significant amounts of feldspar are termed **feldspathic arenites** or **feldspathic wackes**. Feldspar-rich sandstone is also called **arkose**. **Lithic**

CLASSIFICATION OF ARENITES

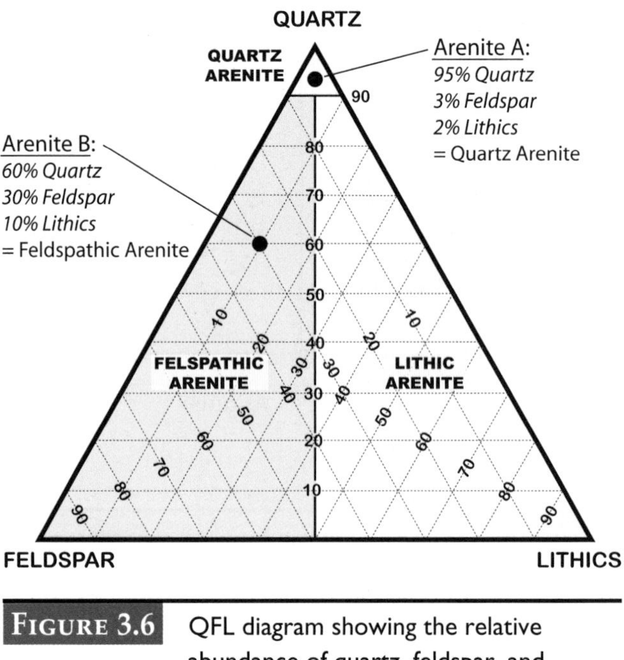

FIGURE 3.6 QFL diagram showing the relative abundance of quartz, feldspar, and lithic (rock) fragments in two selected sandstone samples.

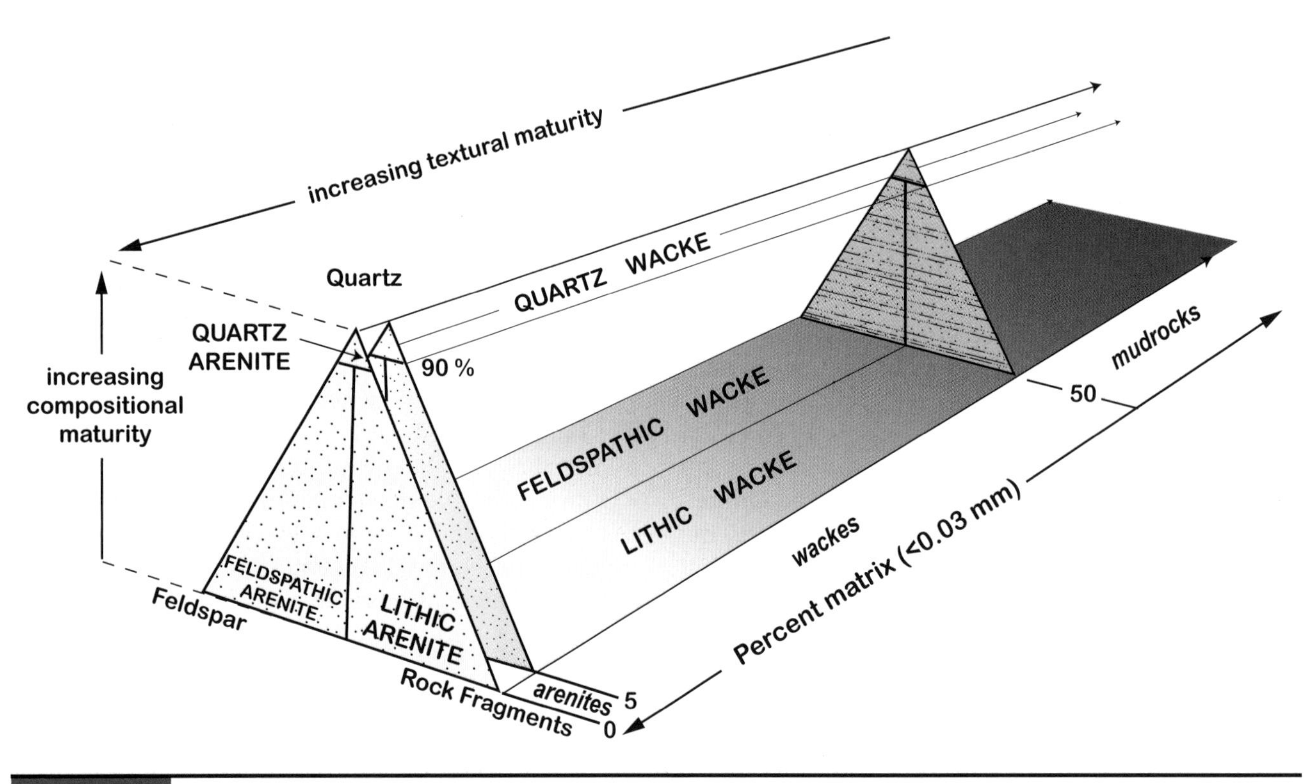

FIGURE 3.5 Classification scheme for clastic sedimentary rocks proposed by Dott (1964) based upon composition (quartz, feldspar, lithics) and texture (percent matrix). The term arenite is applied to sandstones containing less than 5% matrix. Wackes contain between 5 and 50% matrix. Rocks containing greater then 50% matrix are classified as mudrocks and shale (fissile mudrocks).

arenites and **lithic wackes** are sandstones with a significant percentage of rock fragments (**lithics**).

In practice, sandstone classification proceeds by first determining the percentage of matrix in the rock. This may take some practice. The relative abundance of matrix is best determined by performing a point-count on a thin section of the rock in question. This procedure can be quite expensive and time-consuming. You will approximate the relative abundance of fine-grained matrix with a hand lens. If the rock has 0 to 5% matrix it is an arenite. If the rock contains 5 to 50% matrix it is a wacke. If it has greater than 50% mud-size matrix it is classified as a mudrock. Once you have determined if the sample is in fact an arenite or a wacke, further classification requires that you determine the relative abundance of quartz, feldspar, and lithics in the sand-size fraction. Again, the most accurate procedure for determining the relative abundance of these three constituents is the point-count method. In this exercise you will need to estimate the relative abundance of these constituents using a hand lens. This technique is not as accurate but should permit you to estimate the composition within allowable limits. Once you have estimated the relative abundance of sand grains for a given sample, plot the point that represents this ratio on the corresponding QFL diagram (siliciclastic rock data sheet). To illustrate how this is done, we have plotted the composition of two arenites, A and B, on figure 3.6. Arenite A contains 95% quartz, 3% feldspar, and 2% lithics (figure 3.6). The point representing this sandstone falls within the triangle at the top of the QFL diagram, indicating that it is classified as a quartz arenite. Arenite B contains 60% quartz, 30% feldspar, and 10% lithics (figure 3.6). This rock plots in the portion of the triangular diagram reserved for feldspathic arenite. If examples A and B above contained between 5 and 50% matrix they would have been classified as quartz wacke and feldspathic wacke, respectively.

Describing Carbonate Rocks

Carbonate rocks are comprised of carbonate minerals, chiefly **aragonite** ($CaCO_3$) and **calcite** ($CaCO_3$) precipitated from seawater by the shell-secreting activities of algae and invertebrate animals, and to a lesser degree by abiotic precipitation. Compaction and cementation of these biogenically produced sediments (skeletons, shells, and mud) results in the formation of limestone, prompting the sedimentologist Noel James to quip that "Limestones are born, not made." Partial replacement of

Sketches showing the appearance of A. a quartz arenite and B. a feldspathic arenite in thin section. Note scale.

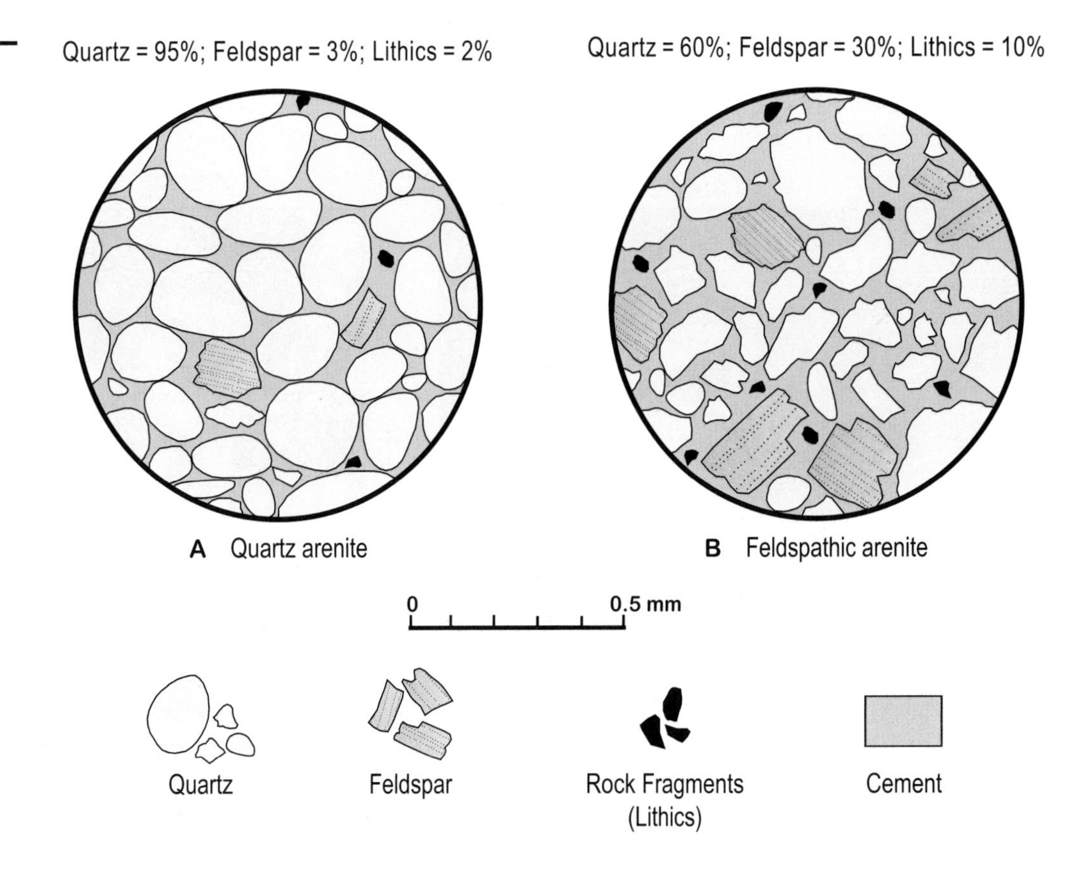

A. Thin section of subrounded quartz (q) grains embedded in blue epoxy resin.

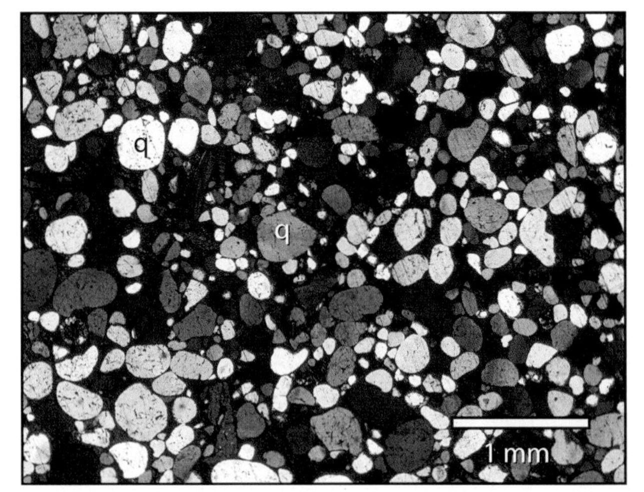

B. Thin section of fine-grained quartz arenite. Note rounded quartz grains.

C. Thin section of felspathic arenite. Angular quartz (q) and feldspar (f) grains floating in calcite cement (c; brown).

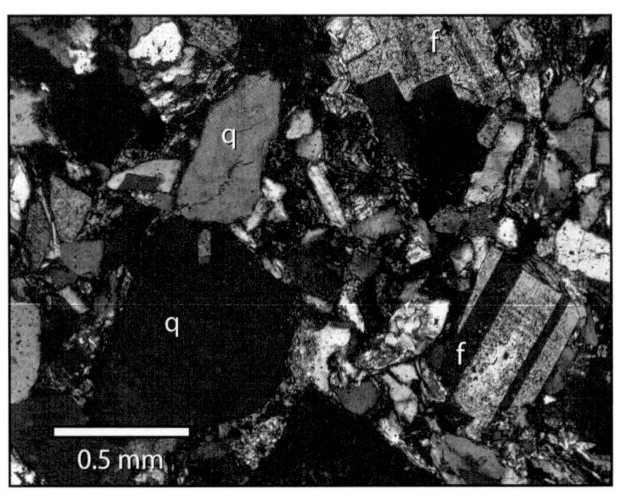

D. Thin section of feldspathic wacke. Grains consist of quartz (q) and felspar (f).

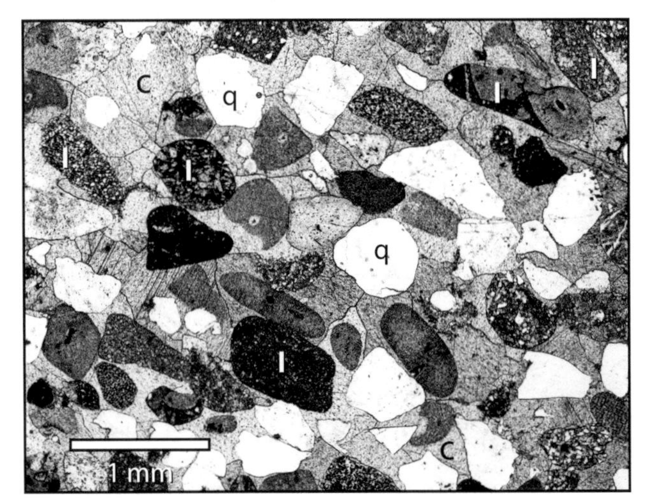

E. Thin section of lithic arenite. Subrounded to subangular quartz (q) and rock (lithic) fragments (l) surrounded by calcite (c) cement (gray).

F. Thin section of a lithic wacke. Quartz (q), lithic fragments (l) and clay (cl) cement.

FIGURE 3.8 Images of selected sandstones showing the appearance of quartz, feldspar, and lithic fragments in thin section. Also note differences in shape and sorting.

calcium with magnesium through interaction with Mg-bearing water, shortly or long after deposition, may alter calcite in limestone to the mineral **dolomite** ($CaMg(CO_3)_2$). A limestone that has been largely or completely altered to dolomite is called a **dolostone**. Together, limestone and dolostone comprise the suite of rocks known as **carbonates**.

Composition of Carbonate Rocks

Sedimentary carbonate rocks are comprised of three main constituents: mud, grains, and cement/pore space. The matrix of most limestones consists of mud-size (< 4 microns) calcite crystals called **micrite** or simply **carbonate mud**. This dense, fine-grained carbonate material results from several processes, including inorganic precipitation of calcite from saturated seawater, disaggregation of calcareous green algae into mud-size crystals, breakdown of skeletons of marine invertebrates, as well as precipitation of mud-grade cements between and within grains. Other processes resulting in production of carbonate mud have also been described, but for purposes of classification it is more important to recognize the presence and relative abundance of primary carbonate mud in a rock sample than to determine the process or processes responsible for its presence. The amount of mud in carbonate rocks ranges from 0 to 100%.

Carbonate grains are commonly separated into four major categories: **skeletal remains**, **ooids**, **intraclasts**, and **peloids** (figure 3.9). As the name implies, skeletal grains (figure 3.9A and 3.9B) are the whole or partial remains of carbonate-secreting plants and animals. Marine algae and invertebrates (snails, clams, trilobites, sand dollars, etc.) furnish the bulk of sediment deposited in carbonate environments and are sensitive indicators of conditions that prevailed during limestone deposition. Because life has evolved through time, the types of organisms represented in a limestone also provide clues to the age of the rock. For example, the presence of the trilobite *Elrathia kingi* in a limestone indicates that the limestone was deposited during Middle Cambrian time.

Ooids, defined as coated grains with a nucleus (shell fragment or sand grain) and concentrically laminated calcareous cortex (figure 3.9E), are unique to carbonate rocks. These grains are typically spherical to elliptical in shape and are less than 2 mm in diameter. They are common in carbonate rocks dating back to the Late Proterozoic and can be seen forming in selected marine settings today. Modern ooids form on shallow, high-energy carbonate shelves, such as Joulter's Cay in the Bahamas, where the ooids are alternatively buried on the shallow seafloor and then reintroduced to seawater by storm- or tide-induced erosion. Upon excavation, a layer of calcium carbonate rapidly precipitates on the surface of the ooid. Crystal poisoning prevents further accretion until the grain has been buried for a period of time and re-excavated. Hence, a well-formed ooid must be buried and eroded in a high-energy, shallow-marine setting for a considerable amount of time. Evidence suggests that ancient ooid-bearing carbonates were likewise formed in shallow, high-energy settings. Large amounts of oil and gas are produced from pore spaces in oolitic limestones throughout the world.

Intraclasts (figures 3.9C and D) are fragments of weakly consolidated carbonate sediment that have been eroded from the seafloor or from tidal flats by waves, storms, or tidal currents. These ellipsoidal to discoidal grains range in size from fine sand to gravel, depending on the degree of sediment consolidation and the nature of the currents. By definition, intraclasts are eroded from within the basin of deposition and redeposited in close proximity to their point of origin.

A peloid is a sand-sized, pellet-shaped grain of carbonate mud that generally lacks internal structure (figure 3.9F). Many peloids originate as fecal pellets, some as small intraclasts, some by alteration of skeletal fragments, and some as small aggregates of mud-size cement crystals. Since it is often difficult to discern which process formed a given peloid, the term should be used descriptively for all sand-sized, elliptical to pellet-shaped particles comprised of carbonate mud. In the Bahamas, peloids comprise a significant portion of sediment deposited in the shallow interior of the Grand Bahama Bank, where tidal and wave currents are weak. Ancient peloid-rich carbonate rocks appear to have formed in lagoons and restricted bank tops as well. Unlike skeletal grains, ooids, and intraclasts, all of which can be seen with the unaided eye, peloids may be difficult to detect without a microscope, especially in rocks with a muddy matrix (wackestone, packstone).

Classification of Carbonate Rocks

As with siliciclastic sedimentary rocks, texture is an important descriptor of carbonate rocks. However, instead of focusing on the size, shape, and sorting of constituent particles, the widely used Dunham (1962) classification (figure 3.10) is based on the particle fabric (mud support versus grain support) and whether the particles were biologically bound during sedimentation. In the former, four classes reflecting decreasing mud content are termed **mudstone**, **wackestone**, **packstone**, and **grainstone**.

A. Slab of Ordovician limestone with abundant skeletal grains. B = brachiopods, BR = bryozoans

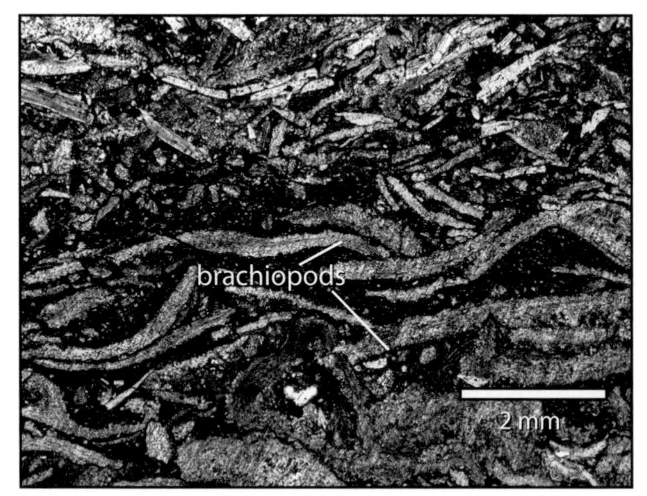

B. Thin section image of a skeletal packstone.

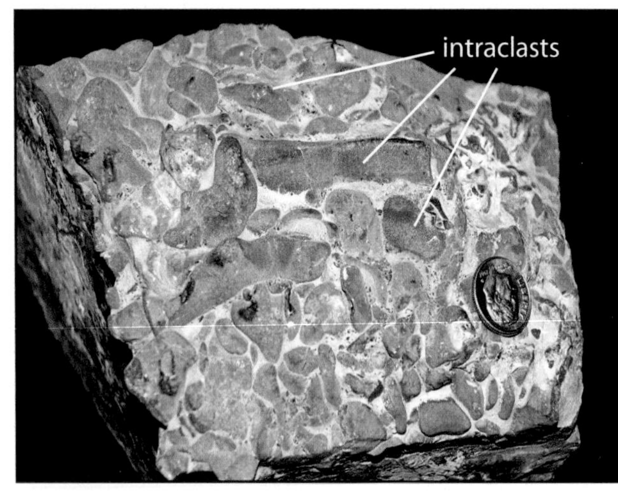

C. Hand sample of Ordovician-age intraclast-rich limestone from western Utah. Note coin for scale.

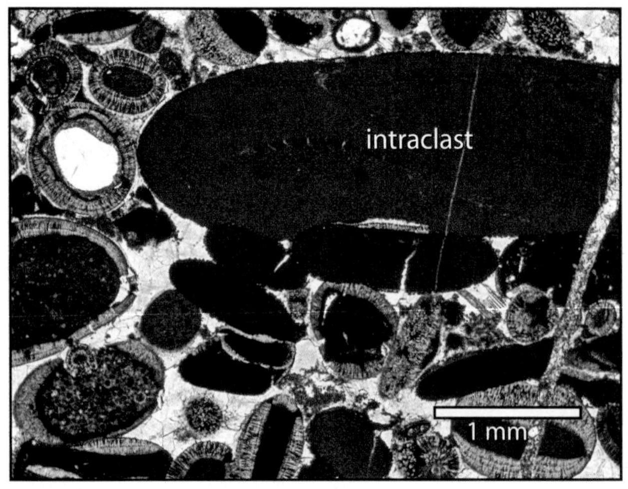

D. Thin section of an intraclast-rich grainstone. Note that some intraclasts (dark grains) possess oolitic coatings.

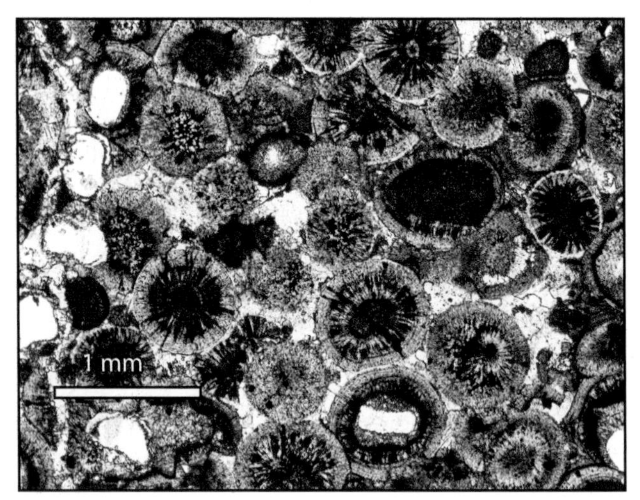

E. Thin section of an oolitic grainstone.

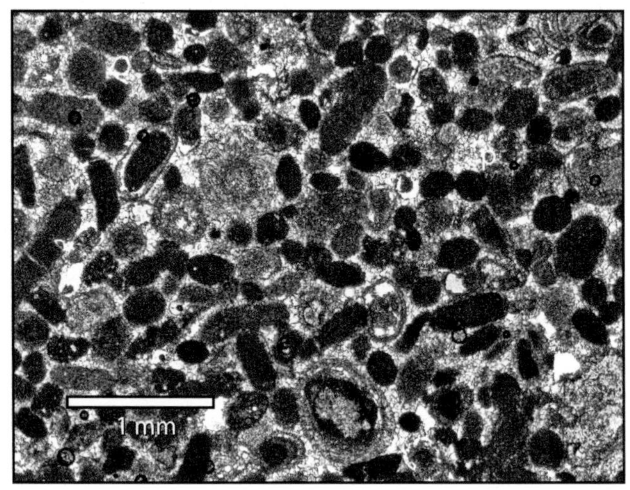

F. Thin section of a peloid-rich grainstone. The peloids are the dark, oval grains.

FIGURE 3.9 Thin section images of common types of carbonate grains.

Mudstones (< 10% grains) and wackestones (> 10% grains) display mud-supported fabrics, meaning that the grains (skeletal fragments, ooids, etc.) generally do not touch each other, but "float" in the mud matrix. Packstones and grainstones display grain-supported fabrics where the grains support each other. Packstones are characterized by the complete or partial filling of interparticle pore spaces by carbonate mud. Grain-supported rocks lacking interstitial mud are classified as grainstones. Mud-rich carbonates indicate deposition below the energy threshold needed to remove fine-grained sediment from the environment of deposition. Increasingly grain-rich rocks indicate an increase in depositional energy. Figure 3.11 shows the textural spectrum (low-energy mudstone to high-energy grainstone) for a suite of skeletal-rich limestones. Carbonate rocks reflecting organic binding (by algae and/or encrusting organisms) or construction of a rigid carbonate framework (e.g., corals) during deposition are called **boundstones** (figure 3.12). Boundstones reflect a broad spectrum of depositional conditions dictated by the ecological requirements of the sediment-binding and frame-building organisms.

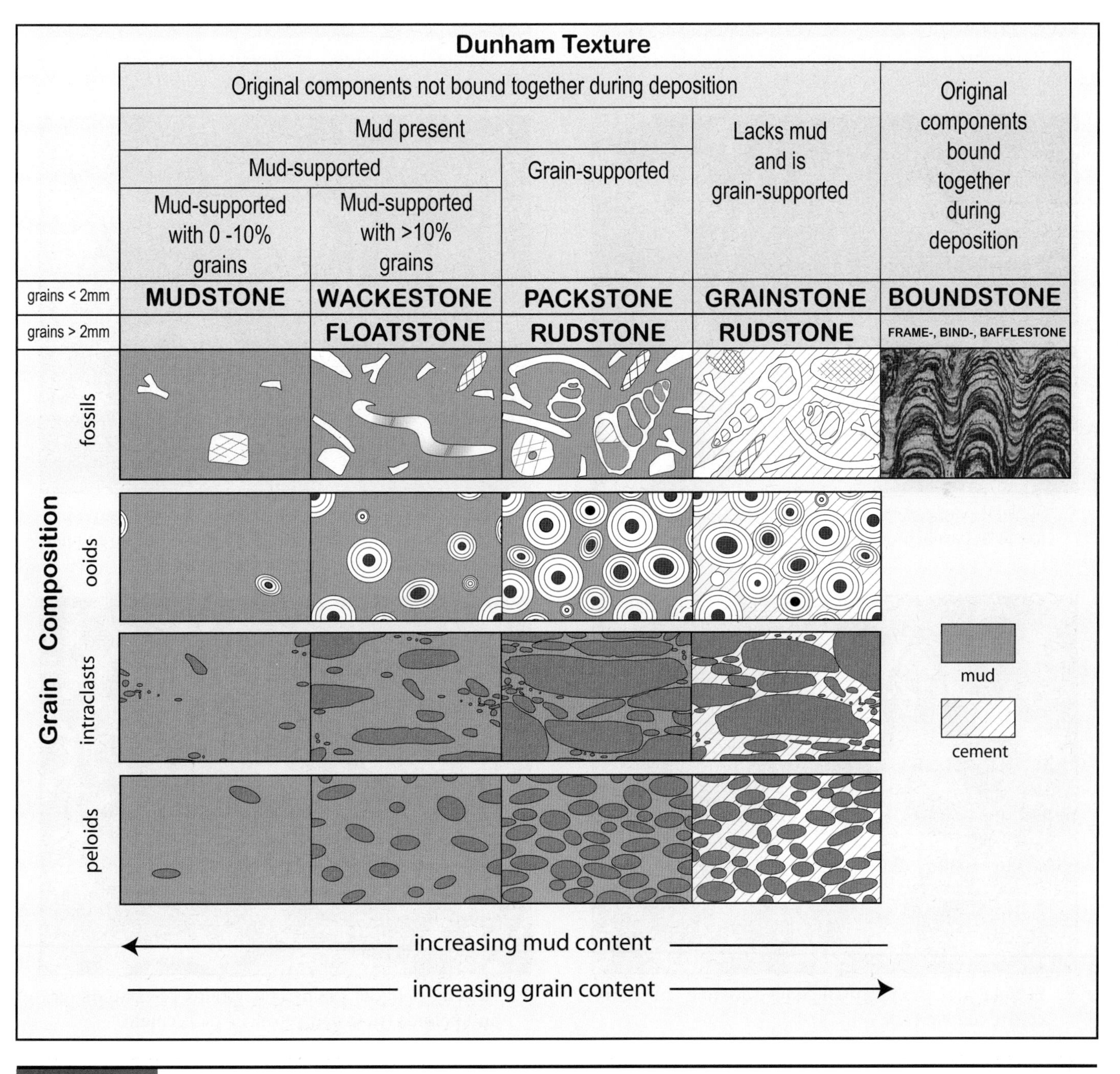

FIGURE 3.10 Limestone classification proposed by Dunham (1962) based on the presence or absence of mud and the presence or absence of grain support.

A. Mudstone.

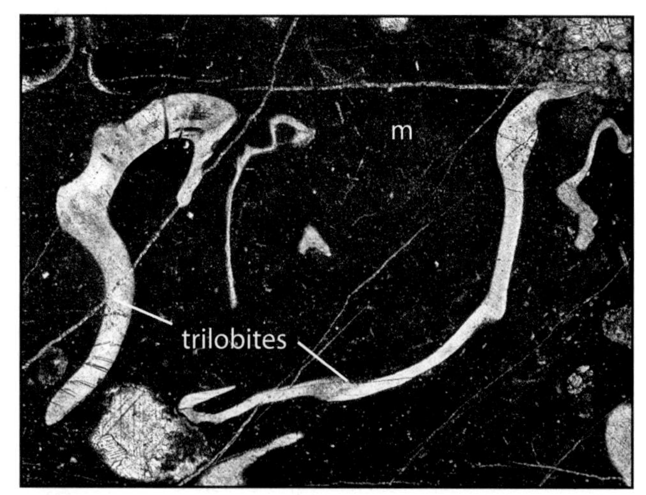

B. Skeletal wackestone. Fragments of fossils (trilobites) "floating" in carbonate mud matrix (m).

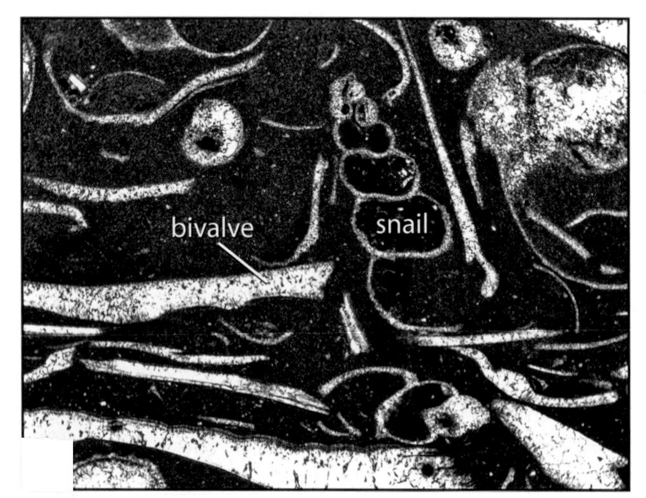

C. Skeletal wackestone to packstone. Whole and broken fossils in partial mud support and partial grain support.

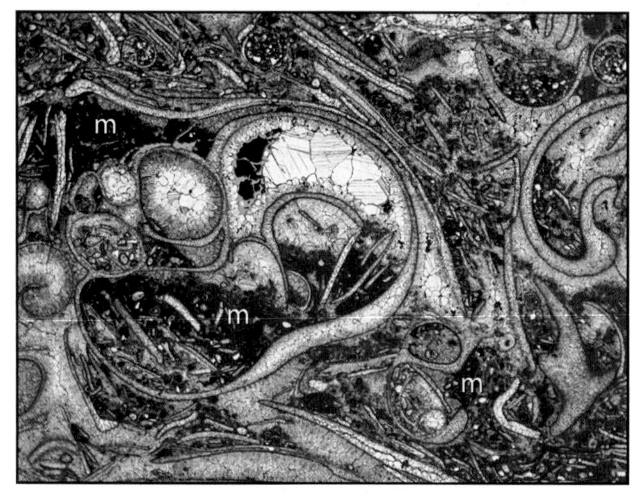

D. Skeletal packstone with whole and broken snail shells. Note mud (m) filling lower part of large snail as well as areas between smaller grains.

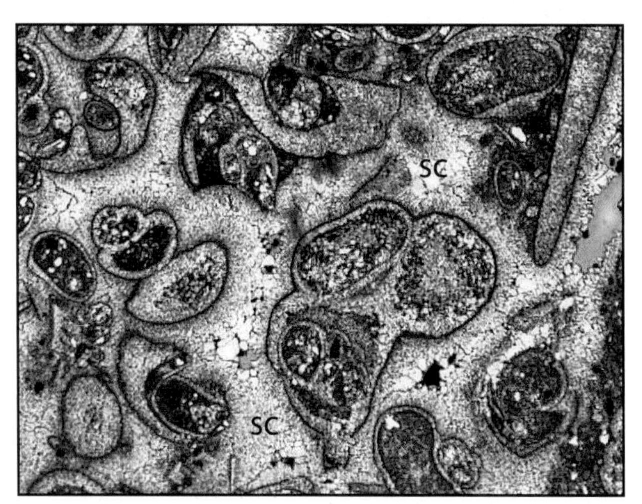

E. Skeletal grainstone. Fossil snails "floating" in sparry calcite cement (sc).

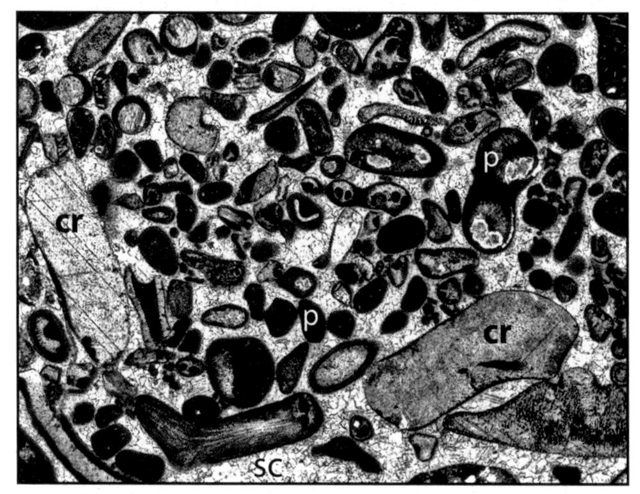

F. Skeletal-peloidal grainstone. Grains of crinoids (cr), and peloids (p) in sparry calcite (sc) cement.

Figure 3.11 Thin section images showing major limestone types according to the classification scheme of Dunham (1962).

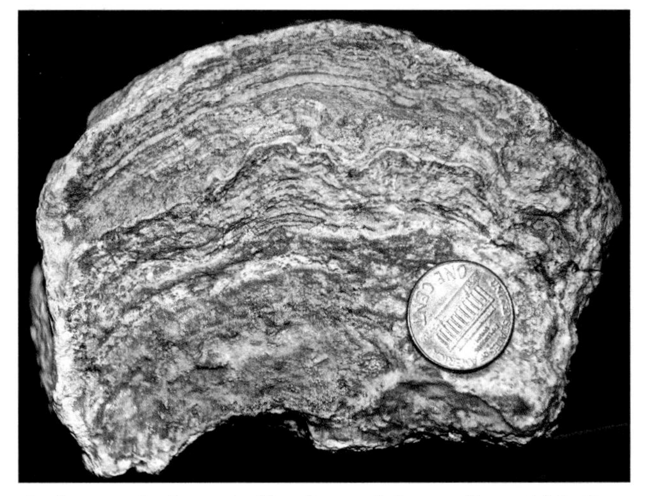

A. Layered stromatolite formed by sediment-binding activity of blue-green algae in the Eocene Green River Formation, Utah.

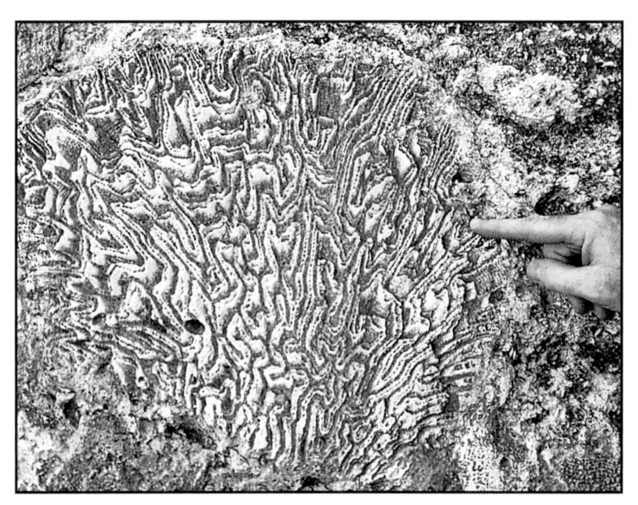

B. Skeletal framestone of the colonial coral *Diploria* in the Pleistocene Key Largo Limestone, Florida.

FIGURE 3.12 Images of limestones classified as boundstones by Dunham (1962).

The most informative rock names result from combining compositional and textural terms. For example, a grain-supported carbonate rock comprised of unbroken gastropod shells with interstitial mud would be called a skeletal packstone, or better yet, a whole-gastropod packstone. Oolitic grainstone, peloidal packstone, and mixed skeletal-peloidal packstone are examples of other informative compositional-textural combinations. If the rock has been dolomitized, simply add the prefix dolo- to the rock. For example, a skeletal-rich wackestone that has been dolomitized would be classified as a skeletal dolowackestone. A dolomitized packstone would be classified as a dolopackstone, etc.

Since calcite and dolomite crystal are both rhomb-shaped, it may be difficult to determine the presence of dolomite with the unaided eye. The simplest method for differentiating between a limestone and a dolostone is the "acid test." A drop of dilute hydrochloric acid will cause limestone to fizz vigorously. Dolostone will fizz only weakly, but for a longer period of time.

Evaporites

Most of the water on Earth is found in the global ocean. The salinity of 90% of the total volume of seawater ranges between 34.61 and 34.70 ppt. As seawater evaporates, a variety of minerals precipitate, and usually in a predictable sequence beginning with aragonite ($CaCO_3$). When 50 to 75% of seawater in a given basin has evaporated, **gypsum** ($CaSO_4$) will precipitate. **Halite** (NaCl) precipitates from highly concentrated and evolved waters when 90% of the original ocean water has evaporated. The most common and commercially important chemical sedimentary rocks that result from evaporation of seawater are **rock gypsum** and **rock salt**. Extensive evaporite deposits are characteristic of the Silurian System of the Michigan basin, the Pennsylvanian System of the Paradox basin, the Permian System in central Europe and the southwestern United States, and the Jurassic System bordering the Atlantic Ocean basin.

PROCEDURE

Complete the rock data worksheet for each specimen of siliciclastic and carbonate rock provided by your instructor. Describe all textural and compositional features that apply to a particular sample. Plot the composition of each sandstone sample on the QFL diagram that corresponds to the sample. Once you have thoroughly described each sample, determine the most appropriate name for that rock. Use the Dott scheme to classify sandstones and the Dunham classification for limestones. Although you will provide a name for each specimen, the most important aspect of the exercise is the careful and complete description of each rock. Be sure to use a hand lens when examining the specimens.

Siliciclastic Rock Data Sheet

Sample #	**Texture and Matrix**		**Grain Composition**
Grain size:			% Quartz ___ % Feldspar ___ % Lithics ___
Grain shape: angular ___ subangular ___ subrounded ___ rounded ___	Sorting: poorly sorted ___ moderately ___ well sorted ___ very well sorted ___	% Matrix: Color:	**Quartz** **Feldspar** **Lithics**
Arenite ___ or Wacke ___			
Textural maturity:			
Compositional maturity:			Dott rock name:

Sample #	**Texture and Matrix**		**Grain Composition**
Grain size:			% Quartz ___ % Feldspar ___ % Lithics ___
Grain shape: angular ___ subangular ___ subrounded ___ rounded ___	Sorting: poorly sorted ___ moderately ___ well sorted ___ very well sorted ___	% Matrix: Color:	**Quartz** **Feldspar** **Lithics**
Arenite ___ or Wacke ___			
Textural maturity:			
Compositional maturity:			Dott rock name:

Sample #	**Texture and Matrix**		**Grain Composition**
Grain size:			% Quartz ___ % Feldspar ___ % Lithics ___
Grain shape: angular ___ subangular ___ subrounded ___ rounded ___	Sorting: poorly sorted ___ moderately ___ well sorted ___ very well sorted ___	% Matrix: Color:	**Quartz** **Feldspar** **Lithics**
Arenite ___ or Wacke ___			
Textural maturity:			
Compositional maturity:			Dott rock name:

Sample #	**Texture and Matrix**		**Grain Composition**
Grain size:			% Quartz ___ % Feldspar ___ % Lithics ___
Grain shape: angular ___ subangular ___ subrounded ___ rounded ___	Sorting: poorly sorted ___ moderately ___ well sorted ___ very well sorted ___	% Matrix: Color:	**Quartz** **Feldspar** **Lithics**
Arenite ___ or Wacke ___			
Textural maturity:			
Compositional maturity:			Dott rock name:

Siliciclastic Rock Data Sheet

Sample #	Texture and Matrix		Grain Composition
Grain size:			% Quartz ___ % Feldspar ___ % Lithics ___
Grain shape: angular ___ subangular ___ subrounded ___ rounded ___	Sorting: poorly sorted ___ moderately ___ well sorted ___ very well sorted ___	% Matrix: / Color:	Quartz / Feldspar / Lithics
Arenite ___ or Wacke ___			
Textural maturity:			
Compositional maturity:			Dott rock name:

Sample #	Texture and Matrix		Grain Composition
Grain size:			% Quartz ___ % Feldspar ___ % Lithics ___
Grain shape: angular ___ subangular ___ subrounded ___ rounded ___	Sorting: poorly sorted ___ moderately ___ well sorted ___ very well sorted ___	% Matrix: / Color:	Quartz / Feldspar / Lithics
Arenite ___ or Wacke ___			
Textural maturity:			
Compositional maturity:			Dott rock name:

Sample #	Texture and Matrix		Grain Composition
Grain size:			% Quartz ___ % Feldspar ___ % Lithics ___
Grain shape: angular ___ subangular ___ subrounded ___ rounded ___	Sorting: poorly sorted ___ moderately ___ well sorted ___ very well sorted ___	% Matrix: / Color:	Quartz / Feldspar / Lithics
Arenite ___ or Wacke ___			
Textural maturity:			
Compositional maturity:			Dott rock name:

Sample #	Texture and Matrix		Grain Composition
Grain size:			% Quartz ___ % Feldspar ___ % Lithics ___
Grain shape: angular ___ subangular ___ subrounded ___ rounded ___	Sorting: poorly sorted ___ moderately ___ well sorted ___ very well sorted ___	% Matrix: / Color:	Quartz / Feldspar / Lithics
Arenite ___ or Wacke ___			
Textural maturity:			
Compositional maturity:			Dott rock name:

Carbonate Rock Data Sheet

Sample # — **Texture and Matrix**	**Grain Composition**
Mud Matrix: present _____ absent _____ Grain support _____ Mud support _____ < 10% grains _____	Ooids: % ______ Peloids: % ______
Color:	Skeletal Grains: % ______ types of fossils: ____________________ ____________________
Limestone _____ or Dolostone _____	diversity of fossils: high __ medium __ low __ preservation: whole __ broken __ both __
Dunham rock name:	Intraclasts: % ____ shape ____ size ____

Sample # — **Texture and Matrix**	**Grain Composition**
Mud Matrix: present _____ absent _____ Grain support _____ Mud support _____ < 10% grains _____	Ooids: % ______ Peloids: % ______
Color:	Skeletal Grains: % ______ types of fossils: ____________________ ____________________
Limestone _____ or Dolostone _____	diversity of fossils: high __ medium __ low __ preservation: whole __ broken __ both __
Dunham rock name:	Intraclasts: % ____ shape ____ size ____

Sample # — **Texture and Matrix**	**Grain Composition**
Mud Matrix: present _____ absent _____ Grain support _____ Mud support _____ < 10% grains _____	Ooids: % ______ Peloids: % ______
Color:	Skeletal Grains: % ______ types of fossils: ____________________ ____________________
Limestone _____ or Dolostone _____	diversity of fossils: high __ medium __ low __ preservation: whole __ broken __ both __
Dunham rock name:	Intraclasts: % ____ shape ____ size ____

Sample # — **Texture and Matrix**	**Grain Composition**
Mud Matrix: present _____ absent _____ Grain support _____ Mud support _____ < 10% grains _____	Ooids: % ______ Peloids: % ______
Color:	Skeletal Grains: % ______ types of fossils: ____________________ ____________________
Limestone _____ or Dolostone _____	diversity of fossils: high __ medium __ low __ preservation: whole __ broken __ both __
Dunham rock name:	Intraclasts: % ____ shape ____ size ____

Carbonate Rock Data Sheet

Sample # **Texture and Matrix**	**Grain Composition**
Mud Matrix: present _____ absent _____ Grain support _____ Mud support _____ < 10% grains _____	Ooids: % _______ Peloids: % _______
Color:	Skeletal Grains: % _______ types of fossils: _______________________
Limestone _____ or Dolostone _____	diversity of fossils: high __ medium __ low __ preservation: whole __ broken __ both __
Dunham rock name:	Intraclasts: % ____ shape ____ size ____

Sample # **Texture and Matrix**	**Grain Composition**
Mud Matrix: present _____ absent _____ Grain support _____ Mud support _____ < 10% grains _____	Ooids: % _______ Peloids: % _______
Color:	Skeletal Grains: % _______ types of fossils: _______________________
Limestone _____ or Dolostone _____	diversity of fossils: high __ medium __ low __ preservation: whole __ broken __ both __
Dunham rock name:	Intraclasts: % ____ shape ____ size ____

Sample # **Texture and Matrix**	**Grain Composition**
Mud Matrix: present _____ absent _____ Grain support _____ Mud support _____ < 10% grains _____	Ooids: % _______ Peloids: % _______
Color:	Skeletal Grains: % _______ types of fossils: _______________________
Limestone _____ or Dolostone _____	diversity of fossils: high __ medium __ low __ preservation: whole __ broken __ both __
Dunham rock name:	Intraclasts: % ____ shape ____ size ____

Sample # **Texture and Matrix**	**Grain Composition**
Mud Matrix: present _____ absent _____ Grain support _____ Mud support _____ < 10% grains _____	Ooids: % _______ Peloids: % _______
Color:	Skeletal Grains: % _______ types of fossils: _______________________
Limestone _____ or Dolostone _____	diversity of fossils: high __ medium __ low __ preservation: whole __ broken __ both __
Dunham rock name:	Intraclasts: % ____ shape ____ size ____

Exercise 4

Depositional Environments

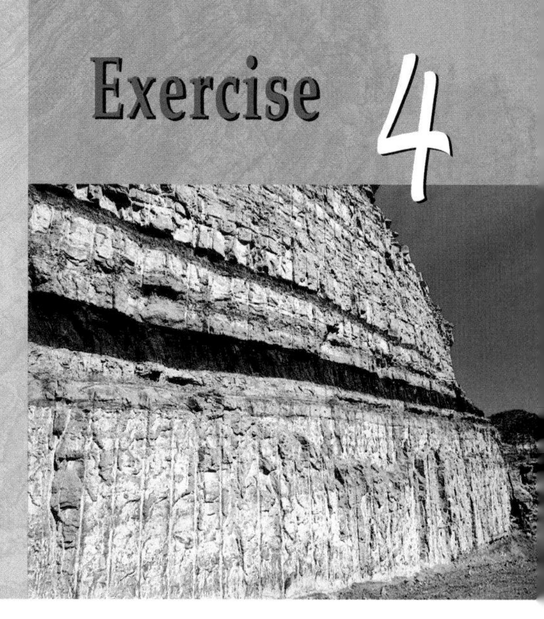

Learning Objectives

After completing this exercise, you will:

1. understand the relationship between modern and ancient depositional environments;
2. be able to describe the three main depositional environments in which sedimentary rocks form;
3. be able to list subenvironments that comprise the continental depositional environment;
4. be able to list subenvironments that comprise the transitional depositional environment;
5. be able to list subenvironments that comprise the marine depositional environment; and
6. be able to diagnose the depositional environment in which a sedimentary rock was deposited by evaluating its textural and compositional attributes.

The principle of uniformity is basic and fundamental to all science. First described by James Hutton, a Scot, in 1785, it is an assumption concerning nature that provides a way to scientifically deal with data about Earth history. The principle simply states that natural law is uniform through time and space. Geologic processes involved in the evolution of Earth have been governed by the same natural laws throughout geological time. Uniformity thus provides a basis for formulating models of the past in terms of our understanding of present geologic processes. Understanding natural laws and their effects on the processes operating on Earth today is a guide to past Earth events, as well as future ones. It does not mean that today's world is an exact replica of either the past or future worlds, but only a sampling of each. Uniformity as a principle does not imply consistency in rate of change. It does not preclude catastrophic events as important in geologic history. Modern catastrophic events such as the 1980 eruption of Mount St. Helens and the 1964 Good Friday earthquake in Alaska are examples of vigorous alteration of the Earth's features. Ancient catastrophic events are well documented in the geologic record, such as the erosion of the scablands of Washington State and the impact event thought by many to be the cause of mass extinction of organisms at the close of the Cretaceous Period. To make that distinction clear, the word actualism has been suggested as an alternative to uniformitarianism.

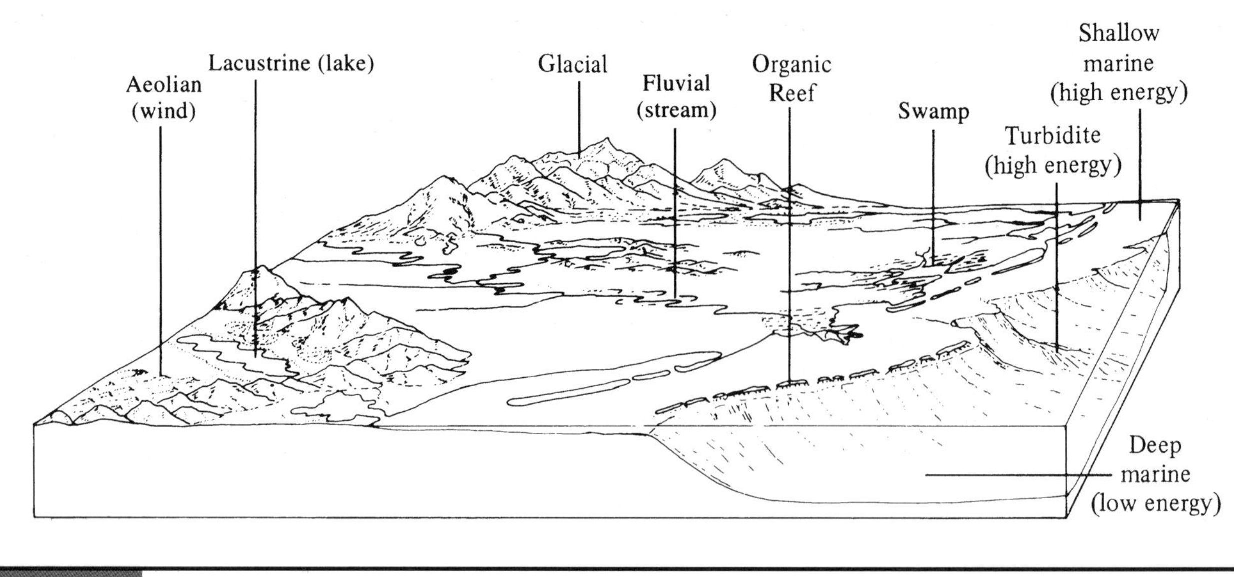

FIGURE 4.1 Block diagram showing typical depositional environments.

Depositional Environments

A key element of the geological history of a given area is the succession of depositional environments that have occupied that area through time. This history is recorded in the local succession of sedimentary strata. Geologists use the concept of uniformitarianism to decipher this archive by comparing features contained in ancient strata with those forming in modern sedimentary depositional environments. The environments where sediments are deposited today (and in the past) are highly varied with respect to physical, chemical, and biological processes. Unique combinations of processes define three main depositional environments: continental, transitional, and marine (figure 4.1). Continental environments include alluvial fans, deserts (eolian), lakes (lacustrine), braided and meandering streams (fluvial), and glaciers. Transitional environments form at the interface between land and sea and include such places as deltas, swamps (paludal), beaches, and tidal flats. Marine environments range from shallow to deep (figure 4.2) and include organic reefs (figure 4.3). Processes and depositional features unique to these settings are summarized in table 4.1. In figures 4.4, 4.5, and 4.6, images of modern depositional environments are matched with images of ancient sedimentary rocks that formed under similar conditions. Pay particular attention to the discussion of rock properties (texture, composition, color, sedimentary structures, and fossils) that are characteristic of each environment.

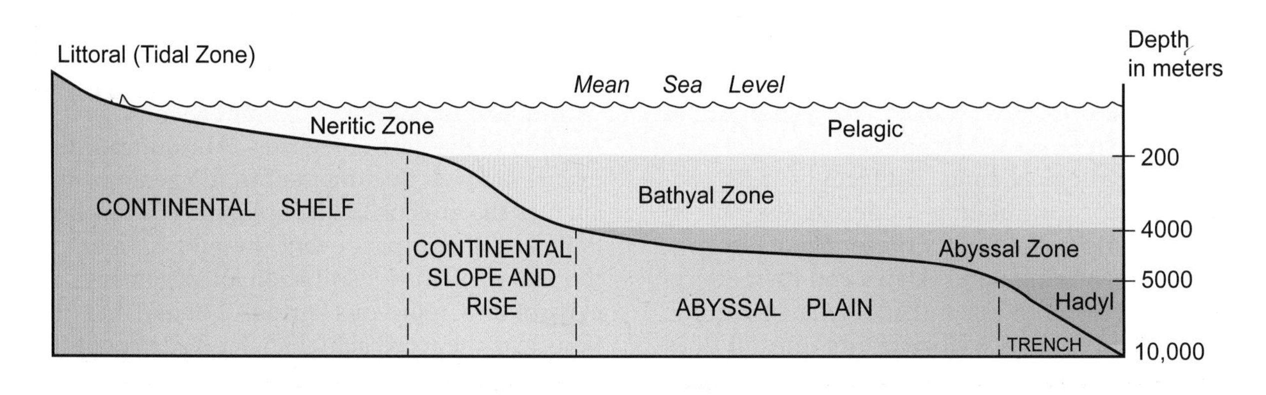

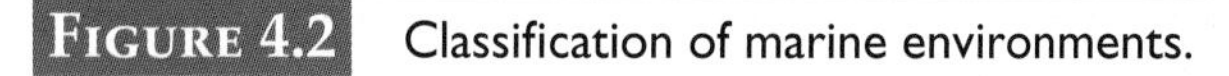
FIGURE 4.2 Classification of marine environments.

Depositional Environment		Environmental Processes	Texture	Organisms	Sedimentary Structures	Sedimentary Rocks
Continental	Alluvial Fan	flash floods, debris flows, high energy events, oxidation	poorly sorted, variable rounding	sparse land plants and animals	inclined bedding planes, channels	conglomerate and breccia, red color
	Glacier	ice transport, grinding, mass wasting of valley walls, outwash	very poorly sorted, angular clasts	few	few	tillite: conglomerate and breccia
	Lake (lacustrine)	low energy, shallow to deep, oxidation or reduction of iron	variable sorting and rounding of grains	freshwater plants and animals; fishes, turtles; well preserved	laminations, cross-bedding, ripple marks	interbedded shale, sandstone, and limestone
	Meandering Stream (fluvial)	high energy channels, deposition on point bar, erosion of cut bank, shifting channel position, oxidation	variable rounding and sorting, variable grain size	vertebrates (dinosaurs and mammals), whole in floodplain shales, broken in channel ss.	cross-beds, cut and fill, ripple marks, lenticular and elongate sand bodies	channels: sandstone and pebble conglomerate floodplain: shale and siltstone, red, tan, green
	Desert (eolian)	variable wind strength and wind direction, arid, oxidation, migrating sand dunes, deflation surfaces	medium to fine sand, well rounded and sorted, frosted	few, occasional trace fossils	small- to large-scale cross-bedding, ripple marks	sandstone, tan to red, may be bleached white by hydrocarbons
	Playa	arid, high evaporation rate, periodic flooding, but otherwise low energy	crystalline or nodular evaporites	none	crystalline or nodular evaporites	gypsum, anhydrite, rock salt, mudrock
Transitional	Beach	high to low energy, waves, tides, wind	well sorted and well rounded, ripple marks	marine invertebrates, generally broken or disarticulated	laminae, horizontal bedding, vertical trace fossils	quartz sandstone, carbonate grainstone, pebble conglomerate
	Delta and Barrier Island	variable environment includes distributary mouth bars, levees, and coal swamps; high to low energy	fine to medium sand, well sorted, well rounded; mud	brackish and marine invertebrates, root molds, trace fossils	laminations, trough cross-bedding, sand lenses in delta plain	sandstone, siltstone gray shale, coal, mud-clast conglomerate
	Tidal Flat	variable ebb and flood tidal currents	fine-grained, well sorted, mudstone to packstone	mollusks, algae, stromatolites	symmetrical ripples, mudcracks, laminae, fenestral fabric, trace fossils	siliciclastic: siltstone, shale, sandstone carbonate: stromatolites, intraclastic conglomerate
Marine	Carbonate Sand Shoals	variable ebb and flood tidal currents, fair-weather and storm waves, photic zone	grainstone, well sorted	marine invertebrates, whole or broken	cross-bedding, ripple marks	skeletal and oolitic grainstone
	Reef	fair-weather and storm waves, tidal currents, abundant light and oxygen	framestone, boundstone, grainstone	diverse assemblage of invertebrates; corals, stromatoporoids, etc.	borings, toppled corals	limestone and dolostone
	Deep Marine	subphotic to aphotic, low energy, turbidity flows, variable oxygen levels	well sorted, fine-grained	trace fossils, pelagic organisms such as cephalopods, etc.	laminae, thin beds, graded bedding	gray to black shale and siltstone, banded chert

TABLE 4.1 Summary of key sedimentary depositional environments, processes, and diagnostic features.

FIGURE 4.3

A typical reef environment with adjacent fore-reef and back-reef lagoon.

A. Modern alluvial fan.

B. Cretaceous conglomerate from northern Utah.

C. Modern valley glacier system in the Swiss Alps.

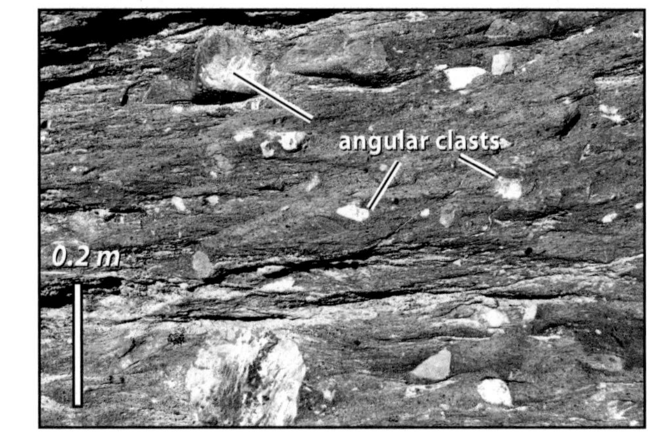

D. Precambrian glacial deposits from Utah. Note the angular nature of the clasts.

FIGURE 4.4 Modern and ancient continental depositional environments. *(continued)*

E. Modern fluvial environment showing meandering channel and wide floodplain.

F. Cretaceous channels (lens-shaped sandstone in the center of the image) surrounded by overbank silts and clays.

G. Modern dune field in southern New Mexico.

H. Cross-bedded sandstone of Jurassic age, Zion National Park, Utah.

FIGURE 4.4 Modern and ancient continental depositional environments.

A. Modern swamp in Louisiana.

B. Cretaceous coal seam in central Utah.

C. Modern beach.

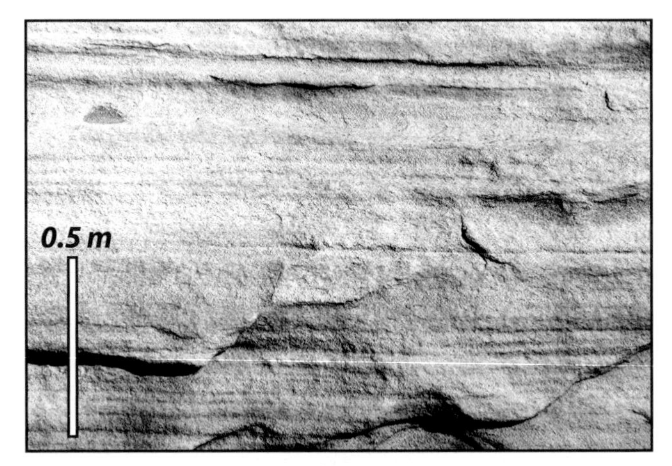

D. Horizontally bedded beach sands of Cretaceous age, central Utah. Note the horizontal bedding typical of beach deposits.

FIGURE 4.5 Modern and ancient transitional depositional environments.

A. Ooid shoal in the Bahamas. Large oolite sand waves in shallow (up to 4 meters deep) marine waters.

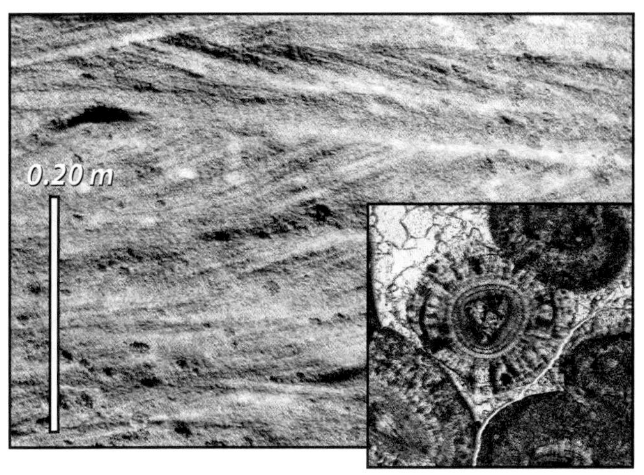

B. Jurassic Carmel Formation, central Utah.

C. Modern reef located on the east side of Andros Island, Bahamas.

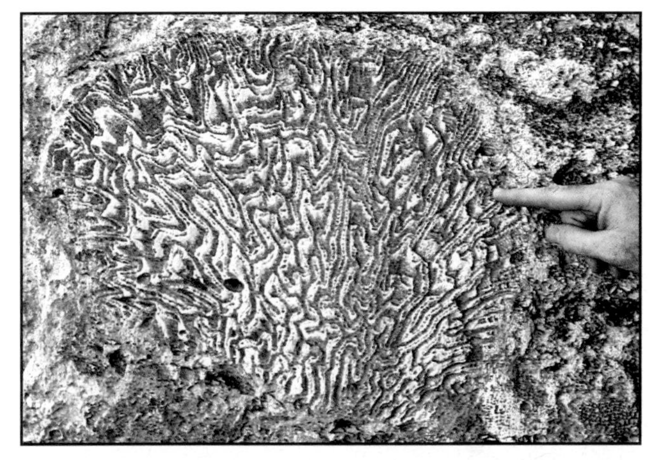

D. Pleistocene coral reef, Key Largo Limestone, south Florida. Note hand for scale.

FIGURE 4.6 Modern and ancient shallow-marine depositional settings.

PROCEDURE

Figures 4.7 through 4.12 show outcrop images of Mesozoic formations from central Utah along with sedimentary data collected at each locality. Carefully study the images and accompanying information. Compare outcrop data with information provided in table 4.1 to determine the depositional environment in which each of the formations was formed. Answer all of the accompanying questions.

PART A

North Horn Formation, Cretaceous, Red Narrows, Utah (Figure 4.7)

1. What is the dominant sedimentary rock type shown in figure 4.7?

2. What does the red color suggest about the environment of deposition?

3. Rocks of the North Horn Formation were most likely deposited in which of the sedimentary environments listed in table 4.1? Explain your answer.

Cretaceous North Horn Formation, Utah

Rock Type: Conglomerate with interbeds of sandstone

Composition: Boulders, cobbles, and pebbles comprised of sandstone and limestone. Matrix comprised of quartz, lithics, and mud

Texture: Poorly sorted, large clasts subrounded, sand-size material angular to subrounded

Color: Red to reddish brown

Sedimentary Structures: Thick-bedded with primary dips of a few degrees to the east; gravel-filled channels

Fossils: None

Regional Trend: Formation is over 2,000 feet thick in central Utah; it becomes thinner and finer-grained to the east

FIGURE 4.7 Cretaceous North Horn Formation, Utah.

Part B

Blackhawk Formation, Cretaceous, Price Canyon, Utah (Figure 4.8)

1. What is the dominant sedimentary rock type shown in figure 4.8?

2. What economically important sedimentary rock occurs in this outcrop?

3. Rocks of the Blackhawk were most likely deposited in which of the sedimentary environments listed in table 4.1? Explain your answer.

Cretaceous Blackhawk Formation, Utah

Rock Type: Sandstone and siltstone with interbeds of coal

Composition: Quartz, lithics, clay-pebble clasts

Texture: Moderately well sorted, subrounded grains

Color: Tan to gray

Sedimentary Structures: Sand-filled channels, cut and fill structures

Fossils: Freshwater plants, dinosaur footprints, root molds

Regional Trend: Underlain by marine sandstone and shale; overlain by fluvial sandstone and shale

FIGURE 4.8 Cretaceous Blackhawk Formation, Utah.

PART C

Cedar Mountain Formation, Early Cretaceous, San Rafael Swell, Utah (Figure 4.9)

1. What two sedimentary rock types predominate in this formation?

2. How would you describe the shape of the sandstone bodies shown in the satellite view of the area (figure 4.9C)?

3. Rocks of the Cedar Mountain Formation were most likely deposited in which of the sedimentary environments listed in table 4.1? Explain your answer.

4. Explain why dinosaur skeletons collected from shales at locality 1 on figure 4.9A are more likely to be complete than dinosaur remains collected from sandstones at locality 2 on figure 4.9A.

Cretaceous Cedar Mountain Formation, Utah

Rock Type: Pebble conglomerate and sandstone encased in siltstone and shale

Composition: Quartz, lithics, pebbles of sandstone chert, and quartzite

Texture: Conglomerate and sandstone poorly to moderately well sorted; siltstone and shale well sorted

Color: Conglomerate and sandstone tan to white; shale and siltstone red, maroon, green, light gray

Sedimentary Structures: Sand-filled channels, cut and fill structures; siltstone and shale is laminated to blocky

Fossils: Dinosaur bones and plants in sandstone lenses are disarticulated and broken; whole dinosaurs and well preserved plant material in shale

Regional Trend: Conglomerate and sandstone confined to elongate bodies surrounded by shale and siltstone; shale is volumetrically dominant

FIGURE 4.9 Cretaceous Cedar Mountain Formation, Utah.

PART D

Mancos Shale, Cretaceous, Road Cuts along I-70 in Central Utah (Figure 4.10)

1. What is the dominant sedimentary rock type shown in figure 4.10?

2. What does the black color suggest about the environment of deposition? What additional information is provided by the types of fossils found in the Mancos Shale?

3. Rocks of the Mancos Shale were most likely deposited in which of the sedimentary environments listed in table 4.1? Explain your answer.

Cretaceous Mancos Shale, Utah

Rock Type: Shale with thin beds of siltstone; bentonite beds are common

Composition: Siliciclastic mud and quartz silt

Texture: Well sorted

Sedimentary Structures: Undisrupted laminae with thin interbeds of siltstone (see figure B above); beds are laterally continuous across the outcrop

Fossils: Cephalopods, fishes, ichthyosaurs, mosasaurs, horizontal trace fossils, rare bivalves (mainly organisms that lived in the water column; few bottom dwellers)

Color: Dark gray to black

Regional Trend: Several thousand feet thick, extends in a north–south belt from Canada to Mexico; extends from central Utah to western Kansas

FIGURE 4.10 Cretaceous Mancos Shale, Utah.

PART E

Moenkopi Formation, Early Triassic, Capitol Reef National Park, Utah (Figure 4.11)

1. How would you describe bedding within the Moenkopi Formation (figure 4.11A)?

2. What do the mudcracks (figure 4.11E) suggest about the depositional setting?

3. What do the ripple marks (figs. 4.11B, 4.11D, and 4.11F) suggest about the direction and intensity of currents operating in the environment of deposition?

4. What do the reptile tracks in figure 4.11C suggest about water depth at the time of deposition?

5. The Moenkopi Formation was most likely deposited in which of the sedimentary environments listed in table 4.1? Explain your answer.

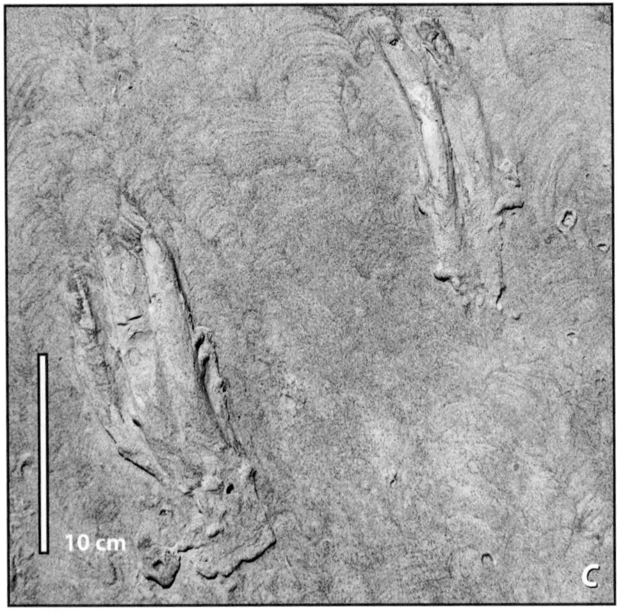

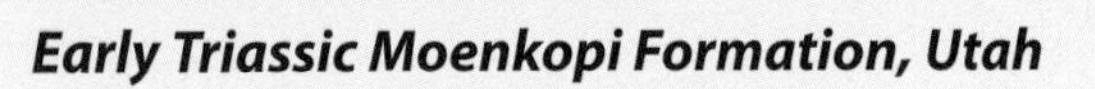

Early Triassic Moenkopi Formation, Utah

Rock Type: Siltstone, shale, sandstone
Composition: Quartz, lithics, siliciclastic mud
Texture: Fine-grained, well sorted
Color: Red to reddish brown, tan
Sedimentary Structures: Thin beds and laminae laterally continuous across outcrop; symmetrical ripple marks, sets of ripple cross-laminated siltstones showing opposite directions of sediment transport, mudcracks
Fossils: Reptile trackways and swim tracks

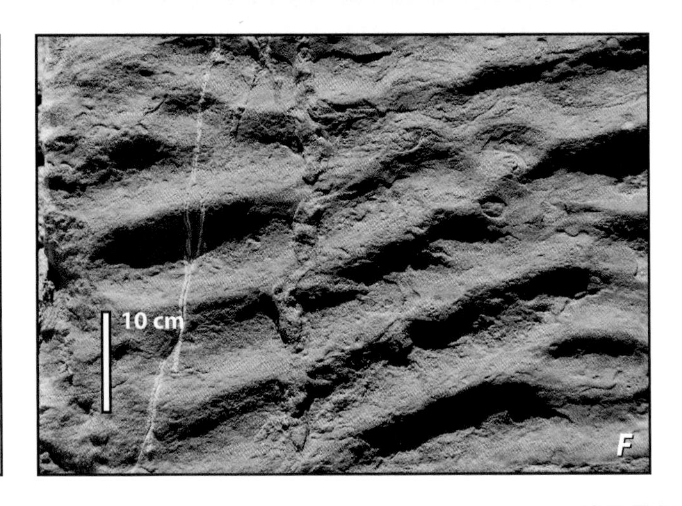

FIGURE 4.11 Early Triassic Moenkopi Formation, Utah.

PART F

Jurassic Carmel Formation, Utah (Figure 4.12)

1. What do the cross-bedding, presence of crinoids, and oolitic nature of the limestone indicate about the environment of deposition?

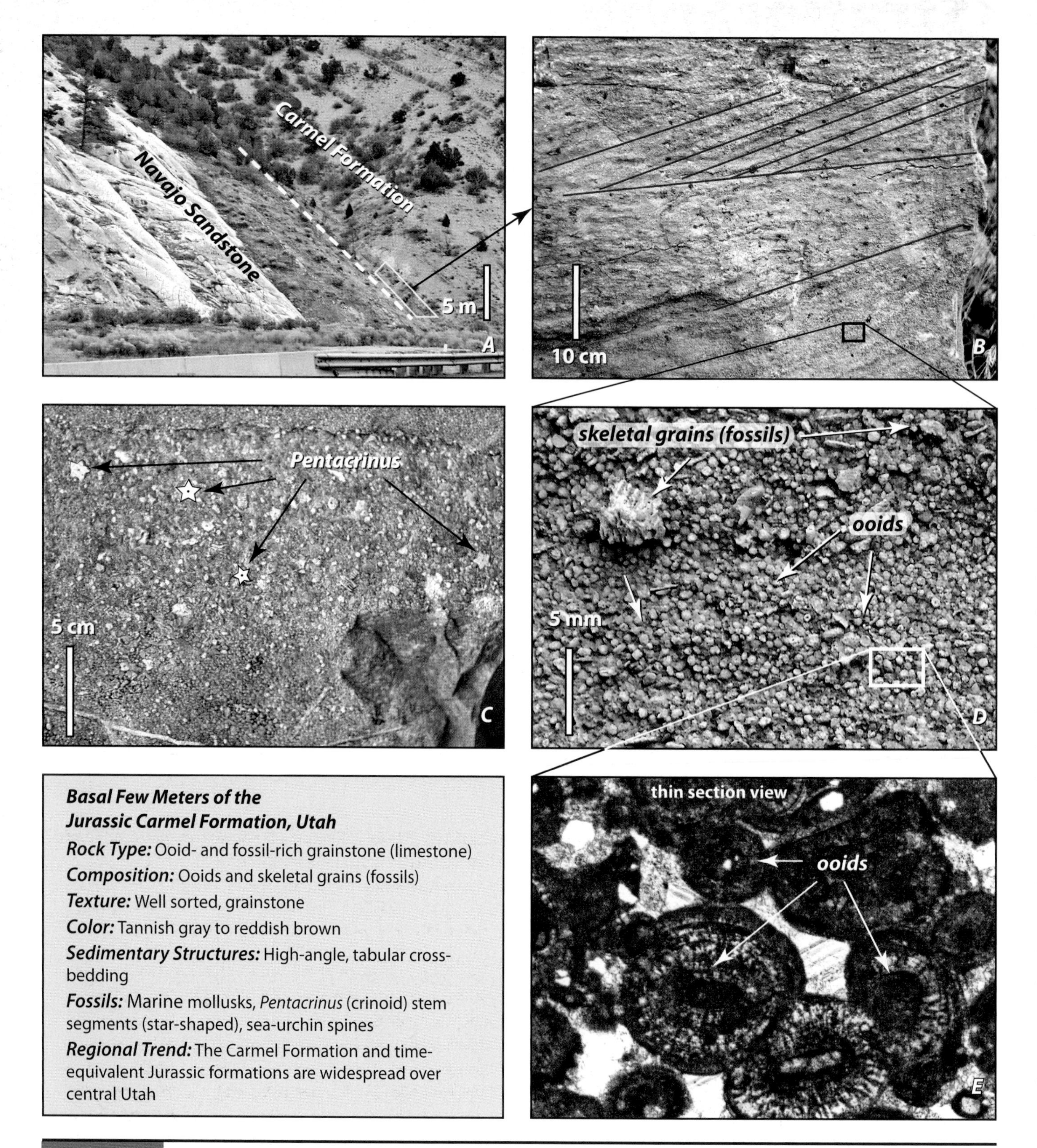

FIGURE 4.12 Jurassic Carmel Formation at Thistle Junction, central Utah.

Stratigraphy

Organizing the Rock and Fossil Record

Exercise 5

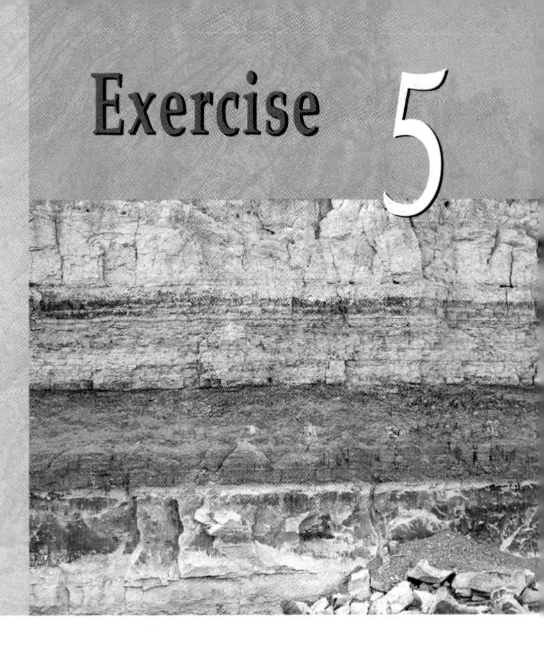

Learning Objectives

After completing this exercise, you will be able to:

1. define stratigraphy and discuss the three subdisciplines of classical stratigraphy (lithostratigraphy, biostratigraphy, and chronostratigraphy);
2. describe the fundamental units of lithostratigraphy;
3. distinguish between the different types of biostratigraphic zones;
4. discuss the difference between geochronology and chronostratigraphy and their fundamental units;
5. correlate two stratigraphic sections using graphic correlation; and
6. find up-to-date stratigraphic data on the web at http://www.stratigraphy.org.

Introduction

As one can see when looking at the side of a mountain or the wall of a canyon, rocks, especially sedimentary rocks, typically show distinctive layering or **stratification** (figure 5.1). The stratification is largely due to lithologic and color differences between adjacent beds or groups of beds. Less obvious are differences in fossil content, magnetic properties, and geochemistry. These visible and not-so-visible differences permit geologists to subdivide a succession of strata into a series of manageable divisions, allowing effective description, classification, and interpretation, as well as communication, concerning them. The subdiscipline of the geological sciences that focuses on the study of stratified rock bodies is known as **stratigraphy**. Stratigraphy provides the organizational framework within which Earth scientists interpret stratigraphic patterns and the formative processes that created them.

Because stratigraphic successions may be described in terms of their rock types, fossils, age, geochemistry, etc., one may subdivide the same stratigraphic succession in a variety of ways, each providing important information about the origin and attributes of the succession. In classical geology, rock type, fossil content, and age were the chief defining criteria for subdividing strata on a global basis. The resulting separate stratigraphic subdisciplines, **lithostratigraphy** (rock type), **biostratigraphy** (fossil content), and **chronostratigraphy** (age of rock), along with radiometric dating, are the basis of the modern geological timescale. In this exercise, you will learn the practice and classification schemes of these three subdisciplines of stratigraphy.

FIGURE 5.1 Outcrop view of Pennsylvanian strata exposed on the cliffs of the San Juan River, southeastern Utah.

Lithostratigraphy

A lithostratigraphic unit is a defined body of sedimentary, metasedimentary, or extrusive igneous rock (basalt flow) that is distinguished on the basis of lithic characteristics and stratigraphic position. The fundamental unit of lithostratigraphy is the **formation**, a homogeneous rock unit or an association of rock units, which is distinct from rock units above and below and can be shown on a geologic map of normal dimensions. A formation may consist almost entirely of cross-bedded sandstone; alternating sandstone and shale; or interbedded sandstone, shale, and limestone, as long as the unit is easily distinguished from rock units above and below.

Formations may be of any mappable thickness. In some areas it is useful to delineate formations that are only a few meters thick, whereas in other areas, because of the great thicknesses of the sedimentary column and its relatively uniform nature, formations may be several hundred meters thick. The accepted lower limit is based on local usefulness and whether the formation is sufficiently thick to appear on a map with a scale of approximately 1 in per mile. Subdivision into units with a thickness less than 20 m is not functional in most regions, although occasionally thinner beds are differentiated if they have distinctive features that can be used to improve regional correlation.

Formation boundaries are known as **contacts** and are normally drawn at horizons showing marked lithologic change, such as where a limestone section changes to sandstone or shale, or where a sandstone bed is succeeded by conglomerate. Contacts may also be placed at some arbitrary marker bed in a transitional lithologic series such as at a thin distinctive conglomerate in the middle of a sandstone and siltstone sequence, at a thick volcanic ash bed in the middle of a shale section, or at a distinctive limestone that separates sandy limestone from calcareous sandstone.

Formations have a formal two-part name reflecting the location of the **type locality** and the lithology. The type locality is the canyon, creek, or geographic region in which the formation is particularly well exposed and where it was first described. The second part of the name is descriptive and refers to the dominant rock type (if homogeneous). If more than one rock type is prevalent the term "formation" is used. For example, the Leadville Limestone of Colorado and adjacent states is predominantly limestone and is named for its type locality near the old mining town of Leadville, Colorado. The Morrison Formation is so named because this Jurassic, dinosaur-bearing unit contains significant amounts of shale, sandstone, and limestone. Since a single lithologic term would not reflect the heterolithic nature of the unit, the term "formation" is used. The type locality of the Morrison Formation is the small town of Morrison, located immediately west of Denver, Colorado.

Formations can be subdivided into smaller **members**, reflecting more subtle lithologic changes than those used to define formations. For example, the Thunder Springs Member of the Redwall Limestone contains distinctive interbeds of dark chert along with the more typical limestone lithology. Conversely, formations may be lumped together to form a more inclusive lithostratigraphic unit known as a **group**. These also have formal two-part names, which give some idea of the geographic occurrence and lithology of the units.

A new formation is measured and described, its contacts are defined, and a specific section is designated for subsequent investigators as the best place to observe the formation and its relation to other formations. This **type section** is usually in the same area as the type locality. If geographic names are limited and exposures are only locally suitable, it is possible to name a formation from one area and designate another exposure as the type section. For example, in Utah, the type locality of the Manning Canyon Shale is in Manning Canyon in the Oquirrh Range, but the type section is in Soldier Canyon, located a few miles to the north, where the entire formation is better exposed.

Biostratigraphy

Fossils are the practical clocks of geology because the process of evolution has produced a unique sequence of life forms through time. Correlation by means of fossils is called **biostratigraphic correlation** or **biostratigraphy**.

William Smith, an English surveyor, discovered in 1799 that various kinds of fossils appear in an orderly and predictable fashion in a vertical succession of rock units. Based on Smith's finding, the **principle of faunal succession** was formulated. The principle states that strata representing successive intervals of Earth history contain unique assemblages of fossils, and therefore rock units can be dated by analysis of their fossil content and through comparison to standard references. Because fossils occur almost exclusively in sedimentary rocks, biostratigraphic correlation is ordinarily not possible in igneous and metamorphic rocks. These rocks, however, can be dated radiometrically in many instances.

Although nearly all kinds of fossils are used for geologic dating on a local scale, only a select few groups, called **index fossils**, are utilized for regional and intercontinental correlation as reference standards in the geologic column. In figure 5.2, the relative biostratigraphic value of the major fossil groups is summarized.

Fossils that are best suited for time correlation over great distances, such as between continents, should have the following characteristics: **abundance** (they must be readily available in order to be of much use), **wide geographic range** (their usefulness is directly proportional to the area throughout which they are found), **rapid evolutionary change** (this results in short geological time ranges and higher-precision dating), **rapid dispersion** (this minimizes differences of age in different parts of the world), and **easy identification** (if only a few experts can distinguish one from another, their value is compromised). Organisms possessing all of these attributes are rare, but planktonic (floating) and nektonic (swimming) organisms are generally best, as their ecological lifestyle allows rapid distribution in many environments.

Examples of various fossils shown in figure 5.2 that are useful for time-stratigraphic determinations include trilobites in the Early Paleozoic, fusulinid foraminifers in the Pennsylvanian and Permian Systems, smaller foraminifers in the Tertiary, graptolites in the Ordovician through earliest Devonian Systems, ammonoids in the Devonian through Cretaceous Systems, and conodonts in the Ordovician through Triassic Systems.

In actual practice, biostratigraphic correlation begins with subdivision of local stratigraphic sections into rock units called **zones** or **biozones**. These are similar to lithostratigraphic formations in the sense that they represent distinct packages of rock layers, or strata, that represent the depositional record of a discrete interval of Earth history. They differ, however, in how they are defined. Instead of using rock type, zones are identified by the presence of a specified individual fossil taxon or by a group of specified taxa, one or two of which serve as name bearers. For example, the *Cheiloceras* Zone is the succession of strata that bear the fossil remains of the ammonoid *Cheiloceras* and/or other Late Devonian ammonoids that are characteristic of that time interval in the physical rock record. Some fossil zones are extremely limited in their geological time range and their presence discriminates a small segment of the geologic column and thus a limited time interval. Within the Jurassic, which William Smith studied in some detail, approximately 60 ammonoid zones are now recognized. Since the Jurassic Period is calculated to be nearly 55 million years long, each of these zones averages less than one million years in duration. Conodonts permit subdivision of Late Devonian strata into zones that may have been deposited in as little as 300,000 years, which is almost instantaneous, geologically speaking. The origin of the zone concept has persisted with little modification, but with great utility, since its original description by the German paleontologist Albert Oppel in 1856.

Recall that zones are defined by the presence of a specified taxon or group of taxa. As such, different types of zones may be specified. The most widely utilized are **taxon range zones**, **concurrent range zones**, and **interval zones** (figure 5.3). The taxon range zone is defined as the body of strata corresponding to the total stratigraphic range of a specified fossil taxon (e.g., genus or species). The zone is named after the defining species. For example, the *Elrathia kingi* Zone comprises all strata

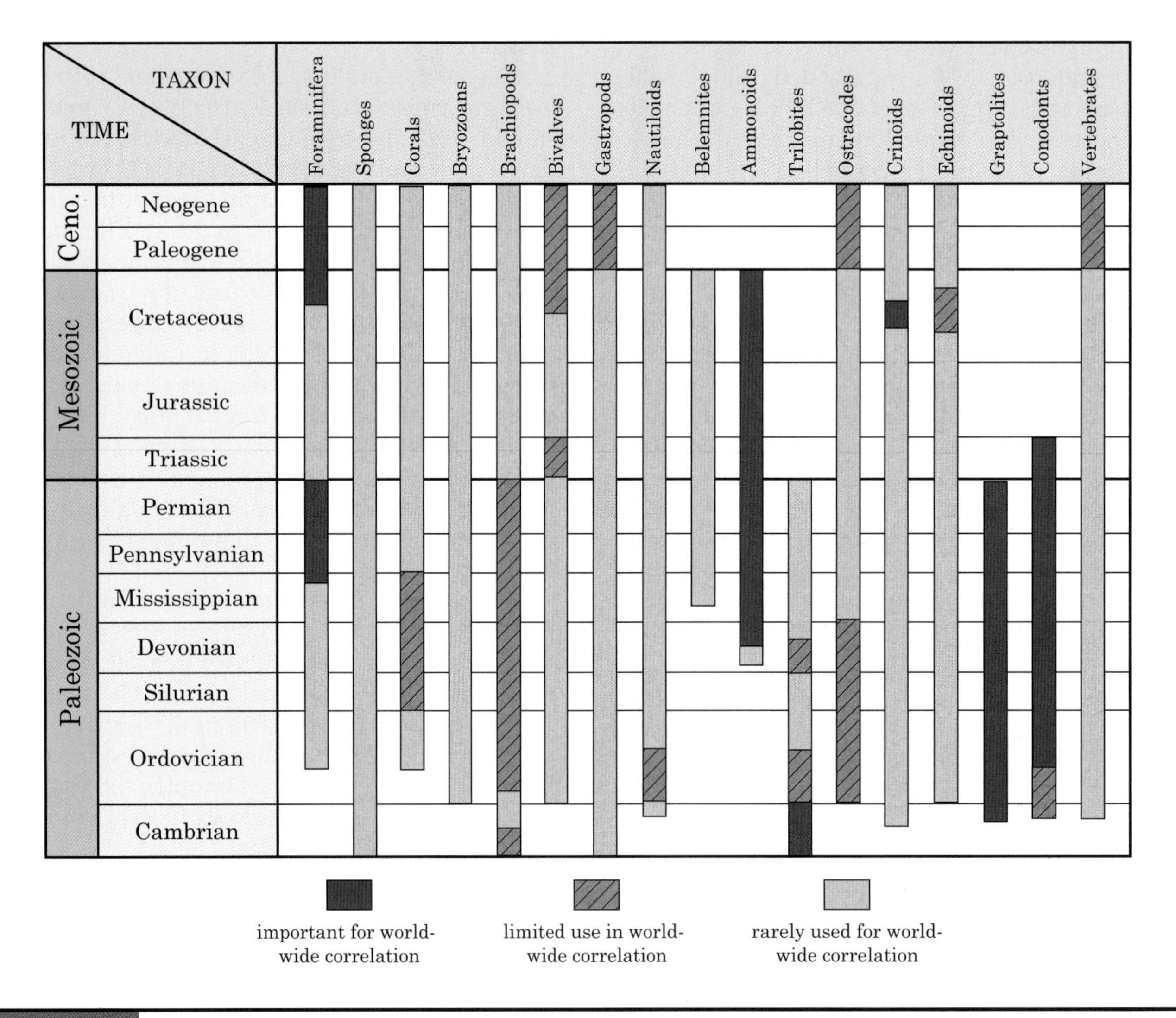

FIGURE 5.2 A chart showing the relative usefulness of various fossil groups as time indicators throughout the geologic column. The dark gray columns indicate common use in worldwide correlation; diagonal lines indicate local use or limited use in worldwide correlation; light gray columns indicate rare use in correlation. (After Teichert, 1958.)

bracketed by the lowest and highest stratigraphic occurrences of the trilobite species *Elrathia kingi* in a given stratigraphic section. The second type of zone is the concurrent range zone (or Oppel zone). This is defined as the stratigraphic interval characterized by the overlapping ranges of two taxa (figure 5.3). Concurrent range zones have higher stratigraphic resolution than taxon range zones because the stratigraphic interval in which the two species co-occur is generally thinner (reflecting a shorter interval of geological history) than the interval of strata containing the entire range of either individual taxon. These are named after the two species whose overlapping ranges delineate the zone. An interval zone is a body of strata corresponding to the interval between any two specified biotic events (e.g., interval between two extinction events or between two origination events) (figure 5.3).

Chronostratigraphy

A chronostratigraphic unit is a body of rock (succession of strata) that was deposited during a given interval of geological time, such as a period, era, or epoch. Chronostratigraphic classification provides a means of establishing age equivalence of rocks on local, regional, and global scales, thereby providing a framework for interpreting Earth history as a whole. The fundamental units of chronostratigraphy are shown in figure 5.4 along with their geochronological referents. It is essential to distinguish the difference between a geochronologic unit (pure time) and its corresponding chronostratigraphic counterpart (body of rock). For example, the Devonian Period is defined as that portion of Earth history that began 416 million years ago and that ended 359.2 million years ago. The Devonian **System** is the body of rock (igneous, metamorphic, and sedimentary) that

GEOCHRONOLOGY	CHRONOSTRATIGRAPHY
Eon	Eonothem
Era	Erathem
Period	System
Epoch	Series
Age	Stage

FIGURE 5.4 Correlation of geochronologic and chronostratigraphic units.

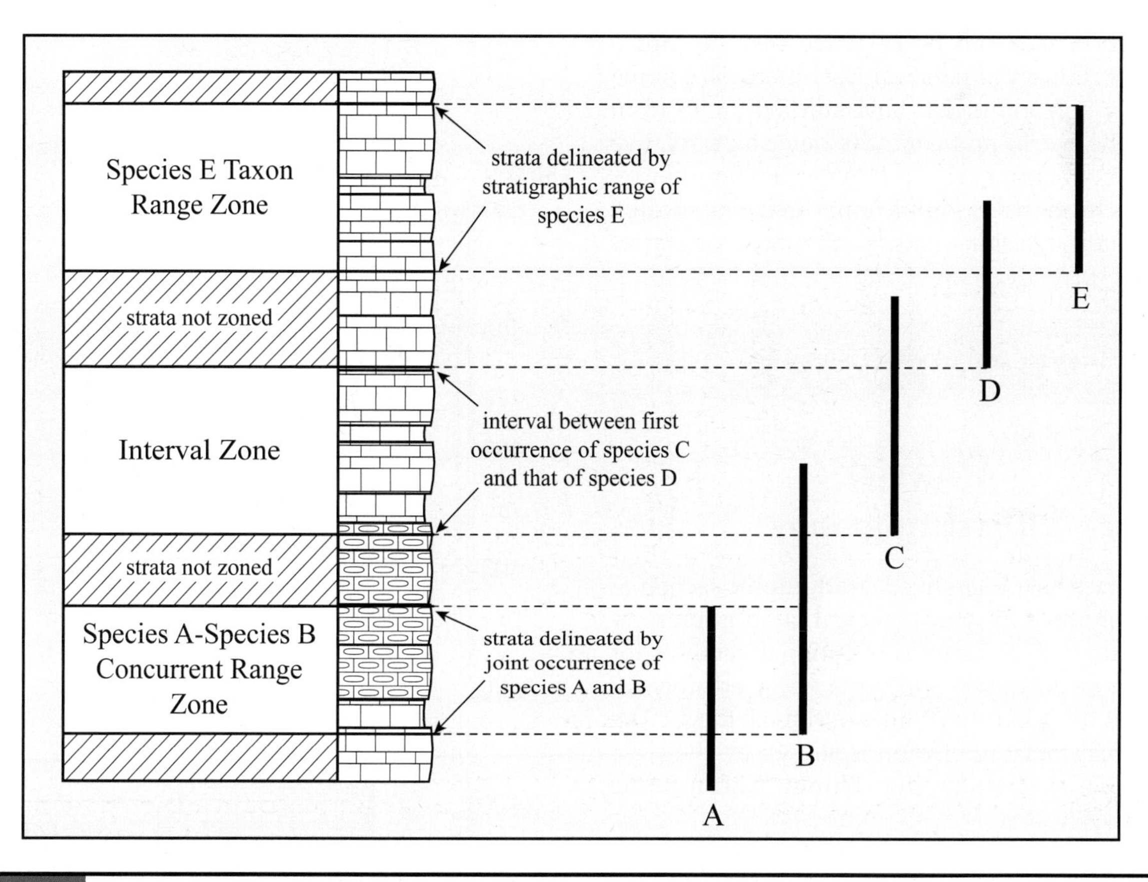

FIGURE 5.3 Subdivision of a hypothetical stratigraphic column into concurrent range, interval, and taxon range zones using stratigraphic ranges of fossils.

formed during that 56.8 million year-long interval of Earth history. Similarly, the Paleozoic **Erathem** is comprised of all rocks formed on Earth during the 291 million year-long span of the Paleozoic Era.

Boundaries between stratigraphic units are defined in a designated stratotype on the basis of observable paleontological or physical features of the rock. Each chronostratigraphic boundary (such as that between the Cambrian and Ordovician Systems) has a single **Global Stratotype Section and Point (GSSP)** that is defined by the **International Commission on Stratigraphy** after painstaking paleontological and sedimentological study of competing candidate sections. The GSSP for the Carboniferous–Permian system boundary is located in northern Kazakhstan at the stratigraphic level where the conodont species *Streptognathodus isolatus* first appears in a section of marine sandstone, siltstone, shale, and limestone. The GSSP for the base of the Ordovician System is located in Scotland. The majority of eonothem, erathem, system, and stage boundaries are now formally defined, with more being added to the list each decade.

Recognition of chronostratigraphic unit boundaries in the field is more difficult than locating formation and biozone boundaries, because one must find fossils or other time-sensitive features in the local section that can be correlated to the global stratotype. Since William Smith's initial study in the late 1700s, paleontologists have studied fossils from around the world and can now determine relative ages of most strata with relative ease. Boundaries between chronostratigraphic units are being located with greater precision as more and more boundaries are being formally defined.

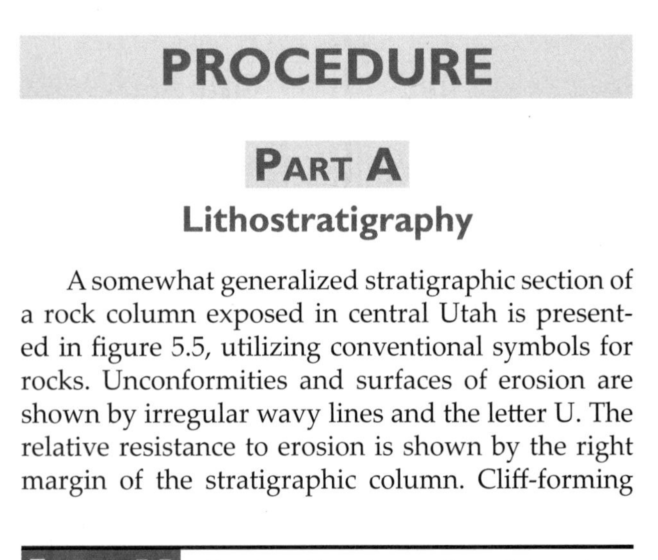

PROCEDURE

PART A

Lithostratigraphy

A somewhat generalized stratigraphic section of a rock column exposed in central Utah is presented in figure 5.5, utilizing conventional symbols for rocks. Unconformities and surfaces of erosion are shown by irregular wavy lines and the letter U. The relative resistance to erosion is shown by the right margin of the stratigraphic column. Cliff-forming

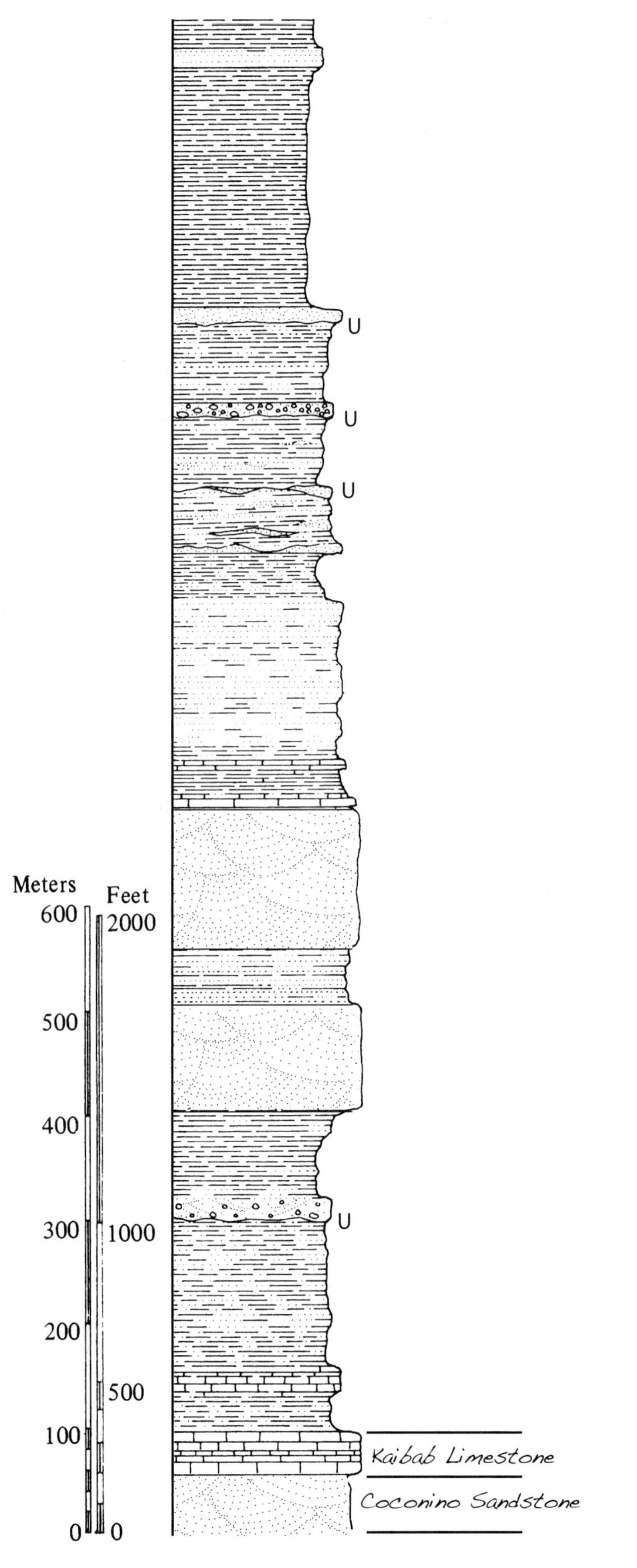

FIGURE 5.5 Stratigraphic column of Late Paleozoic and Mesozoic strata in central Utah.

units extend to the right; slope-forming members are recessed to the left. In central and eastern Utah, limestone and some well-cemented sandstone and conglomerate formations form cliffs, while finer-grained siltstone and shale beds form slopes and strike valleys.

1. Based upon your analysis of the column, subdivide the rocks in figure 5.5 into formations, using the lithology and erosional characteristics as a guide. Mark your subdivisions or formations with heavy horizontal lines as shown for the lower two formations.

2. Give each formation a formal two-part name (as shown for the Coconino Sandstone and the Kaibab Limestone). Geographic names will necessarily be arbitrary, but the descriptive part of the name should reflect the lithology of each of your units. Use the term formation where the lithology appears to be heterogeneous. Formal stratigraphic names begin with upper case letters.

3. On a separate sheet of paper or on your computer, briefly describe each of your formations in terms of its thickness, lithology, and weathering profile (slope, ledge, or cliff). For example:
   - Coconino Sandstone: sandstone, approximately 250 ft thick, cross-bedded, forms cliff
   - Kaibab Limestone: limestone, approximately 150 ft thick, forms cliff

PART B
Formations as Mappable Units

Formations are mappable rock units. Lines delineating the boundaries between successive formations on a geological map are called **contacts**. On figure 5.6, trace the contacts between the ten formations exposed in the vicinity of Shoshone Point in the eastern Grand Canyon. Assume that the layers are horizontal and that the planar contacts between formations neither gain nor lose elevation across the map area. Create a geological map of the area by drawing the contacts with black lines and by coloring the outcrop areas of the ten formations with pastel colors of your choosing. Label the formations using the symbols shown on figure 5.6. What is the relationship between geological contact lines and topographic contour lines when the formations are oriented horizontally?

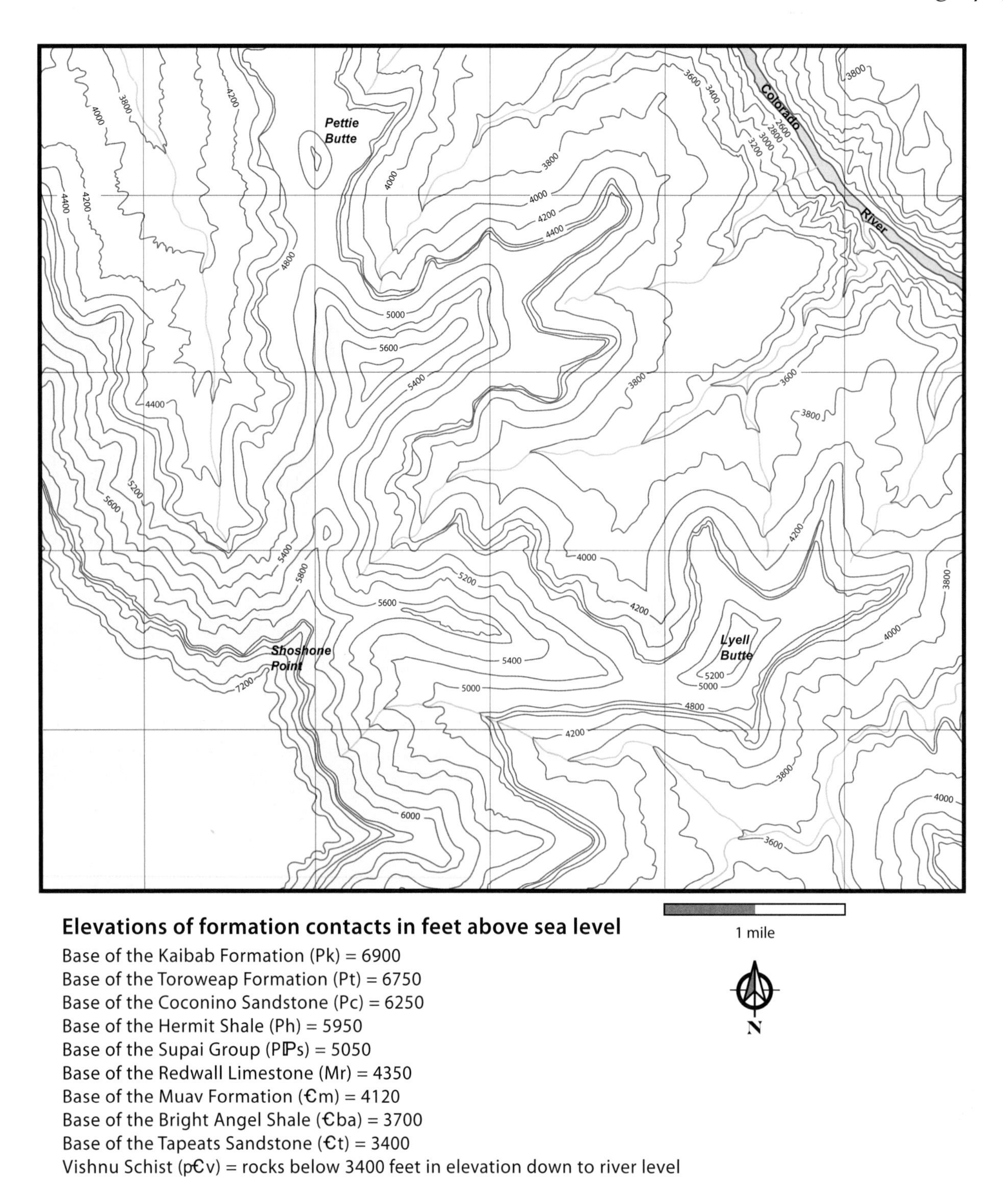

FIGURE 5.6 Topographic map of the Shoshone Point area, eastern Grand Canyon National Park, Arizona.

Part C
Biostratigraphy

1. The Redwall Limestone of the Grand Canyon is 500 ft thick and contains the remains of several different fossil phyla, including brachiopods, crinoids, mollusks, arthropods, bryozoans, foraminifers, and conodonts. Figure 5.7 shows the stratigraphic position and cliff-forming nature of this Mississippian formation. The Redwall was studied in detail by Edwin McKee and Raymond Gutschick in the 1960s, who reported the stratigraphic ranges of several species. Figure 5.8A shows the occurrence of selected species in samples taken at 50-ft intervals through the Redwall Limestone. Using these data, indicate on the left side of the chart shown in figure 5.8B the stratigraphic range of each of the remaining nine species using *Michelenia expansa* as an example.

2. Using the ranges of species that you just drew, subdivide the Redwall Limestone column on the right of figure 5.8B into the *Polygnathus communis–Spirifer granulosus* concurrent range zone and *Endothyra tumula* range zone using the *Michelenia expansa* zone as an example.

3. Color each of these zones a different color on the graphic column.

4. Using colors adopted in item 3 above, trace the outcrop pattern of the three zones across the Redwall cliff depicted in figure 5.7B. Remember that biostratigraphic zones are bodies of sedimentary rock, even though they are named after fossils.

5. What is the average number of species occurring in each zone?

6. Give an example of a zone in which the name-bearing species never occurs outside its zone.

7. Give an example of a zone in which the name-bearing taxa occur in another zone as well.

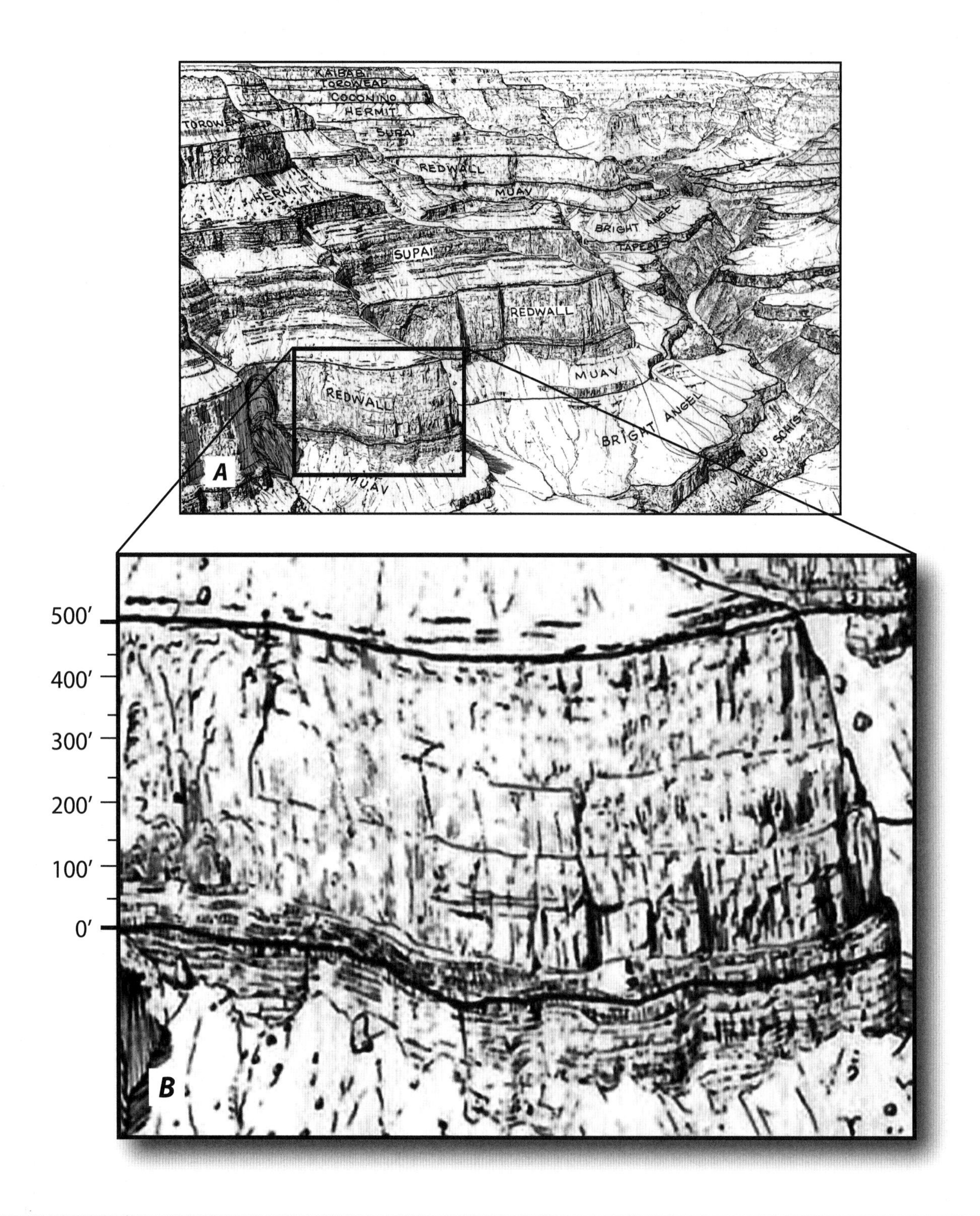

FIGURE 5.7 A. Sketch showing distribution of Paleozoic and Precambrian rock units in the eastern Grand Canyon. B. Inset shows the thickness of the Redwall Limestone. (Sketch by William Chesser.)

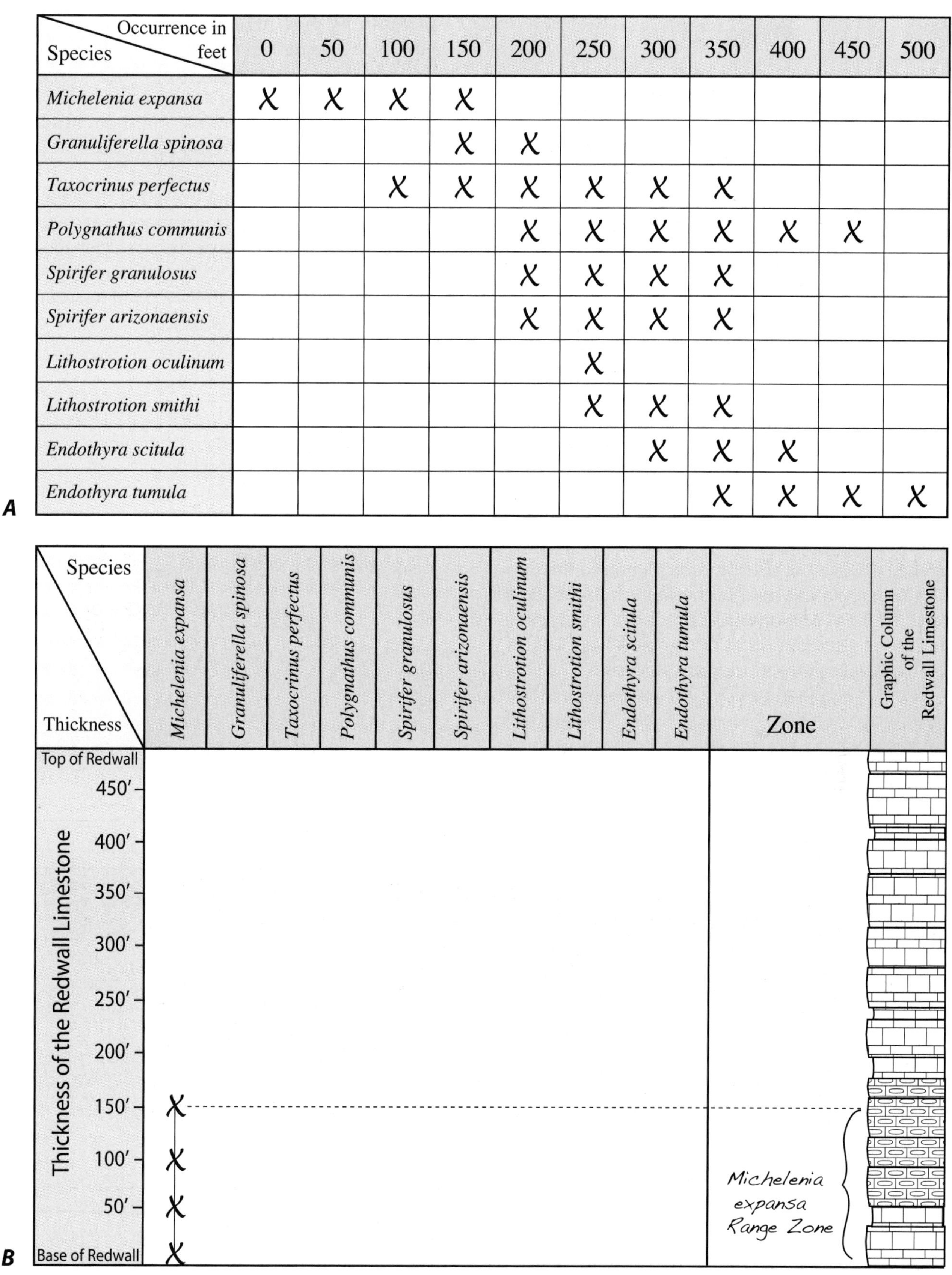

A

Species \ Occurrence in feet	0	50	100	150	200	250	300	350	400	450	500
Michelenia expansa	X	X	X	X							
Granuliferella spinosa				X	X						
Taxocrinus perfectus			X	X	X	X	X	X			
Polygnathus communis					X	X	X	X	X	X	
Spirifer granulosus					X	X	X	X			
Spirifer arizonaensis					X	X	X	X			
Lithostrotion oculinum						X					
Lithostrotion smithi						X	X	X			
Endothyra scitula							X	X	X		
Endothyra tumula								X	X	X	X

FIGURE 5.8 A. Chart showing the stratigraphic distribution of selected Redwall fossils.
B. Stratigraphic column of the Redwall Limestone.

Part D

Fossils and Graphic Correlation

Alan B. Shaw, an innovative American petroleum geologist, devised a graphic system of chronostratigraphic correlation called **graphic correlation**. The procedure is as follows: using one section of fossiliferous rocks as a standard, the measured level of the first appearance and the last appearance of each species of fossil throughout the section is carefully recorded. Similar data for all species in common at another stratigraphic section is also recorded. Data from the standard section are plotted on the horizontal, or X, axis and corresponding data from the second section are recorded on the vertical, or Y, axis of an X-Y plot. The first occurrence of each species is plotted with a dot, and the last appearance of the species is illustrated with a small box. The points will cluster along a line, called the **line of correlation**, that is approximated to fit the points. Once this line has been established, it is possible to correlate precise points from one section to the other. By accumulating data in the standard section as additional local sections are correlated, a composite standard is generated that has great power in correlating new sections as they are studied.

As an example, figure 5.9 illustrates the method of graphic correlation. Eleven species (A through K) have been plotted. On this chart, the line indicates a time correlation between the limestone bed located 50 m (arbitrary pick) above the base of section X and a shale layer situated 70 m above the base of section Y. To correlate any layer in X, simply draw a vertical line from that point to the line of correlation. Then trace a horizontal line from the point of correlation on the line of correlation to the vertical axis.

1. Using stratigraphic data provided in table 5.1A, plot the first and last appearances of each species (A through K) on one of the graph sheets provided in figure 5.10. What stratigraphic level in section X corresponds to level 55 m in section Y?

2. Repeat the plotting procedure in question 1 on one of the graph sheets provided in figure 5.10 using data provided in table 5.1B. What is the explanation for the horizontal part of the line?

Table 5.1A

	Start		End	
Species	**X**	**Y**	**X**	**Y**
A	15	5	49	23
B	21	14	94	53
C	27	12	99	58
D	16	10	80	47
E	32	18	83	54
F	42	28	91	63
G	54	24	68	38
H	53	34	79	58
I	57	32	86	51
J	69	34	85	43
K	43	23	81	51

Table 5.1B

	Start		End	
	X	**Y**	**X**	**Y**
A	14	20	42	50
B	6	6	47	50
C	21	25	36	47
D	15	13	80	50
E	22	33	85	58
F	32	31	90	54
G	7	14	75	50
H	25	26	82	53
I	70	50	90	63
J	36	40	81	52
K	40	47	86	53

Table 5.1C

	Start		End	
	X	**Y**	**X**	**Y**
A	5	14	68	71
B	4	5	40	50
D	15	24	50	58
E	25	34	46	53
F	26	38	77	73
G	33	47	56	59
H	32	42	70	68
I	37	50	58	64
J	45	55	77	77

Table 5.1 Data to plot in answering questions in part D of exercise 5.

3. Repeat the procedure again on one of the graph sheets provided in figure 5.10 using data provided in table 5.1C. What does the pattern of points (and line of correlation drawn through the points) suggest about the relative sedimentation rates between section X and Y?

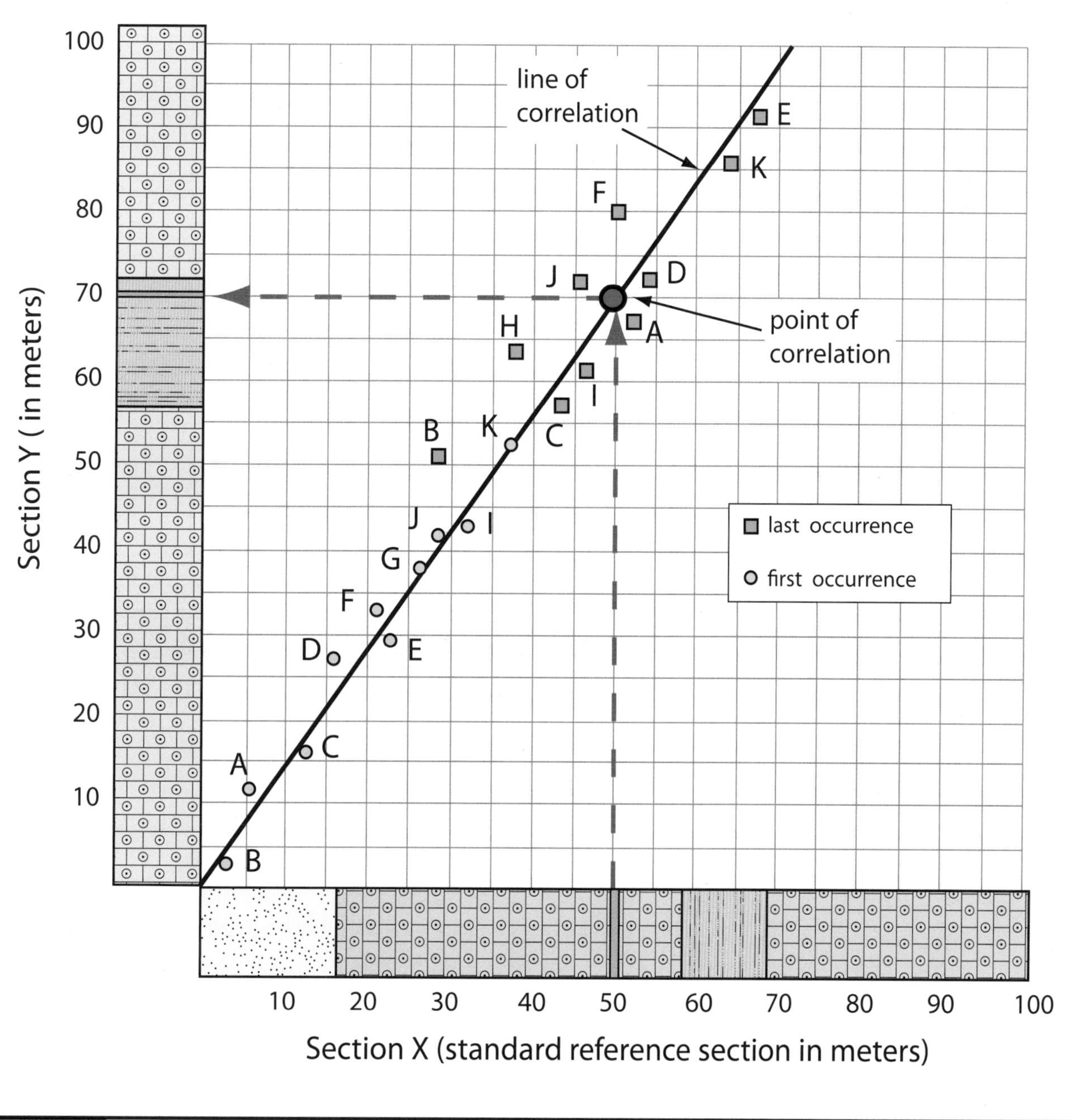

FIGURE 5.9 An illustration of the graphic correlation method of establishing chronostratigraphic equivalence between stratigraphic section X (horizontal axis) and Y (vertical axis) using hypothetical fossil data.

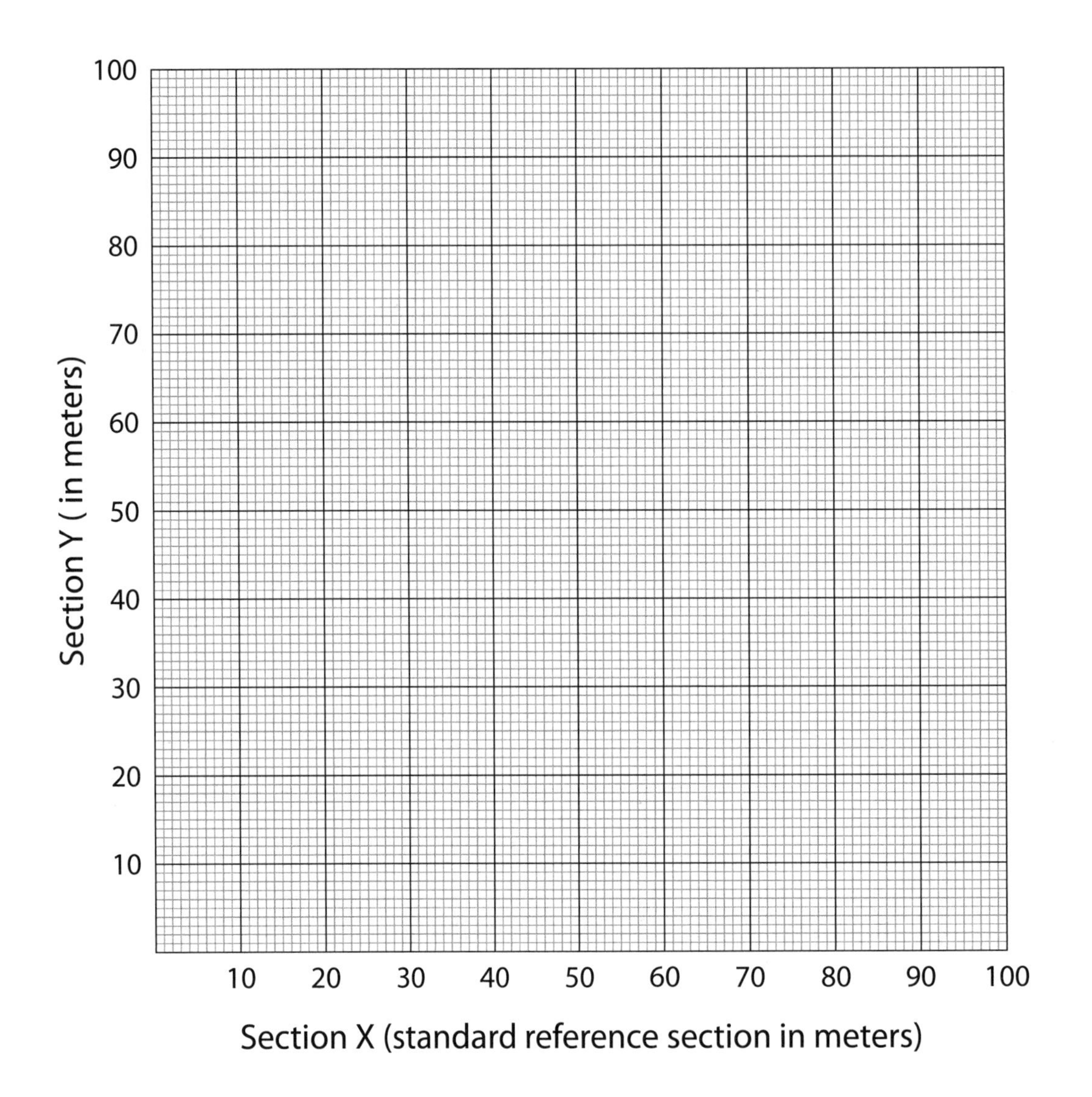

FIGURE 5.10 Charts to plot your results for part D.
Draw first occurrences using dots and last occurrences using plus signs.

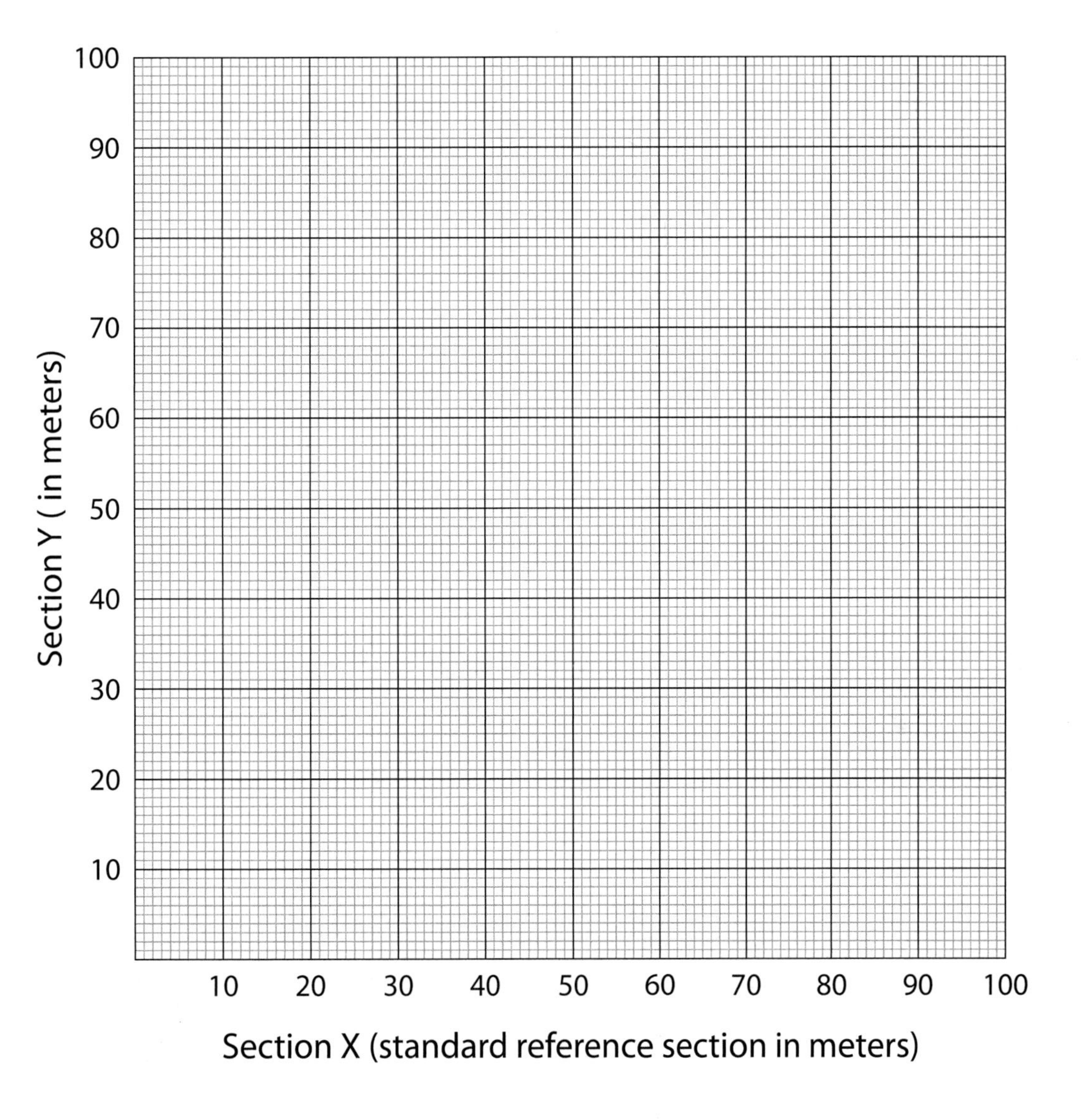
100
90
80
70
60
50
40
30
20
10
Section Y (in meters)
10 20 30 40 50 60 70 80 90 100
Section X (standard reference section in meters)

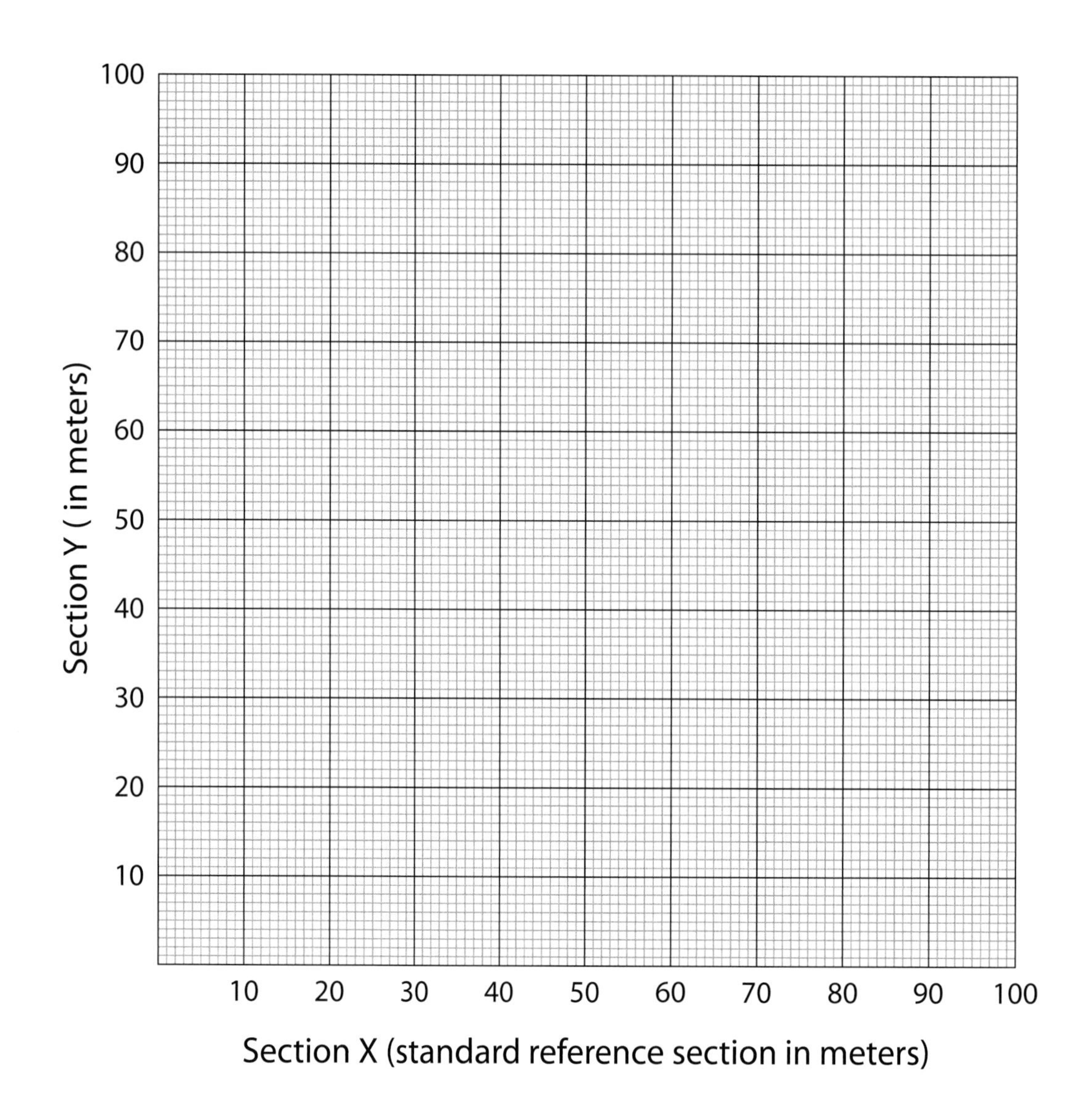
100
90
80
70
60
50
40
30
20
10
Section Y (in meters)
10
20
30
40
50
60
70
80
90
100
Section X (standard reference section in meters)

Part E
Chronostratigraphy

To underscore the concept that chronostratigraphic units are comprised of rock, draw the boundary between the "Precambrian" Eonothem and Paleozoic Erathem in the Grand Canyon using elevation data provided below the map on figure 5.11. Also draw the trace of the contacts between the Cambrian, Mississippian, Pennsylvanian, and Permian Systems. Color each of these systems a different color. Compare this map with the geological map that you drew previously in part B.

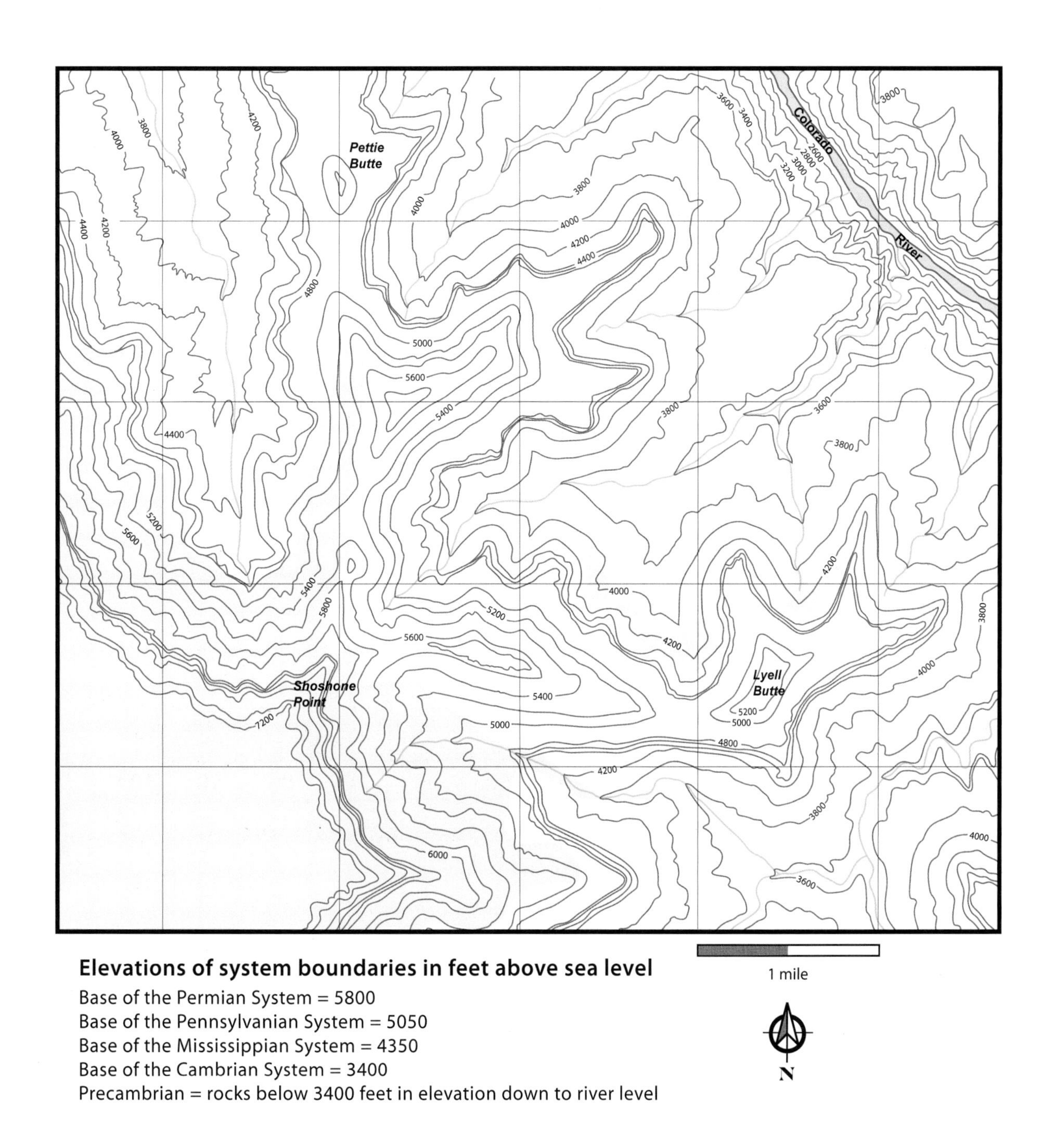

Figure 5.11 Topographic map of the Shoshone Point area, eastern Grand Canyon, with topographic positions of system boundaries listed.

PART F

Chronostratigraphy: Global Stratotype Section and Point

The International Commission on Stratigraphy maintains a website at http://www.stratigraphy.org. The stated objective of the commission is to precisely define global units (systems, series, and stages) on the International Chronostratigraphic Chart, thus setting global standards for the fundamental scale for expressing Earth history. The site defines the nature and composition of the commission and provides up-to-date information on progress toward the stated goals. Take a few minutes to familiarize yourself with the website, then click on the link to GSSPs, which will bring up the GSSP Table—All Periods with the following headings: Stage, Numerical Age (Ma), GSSP Location, Latitude and Longitude, Boundary Level, Correlation Events, Status, and Reference.

Using the GSSP Table, find the data that is missing from table 5.2 for this exercise. The purpose of this activity is to familiarize you with this informative website and to help you gain a better understanding of how and where chronostratigraphic boundaries are defined. For example, the base of the Devonian System is defined by the **Graptolite FAD** of *Monograptus uniformis*. FAD is the acronym for first appearance datum, or the first stratigraphic occurrence of the graptolite species of that name. Also note that the base of the Devonian System is the same as the base of the Lochkovian Stage (lowest stage of the Devonian System). Likewise, the base of each system is identical to the base of its lowest stage and is defined by the same fossil species.

refer to www.stratigraphy.org	Geographic and stratigraphic location of GSSP at base of system	Year in which the GSSP was established	Fossil species or feature that defines the base of the system
Neogene			
Paleogene			
Cretaceous			
Jurassic			
Triassic			
Permian			
Pennsylvanian			
Mississippian			
Devonian			
Silurian			
Ordovician			
Cambrian			

TABLE 5.2 Stratigraphic table for part F.

Exercise 6

Physic l Correlation

Learning Objectives

After completing this exercise, you will be able to:

1. define the meaning of geological correlation;
2. explain the difference between chronostratigraphic and lithostratigraphic equivalence; and
3. discuss a number of techniques used by geologists to correlate geologic data from one area to another.

To **correlate** means to make equal or to establish comparable relationships. To correlate in a geologic sense is to establish either contemporaneity or continuity; that is, rock units may be equivalent in terms of time (chronostratigraphy) or in terms of being part of the same rock body (physical correlation or lithostratigraphy). Under unusual circumstances, rocks may be equivalent in both time and lithology, but, generally speaking, one correlates rock units either in terms of time or lithology, not both.

Correlation is critical to understanding historical geology, because with it we can establish events in relationship to one another. Thus, one can gradually build a geologic column or a sequence of events and ultimately develop a geologic history. This exercise concerns correlation of rock units to establish lateral continuity or change.

Physical correlation denotes the use of physical criteria to match rock units. The goal of physical correlation is to establish the geographic extent, or continuity, of particular rock units. Physical correlation can be accomplished by any one or more of the following methods:

1. Lateral continuity
2. Lithologic similarity
3. Sequence of beds
4. Geophysical characteristics (seismic, electrical, sonic, radioactive, or magnetic properties)
5. Sedimentary sequences

We can demonstrate physical correlation by **lateral continuity** of single laminae or beds that can be traced from area to area. Single beds might be typified by ash falls, which occur geologically at a single instant over a wide area and hence offer ideal evidence of contemporaneity. The 1980 eruption of Mount St. Helens is a modern example of such an event. Volcanic ash was essentially instantaneously deposited over an enormous area extending from Washington to, after four days, the East Coast. We may correlate on **lithologic similarity** where unique kinds of rocks may be recognized in various areas and correlated. In the Colorado Plateau region of western North America, for example, the Navajo Sandstone is a distinctive, white, cross-bedded sandstone that retains many of its general characteristics over a vast area, even though the rocks above and below change somewhat in their composition.

We may correlate on the basis of a **sequence of beds** of distinctive lithologic units. Such correlations are more reliable than correlating on lithologic similarity alone. Statistically a sequence of beds is more reliable as an indicator of lateral continuity, particularly if the sequence exhibits a distinctive relationship clearly separating it from other rock units in the local area.

We may also correlate on various **geophysical characteristics** of rock sequences such as the electrical, radioactive, or sonic characteristics of any given sequence. In many areas where subsurface geology is the primary exploration tool for petroleum, the rock sequence in most of the wells has been correlated by electrical, sonic, or radioactive properties. For example, a very weak current is generated where drilling fluid comes into contact with various formations. One can plot, as a curve or log against depth, the strength of this spontaneous-potential or self-generated current. This curve is distinctive in many instances. A curve also can be plotted of the depth against the resistivity of the various formations to an induced current. These curves can be compared and correlated even though the lithology of the sequence may be poorly known. A curve of natural radioactivity of the sequence is easily compiled. Such curves have proven very useful in many areas. Sonic logs, or curves that record variations in velocity of sound transmitted by individual rock layers, are also useful tools in correlation. Logs of each of these various physical properties, as well as others, are commonly composed of distinctive curves and may be correlated from region to region, based on unique points in the patterns.

Paleomagnetic effects offer still another method of geophysically based physical correlation. The discovery of variations in the intensity or strength of the Earth's magnetic field over the past 160 million years has allowed determination of a series of major epochs of normal and reversed polarity that permit correlation on a worldwide scale. When magnetic, iron-bearing minerals crystallize, or when they settle slowly out of the transporting medium, individual mineral grains align themselves with the existing magnetic field of the Earth. This produces a differential of direction of magnetism in various rocks, which is termed **remanent magnetism**. By measuring polarity, variations in intensities, and directions of inclination through a sequence of rocks or sediments, the sequences can be correlated on the basis of their paleomagnetic history. Remanent magnetic characteristics are particularly useful because they allow correlation between marine and nonmarine sediments, between volcanic rocks and sedimentary rocks, between very unlike sequences in widely separated areas, and, owing to their frequency, they provide a means of worldwide, nearly instantaneous, correlation. Figure 6.1 is a graph showing the history of polarity reversals in the intensity or strength of Earth's magnetic field for approximately the last 5 million years. The bar graph shows four major epochs of reversed and normal polarity beginning with the Gilbert Reversal Epoch approximately 5 million years ago. During the following Gauss Normal Epoch the polarity was the same as that of modern times. The Matuyama Reversal Epoch was of long duration and preceded the present Brunhes Normal polarity.

During the 1960s, paleomagnetists unraveled a 4-million-year record of a succession of reversals in the polarity of the Earth's magnetic field. The discovery of this phenomenon dates back to 1929 with the work of Professor Motonori Matuyama at the Kyoto Imperial University in Japan. With refinement in the potassium-argon dating method, precise dates could be obtained for relatively young (Pliocene and Pleistocene) volcanic rocks, allowing the further development of the geomagnetic-reversal time scale. More recent work has extended

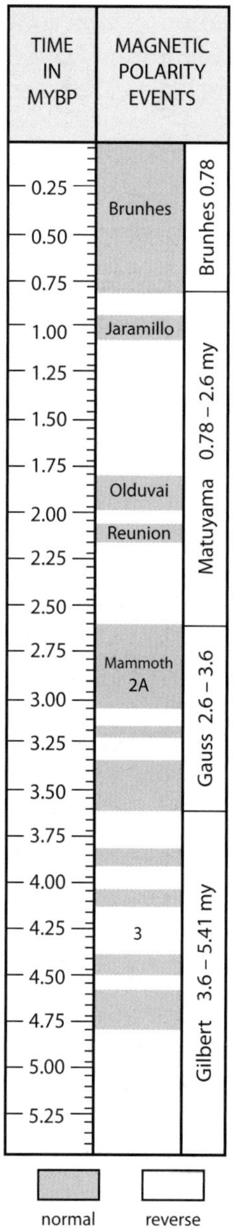

TIME IN MYBP
MAGNETIC POLARITY EVENTS
0.25
0.50
0.75
1.00
1.25
1.50
1.75
2.00
2.25
2.50
2.75
3.00
3.25
3.50
3.75
4.00
4.25
4.50
4.75
5.00
5.25
Brunhes
Jaramillo
Olduvai
Reunion
Mammoth 2A
3
Brunhes 0.78
Matuyama 0.78 – 2.6 my
Gauss 2.6 – 3.6
Gilbert 3.6 – 5.41 my
normal
reverse

FIGURE 6.1

A graph of paleomagnetic polarity reversals for the past 5 million years. *Four* major epochs can be recognized along with *four* minor events.

the polar reversal chronology record back 160 million years, using fossils recovered in deep-sea cores as the basis for the ages of the rock.

Three short-duration events interrupt the major oscillation of the epochs. The Mammoth Event took place approximately 3 million years ago, the Olduvai Event approximately 2 million years ago, and the Jaramillo Event approximately 1 million years ago. Utilizing the paleomagnetic history, a polar reversal chronology has been developed that can be utilized worldwide, even in sequences that greatly contrast in lithology.

Superimposed cyclic patterns in the stratigraphic record of marine rocks seem to reflect global climatic changes recorded as eustatic, or worldwide, synchronous, short-term changes in sea level, modified by longer-term tectonic changes, sediment supply, and climate. The interpretation and application of this concept leading to a new method of subdividing, correlating, and mapping sedimentary rocks is called **sequence stratigraphy**. Recognition of unconformity-bounded stratigraphic units called **sequences** has been in practice in geology for many years, largely beginning with the work of Professor L. L. Sloss of Northwestern University in the late 1940s. A renewed interest in this method of interpreting marine strata evolved from **seismic stratigraphy** methods developed by P. R. Vail (a former graduate student at Northwestern) and others at Exxon Research during the 1970s.

Fundamental to the methods of sequence stratigraphy is the assumption that for all practical purposes, the seismic reflectors (figure 6.2) used in identifying sequence units are time lines; thus, sequence stratigraphy is a chronostratigraphic method. A sequence represents one complete transgressive-regressive cycle of sea level, forming a sedimentary deposit representing a single genetic package bounded by unconformities, thus bounded by time lines. As sea level falls, the platform is eroded, and deposition takes place in the basin. When sea level rises abruptly, flooding the platform,

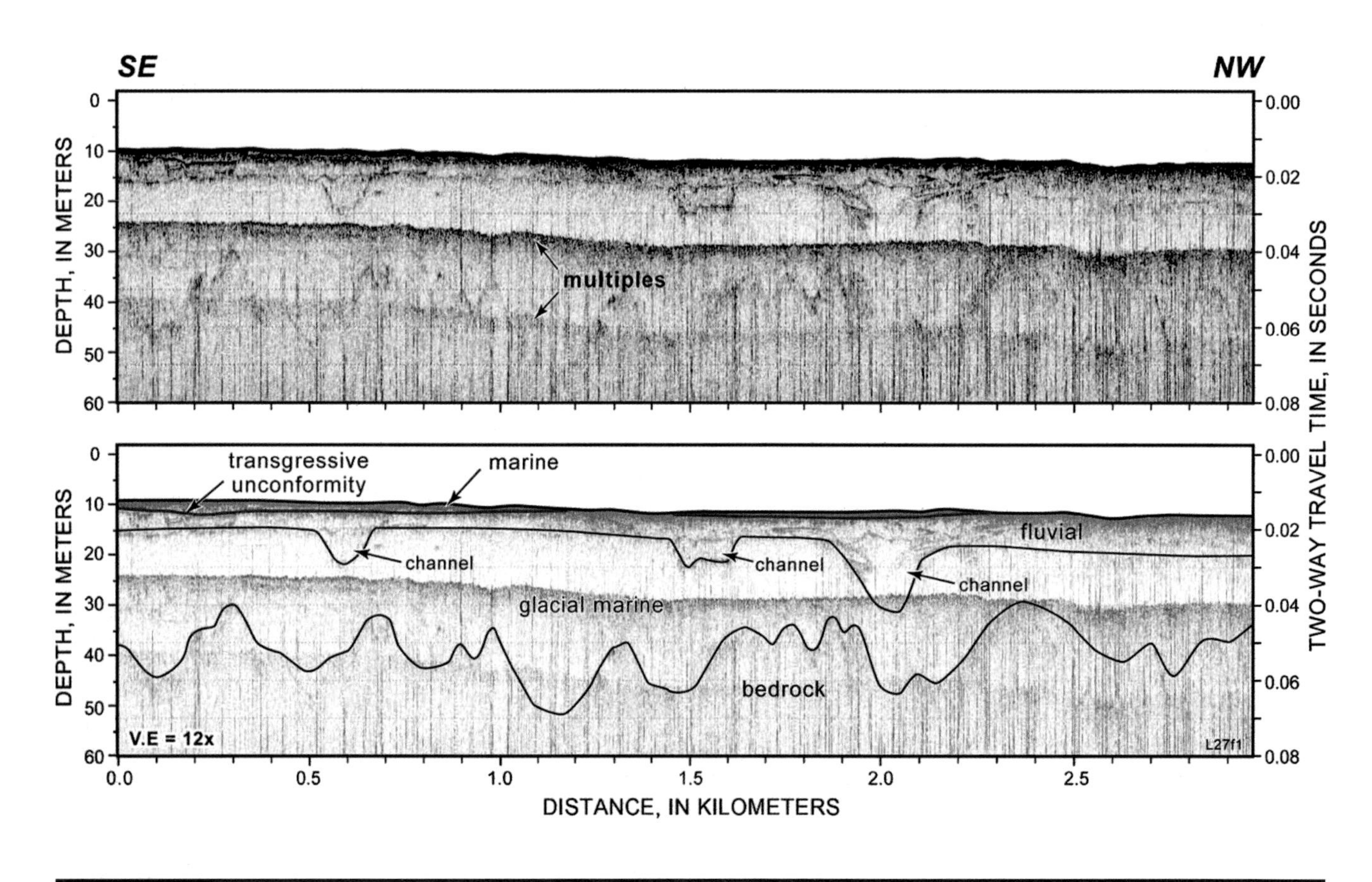

FIGURE 6.2 Seismic-reflection profile showing three small channels cut into glacial-marine sediment in the central part of the study area. The transgressive unconformity (red line) is eroded into the upper surface of fluvial deposits and locally overlain by a sheet of sandy marine sediment generally less than 1 m thick. A constant seismic velocity of 1,500 m/s through water, sediment, and rock was used to convert from two-way travel time to depth. (Barnhardt et al., 2009.)

onlap or transgression occurs as the water progressively deepens and sedimentation takes place more and more landward. At the highstand of flooding, there is practically no sedimentation accumulating in the basin, forming a condensed interval of sediment. As sea level again falls, the sediment progrades toward the basin, forming a downlap surface in the accumulated sediments.

A complex terminology, developed by Peter Vail and his associates, must be mastered by those employing sequence stratigraphic methods. For the sake of introducing the concept of sequence stratigraphy, we will ignore the terminology and observe stratigraphic sequences to explore major sedimentary cycles.

Sedimentary cycles range in duration from approximately 20,000 years to 150 million years and occur as superimposed orders of cyclicity (table 6.1). The mechanism for changing sea level within the interval 20,000 to 500,000 years is thought by some geologists to derive from the Earth's orbital perturbations, the same forces often cited as the mechanisms that drove Pleistocene cyclic patterns.

Vail and his associates have published what have become known as "Vail Curves," which illustrate changes in sea level through time. Figure 6.3 is an example of a Vail Curve for part of the Late Cenozoic (Upper Eocene through Holocene).

PROCEDURE

PART A

The Colorado Plateau of southwest Colorado, southeast Utah, northwest New Mexico, and southeast Arizona contains some of the most spectacular exposures of sedimentary strata in the world. Outcrops and landforms sculpted from Paleozoic and Mesozoic strata by the Colorado River system are on display at Grand Canyon, Zion, Bryce Canyon, Canyonlands, and Arches National Parks as well as at Glen Canyon National Recreation Area and several state and local parks. Figure 6.4 shows a representative sample of these world famous Late Paleozoic and Mesozoic formations at four Utah localities shown on the index map.

1. Using the formation contacts and formation names provided, construct lines of lithostratigraphic (formation) equivalence between the four sections. After correlating the four sections, complete the following questions.

2. List the names (or rock types) and ages of rocks resting directly below the Chinle Formation in each of the stratigraphic sections.

Sequence Terminology	Approximate Duration	Examples	Amplitude (m)	Rise/Fall Rate (cm/1,000 yr)
1st Order	> 100 million years	Sequences of Sloss	—	< 1
2nd Order	10–100 million years	Supersequences	50–100	1–3
3rd Order	1–10 million years	Basic Sequence	50–100	1–10
4th Order	100,000–1 million years	—	1–150	40–500
5th Order	10,000–100,000 years	Parasequences	1–150	600–700

TABLE 6.1 A classification of the units used in sequence stratigraphic studies.

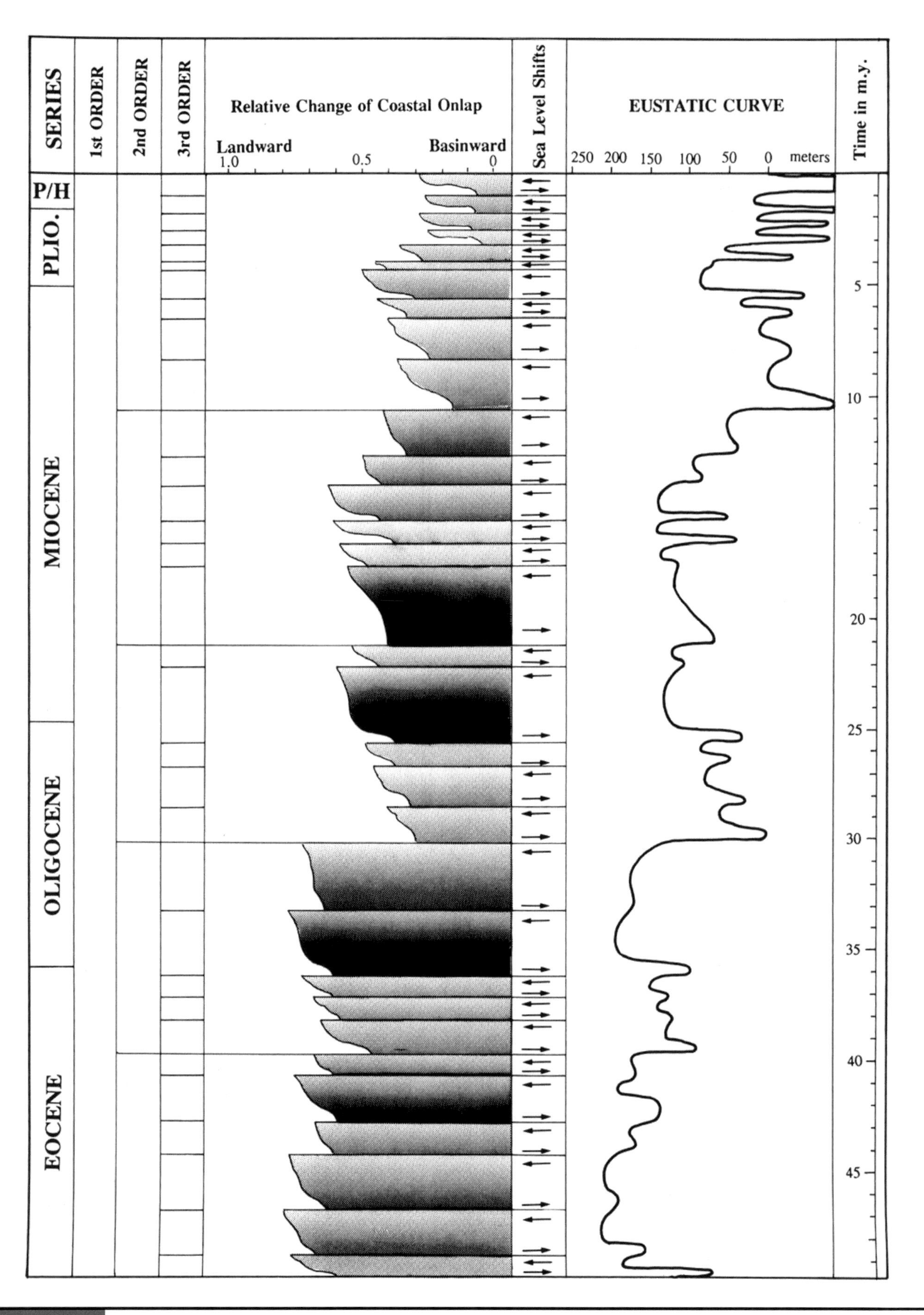

FIGURE 6.3 Eustatic cycle chart (Vail Curve) for the Late Cenozoic (Upper Eocene through Holocene). (After Haq et al., 1988.)

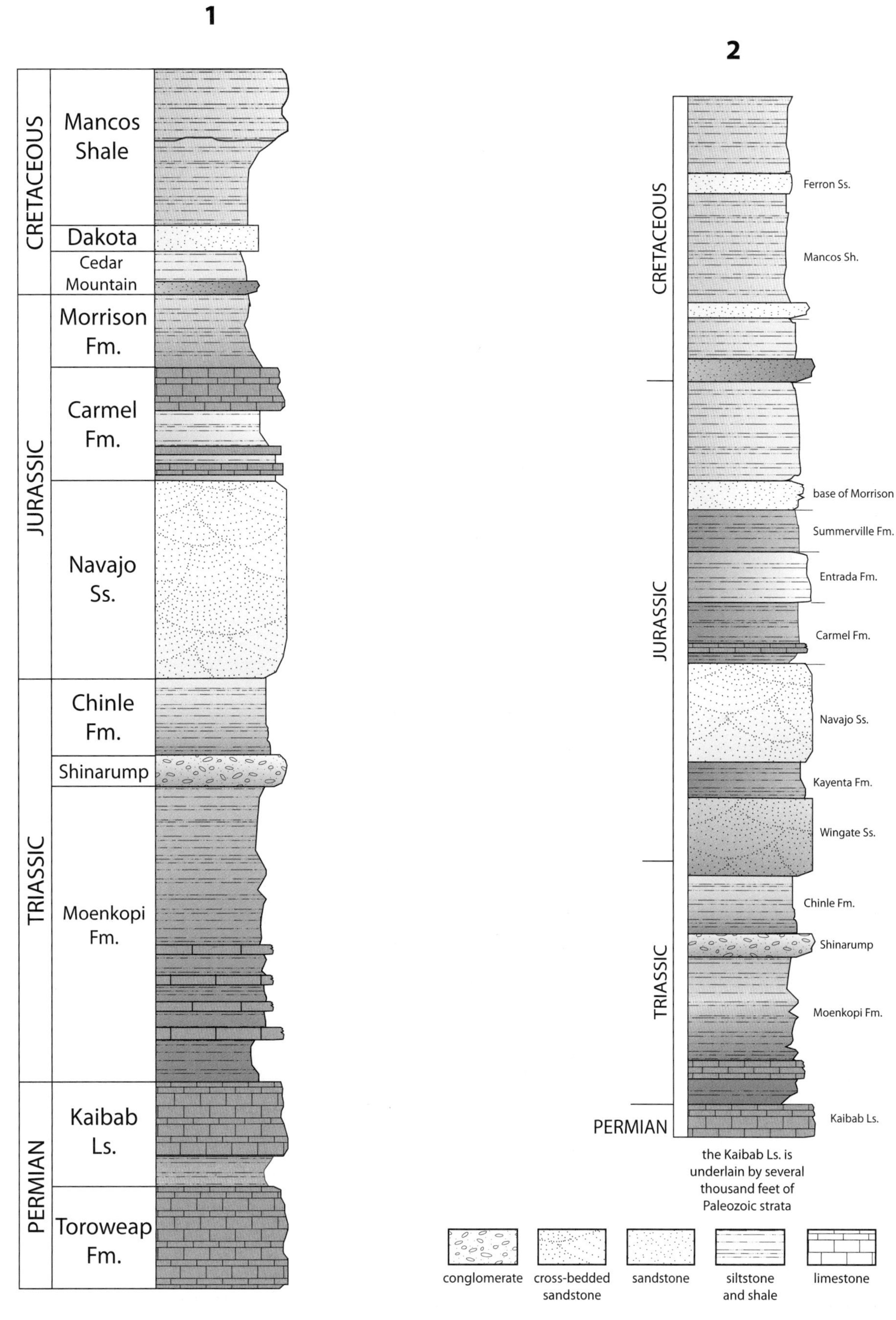

FIGURE 6.4 Four columns of sedimentary rocks exposed in the Wasatch Mountains (1) and Colorado Plateau (2, 3, 4), Utah. *(continued)*

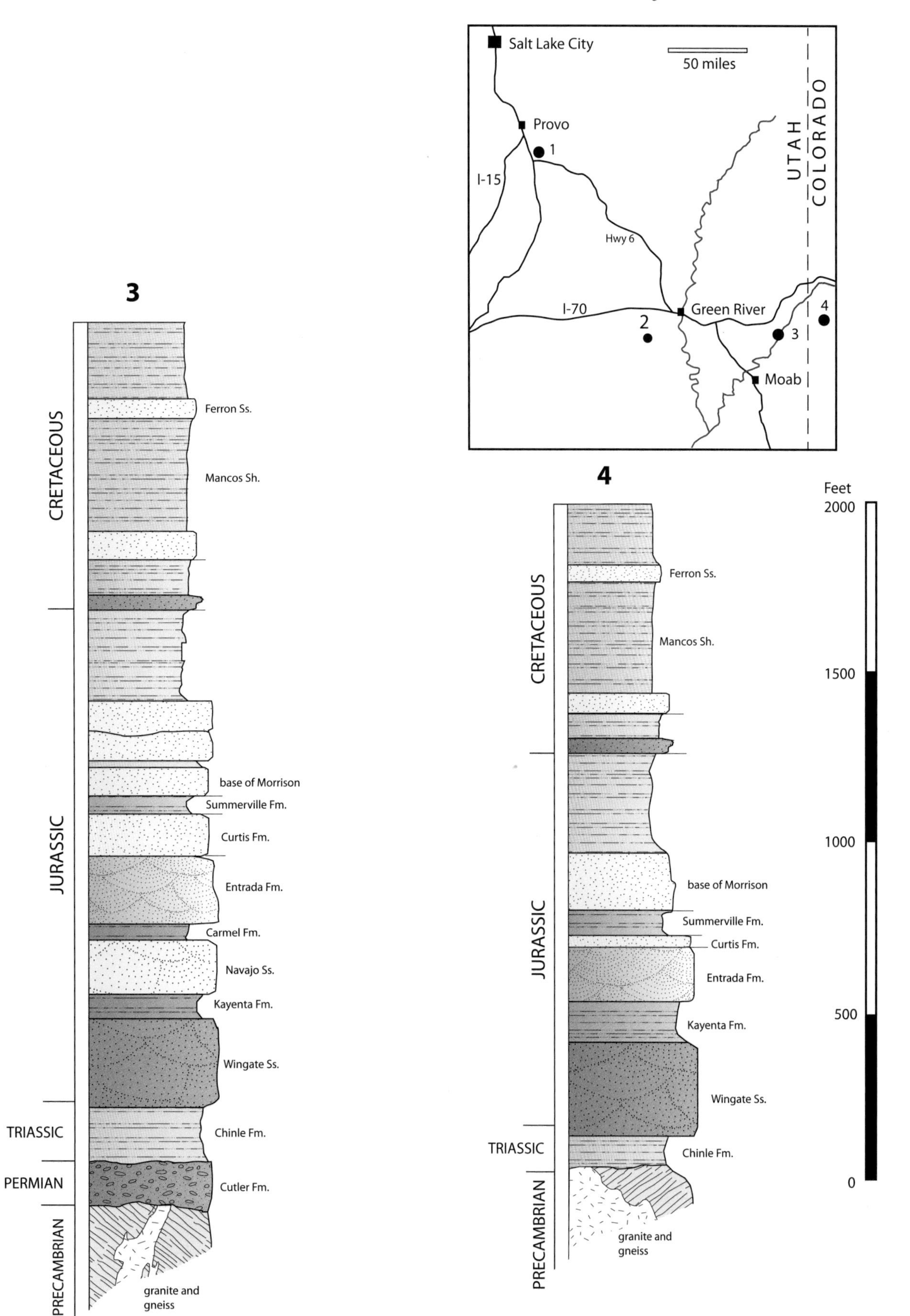

FIGURE 6.4 Four columns of sedimentary rocks exposed in the Wasatch Mountains (1) and Colorado Plateau (2, 3, 4), Utah.

3. How do you account for the differences in thickness and lithology of rocks below the Chinle Formation across the region?

4. Which of the four sections were subjected to significant uplift and erosion during Late Paleozoic and Early Triassic time? Cite evidence.

5. How do you account for regional differences in thickness and/or lithology of the

   a. Navajo Sandstone?

   b. Entrada Formation?

   c. Carmel Formation?

PART B

Figure 6.5 illustrates three electric logs of wells drilled in southern Arkansas through Upper Cretaceous and Tertiary beds. The datum, or reference plane, is a level plane 1,800 ft below sea level. The lithology and the names of the formations to be correlated are shown in the left-hand section. The two curves of electrical characteristics are the principal means of correlation. They can be correlated on the basis of very distinctive "kicks" or irregularities on the electric log. The curve of spontaneous-potential (S.P.) is to the left and the curve of resistivity (Res.) is on the right. Spontaneous-potential and resistivity curves for shale sections, in general, fall close to the median lines. Sandstone beds are more permeable and porous and contain freshwater, salt water, or natural hydrocarbons. Their curves swing away from the center lines on both resistivity and spontaneous-potential curves.

Marls or limestones appear much like slightly sandy shales on the spontaneous-potential curve, but swing away from the shale line on the resistivity curves. Footages below the derrick floor are shown by the tic marks and the depths along the left side of each of the logs.

1. Correlate the electric logs of the three wells using as many points as applicable.

2. These wells are approximately 10 km apart. Is this an area of strong folding or tilting of the sedimentary beds? How can you tell?

3. Is there evidence of thinning or wedging of beds at any particular stratigraphic level?

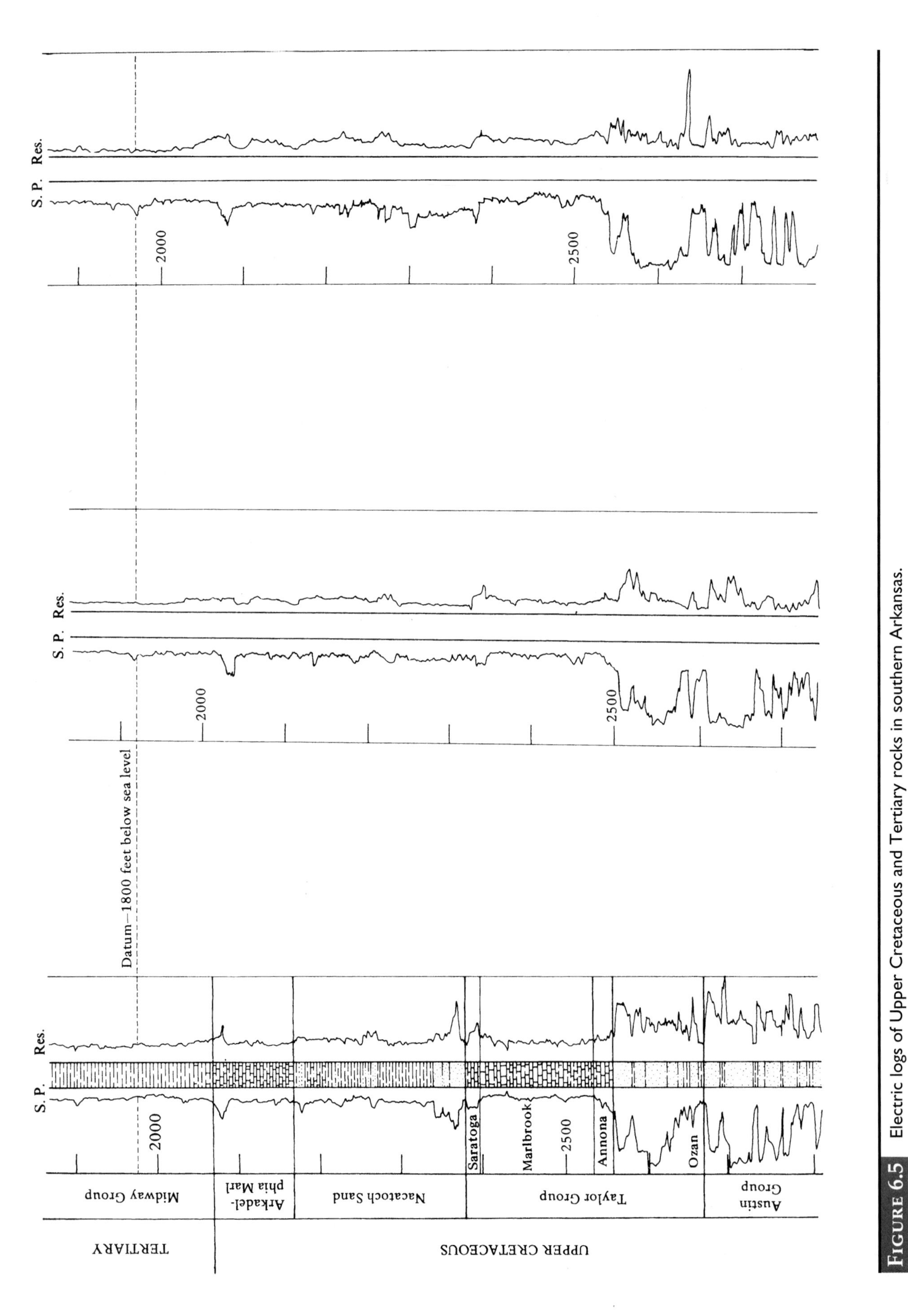

FIGURE 6.5 Electric logs of Upper Cretaceous and Tertiary rocks in southern Arkansas.

PART C

Another method of physical correlation is related to the properties of transmittal and reflection of seismic or earthquake energy through Earth's layered crust. Early in the history of geophysics, it became apparent that energy pulses generated at the surface penetrated the upper crust, with part of the energy being reflected back from horizons where markedly differing seismic properties were encountered. The energy appears much like light reflected through a series of stacked glass plates, with some energy reflected and some transmitted through each of the plates (figure 6.6). If sudden pulses of energy are generated by explosions of dynamite or gas or by vibrating a heavy weight on Earth's surface, the energy is transmitted into Earth's crust and part is reflected back at each reflecting horizon encountered at depth. This energy is picked up by a **seismometer**, an instrument that is sensitive to ground movement, and plotted as a record or curve of Earth's motion on a **seismogram**. Some horizons reflect more energy than others and produce larger motion in the seismometer. Such beds are most useful in seismic studies and appear in records as high-level reflecting horizons. If the elapsed time between initiation of the primary energy source and arrival back at the surface of the reflected energy is plotted, the relative depths and velocity of the energy pulses through the rocks or sediments can be established. If the seismometer is moved, a series of seismograms are produced with distinctive horizons of peaks of high reflectance.

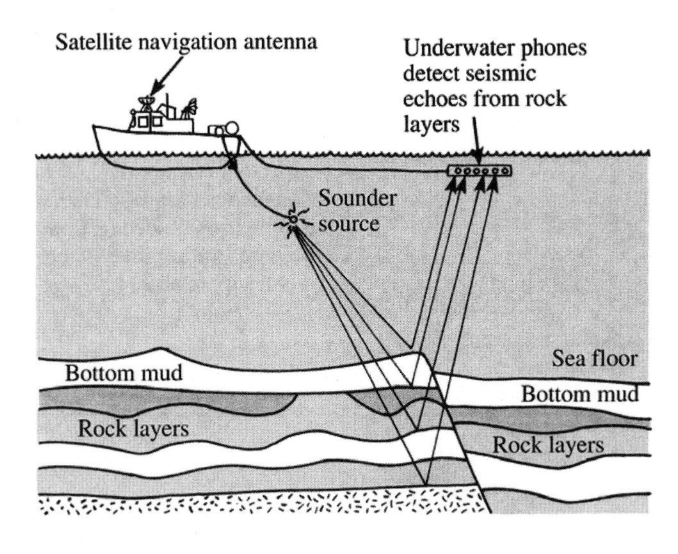

FIGURE 6.6 Seismic data collecting in offshore oil and gas exploration.

This series can be arranged into a **profile** such as the one shown in figure 6.7. In figure 6.7, the horizons of high reflectance are shown as relatively light zones on the numerous seismic curves.

1. On figure 6.7 correlate key reflecting horizons across the seismic profile by highlighting them with a felt-tip pen or colored pencils. Show faults or folds where these are evidenced by offsets or flexures on the various reflecting horizons.

2. What methods of correlation did you use as you worked the problem? (Review the list of methods under physical correlation at the beginning of this exercise.)

3. What is the explanation for the poor definition of the reflecting surfaces in the center of the cross section?

4. Where is the most likely place to drill for oil in this section?

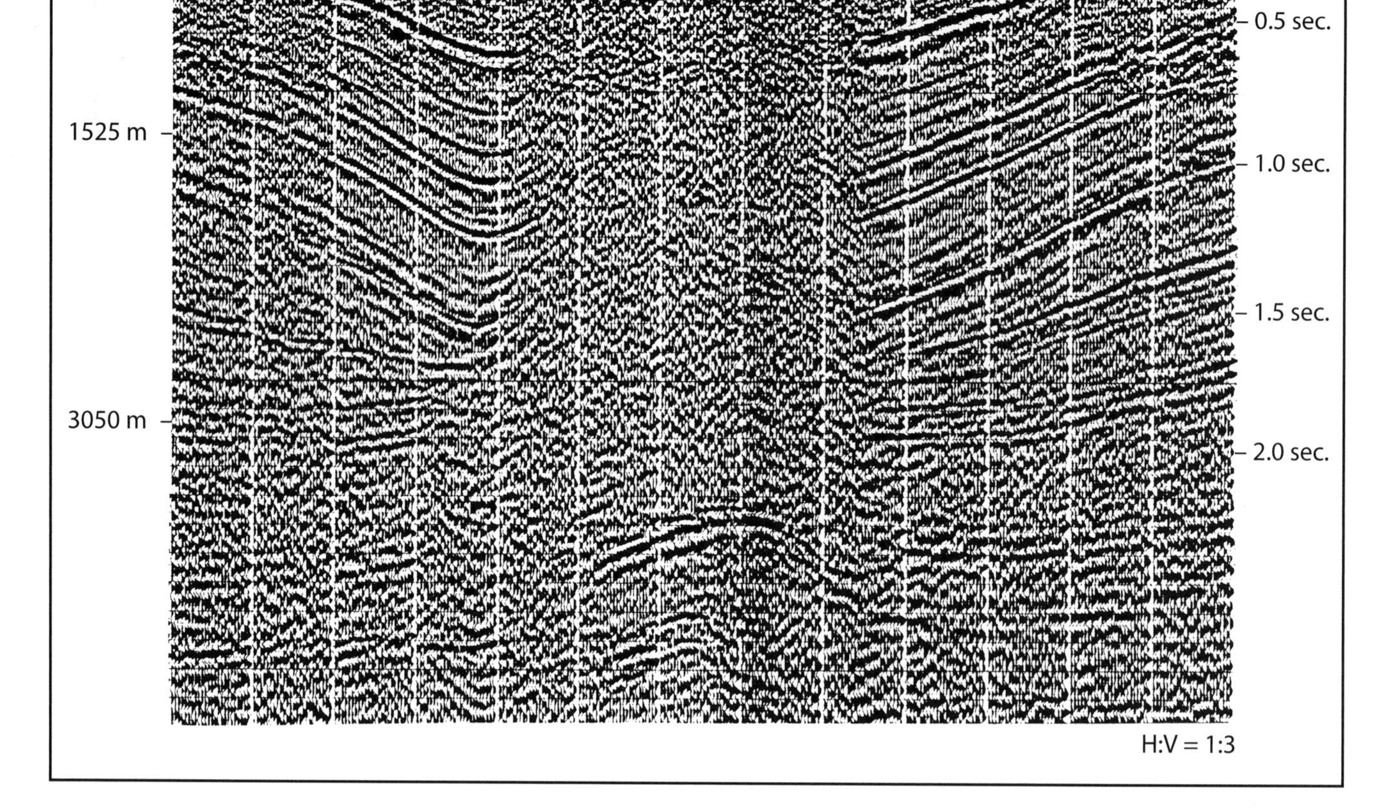

FIGURE 6.7 Seismic reflection profile from the Gulf Coast of Texas.

Facies Relationships and Sea-Level Change

Exercise 7

Learning Objectives

After completing this exercise, you will be able to:

1. define the concept of sedimentary facies;
2. discuss the meaning and significance of Walther's law;
3. understand how changes in sea level cause facies belts to move onshore during transgression and offshore during regression; and
4. correlate sedimentary strata between adjacent outcrops of siliciclastic and carbonate strata.

The term **facies** is used in many ways in geologic literature. In a broad sense, it refers to any aspect of a rock, including appearance, composition, and environment of formation, and any lateral change, or variation, in these attributes over a geographic area, all of which reflect the environment under which the rock was originally deposited. You used facies interpretation in completing exercise 4, the exercise on depositional environments. Facies is also used to denote essentially contemporaneous rocks of different lithology, or type, caused by environmental differences; for example, a rock body of limestone might be the time equivalent of an adjacent and interfingering formation of shale (figures 7.1, 7.2, and 7.3). Such an association is commonly developed in near-shore marine environments.

Facies changes in sedimentary rocks are commonly the result of variations in sedimentary environments during deposition. If one were to look at the continental shelf off any of the present coastlines, one would see considerable variation in the kinds of sediments being deposited: e.g., coarse gravel or sand along the beach zone, grading offshore into silt and clay. In some areas, organisms produce carbonate masses, or reef-like structures. If these same sedimentary belts persist through time, accumulations of similar sediments would be deposited that might be classed as sedimentary facies (figure 7.1), or as **lithofacies** in contrast with the depositional pattern associated with particular biologic groups, which are termed **biofacies**.

Biofacies or lithofacies developments are the result of sediments or organisms responding to a variety of factors in the environment, such as texture of substrate, chemistry (in terms of the amount of oxygen and carbon dioxide available), salinity, water turbulence, and turbidity (the amount of suspended sediment). These as well as other factors influence the characteristics of the sedimentary and biologic association that we see in the geologic record as facies.

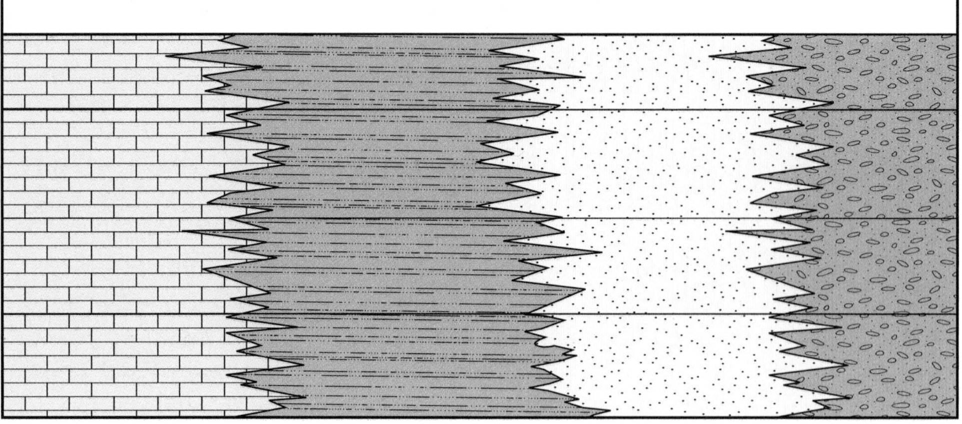

FIGURE 7.1

Schematic cross section showing facies relationships under static conditions of subsidence and sediment supply in a marine environment. The dark horizontal lines delineate synchronous units. Sediment was derived from the area to the right of the depositional site illustrated. The primary cause of sorting, as determined by coarser grain size, is water depth, or distance from shore.

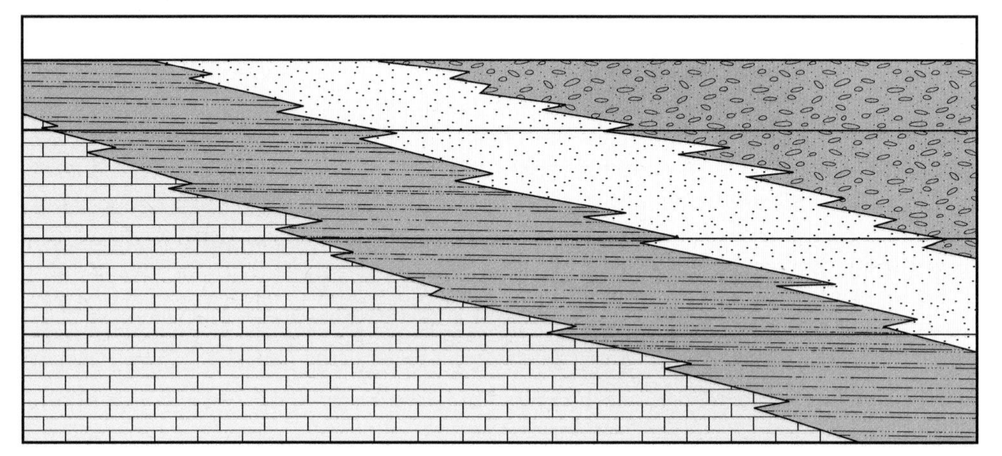

FIGURE 7.2

Schematic cross section showing facies variations produced by fluctuations in the amount and kind of sediment deposited under uniform conditions of subsidence. The sediment source area is to the right of this depositional site. Variation seen here would probably be caused by an uplift of the source area with an attendant increase in particle size and quantity of transported sediment or by a drop in sea level.

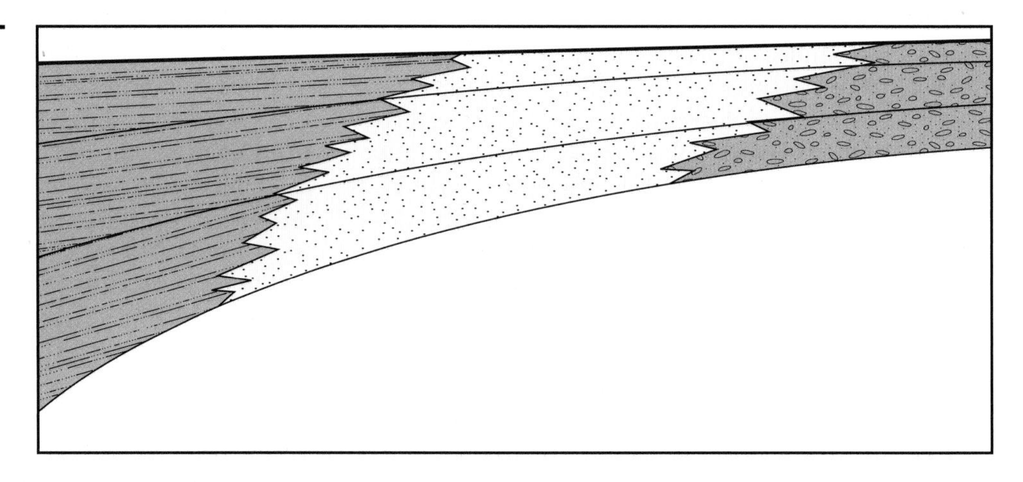

FIGURE 7.3

Schematic cross section showing facies relationships under conditions of varying subsidence and sediment supply. The curved lines represent time lines whose positions were determined by fossils. Source area for these sediments is to the right. Subsidence rates are higher at the left than at the right, creating more accommodation and a thicker accumulation of sediment.

Early geologists who studied the rocks of Europe and North America thought primarily in terms of uniform layers of rocks that extended virtually worldwide and that were deposited from a primeval sea. They had complex interpretations to explain areas where such supposed worldwide uniform patterns were not consistent. Near the turn of the nineteenth century, lateral variations in contemporaneous strata were recognized, and a new study of facies relationships began.

If all aspects of an environment remained constant during some period in geologic time, a series of vertical belts would develop, such as illustrated in figure 7.1. Depth of water, composition and texture of sediments, rate of sedimentation, rate of subsidence, and other factors in this example have been held constant, building a series of sedimentary rocks with relatively constant patterns. Such constancy in the geologic record is rare, however, and generally the environment varies, resulting in a lateral shift in the kinds of rocks that are produced, as illustrated in figure 7.2. In this relationship, the facies belts, or areas of sedimentation of various types of sedimentary rocks, have migrated toward the left through time as a result of increase in the coarseness of the incoming sediment and of the amount of material being transported into the sedimentary basin. The rate of sedimentation was more rapid than the rate of subsidence, resulting in filling of the sedimentary basin on one side and ultimately crowding the shoreline to the left, expanding the landmass at the expense of the area covered by the sea. The thickness of sediment deposited during any one time interval, as shown in figures 7.1 and 7.2, remained relatively constant. There was no development of a thick wedge of clastic sediment toward the source area.

If the rate of sedimentation was influenced by rapid subsidence in the area of fine sediments, a marked thickening of the sediments deposited during any time interval should result, such as shown in figure 7.3. In the region to the left, the rate of sedimentation was more rapid than to the right and a thicker sequence of sediments accumulated. This is shown by a divergence of the time lines. In facies studies, therefore, one can learn something concerning the relative rates of sedimentation, direction of transport, and capacity of the transporting medium.

Figure 7.4 is a restored cross section of Cambrian rocks visible in walls of the Grand Canyon and shows an intertonguing relationship of dolomite and limestone in the west (left) with an eastern belt of shale. Time lines, established by the occurrence of distinctive fossils, are shown as heavily dashed, approximately horizontal lines. The locations of measured stratigraphic sections are shown by the heavily dashed, vertical lines. In the central part of the cross section, limestones can be seen grading laterally into thin tongues of dolomite that interfinger with shales in the upper part of the Bright Angel Shale. Intertonguing relationships such as these suggest very strongly that the limestones are time equivalents of the thin dolomite beds, and that the dolomite beds are the lateral temporal equivalents of the shale.

To the east, beyond the limit of the cross section, even these intermediate and upper beds grade into sandstone. The pattern is much like that shown in the lower part of the measured sections to the west, where Bright Angel Shale is demonstrated to be contemporaneous, or to have been deposited at the same time, as the beds of the Tapeats Sandstone.

In the exercise on physical correlation, it was emphasized that we correlate either on the basis of equivalence in terms of time or in terms of continuous rock bodies. In figure 7.4, the term Tapeats Sandstone is applied to all of the sandstone at the base of the Cambrian section, although beds in the eastern part of the canyon are younger than those in the west. On the basis of lateral continuity, stratigraphic position, and lithology, the Tapeats Sandstone beds in the west are correlated to the east as part of a continuous sandstone body, even though of slightly differing ages. One can speak of the sandstone facies at the base of the Cambrian rocks in contrast to the shale facies that overlies it, and in turn to the dolomite and limestone facies represented in the younger rocks.

To some extent, the Cambrian rocks demonstrate **Walther's law** or **principle**, which is: Those facies that occur adjacent to each other at a given moment in time can be superimposed. For example, the Bright Angel Shale grades laterally into the Tapeats Sandstone and also overlies it. Similarly, in the central part of the section, dolomite and shale units are interbedded and grade laterally into one another. Although not infallible, the general observation holds that various kinds of rocks that are superjacent to one another in a stratigraphic sequence also grade laterally into one another along the outcrop band unless it is interrupted by some external event. This relationship can be seen in shale, sandstone, and conglomerate sequences, as well as in limestone, dolomite, and evaporite sequences.

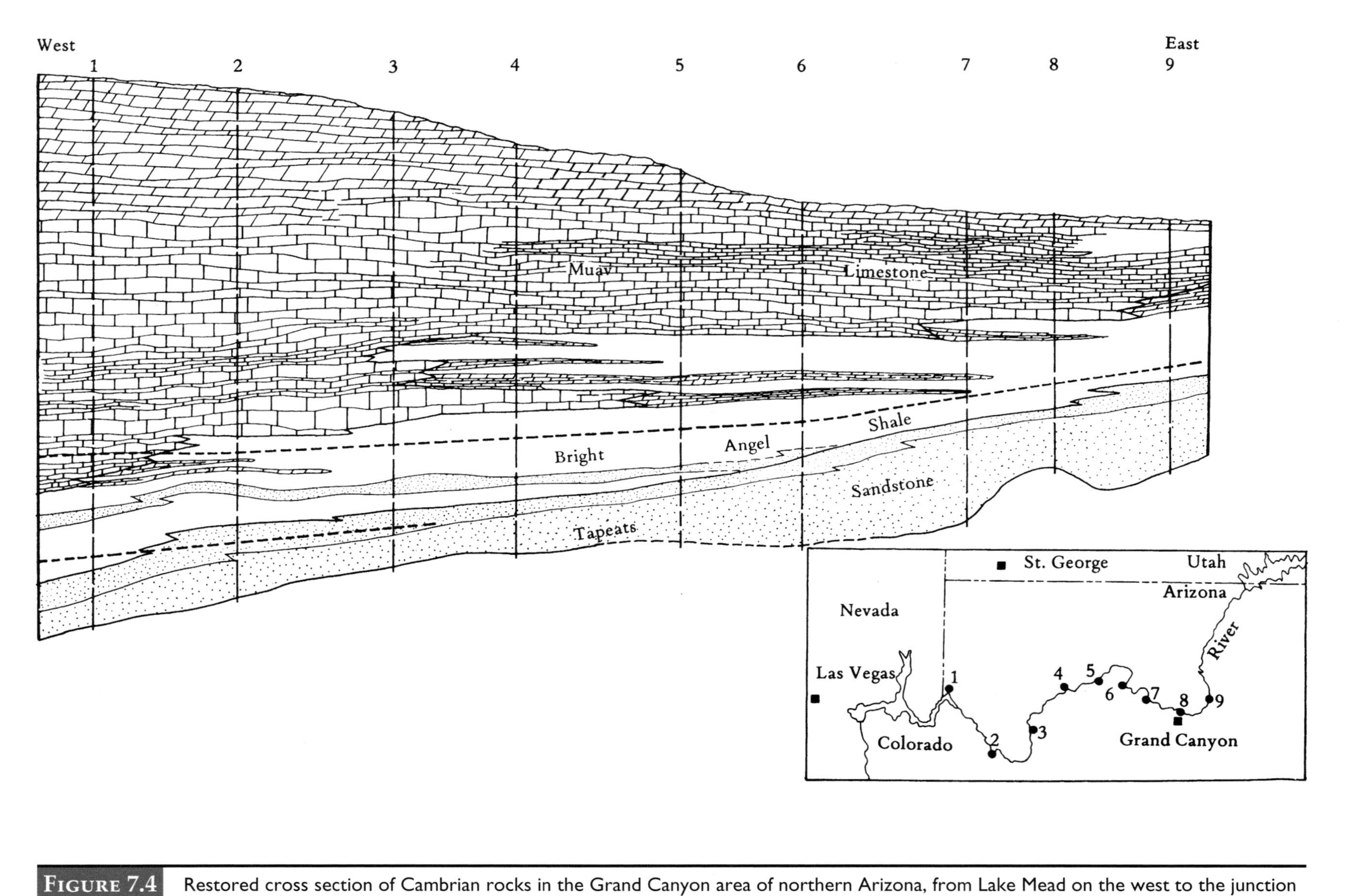

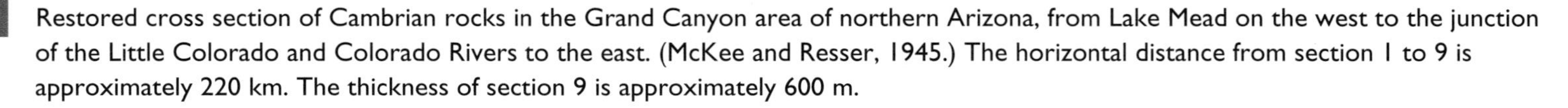
FIGURE 7.4 Restored cross section of Cambrian rocks in the Grand Canyon area of northern Arizona, from Lake Mead on the west to the junction of the Little Colorado and Colorado Rivers to the east. (McKee and Resser, 1945.) The horizontal distance from section 1 to 9 is approximately 220 km. The thickness of section 9 is approximately 600 m.

PROCEDURE

PART A
Correlation of Siliciclastic Rocks

The first part of the exercise utilizes the Devonian rocks in New York and Pennsylvania. These rocks formed the basis for the acceptance of the facies concept in North America. Fifteen somewhat generalized stratigraphic sections, which were measured through the Devonian rocks at localities approximately 32 km apart, are plotted as logs in figure 7.5. Section 1 is toward the west and section 15 is toward the east in a traverse that lies generally along the New York–Pennsylvania border. Symbols of the lithology are those used in previous exercises.

Total thickness of the preserved sections are shown to scale. Various time lines, which have been identified and correlated by the use of fossils, are shown by a series of dots through each column and are marked by small letters. The various time horizons or levels of contemporaneous deposition are shown by the same letters. For example, all the rocks immediately below the dotted line marked "a" in each of the 15 sections were deposited contemporaneously.

1. Construct a restored section for these Devonian rocks (figure 7.5) similar to the example of Cambrian rocks in figure 7.4. Detach both pages of figure 7.5 (left and right). Placing them side by side the long way, tape the two pages together and proceed. With lines and symbols, interconnect various lithologic units and show the facies relationships of the relatively coarse-grained rocks in the east to the fine-grained rocks in the west.

2. Does Walther's law apply to these rocks? Are there exceptions?

3. Are all of the conglomerates the same age?

4. What trend is visible in the sandstone beds as they are traced from east to west?

5. Why do shale beds thin as traced from west to east?

6. What happens to the sandstone that occurs near the base of sections 12, 13, 14, and 15?

7. What is suggested by the small lens of conglomerate near the top of section 3 in the time interval between "f" and "g"?

8. What is the overall textural (sediment size) and thickness pattern displayed by these sedimentary rocks?

9. From which direction were the sediments transported?

10. By what media (e.g., wind, glaciers, streams, marine currents) were the sediments deposited?

11. This "wedge" of sediment is known as the "Catskill Delta," although it is not a delta in the strict sense. Which Paleozoic mountain chain was the source of sediment comprising the Catskill Delta?

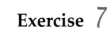

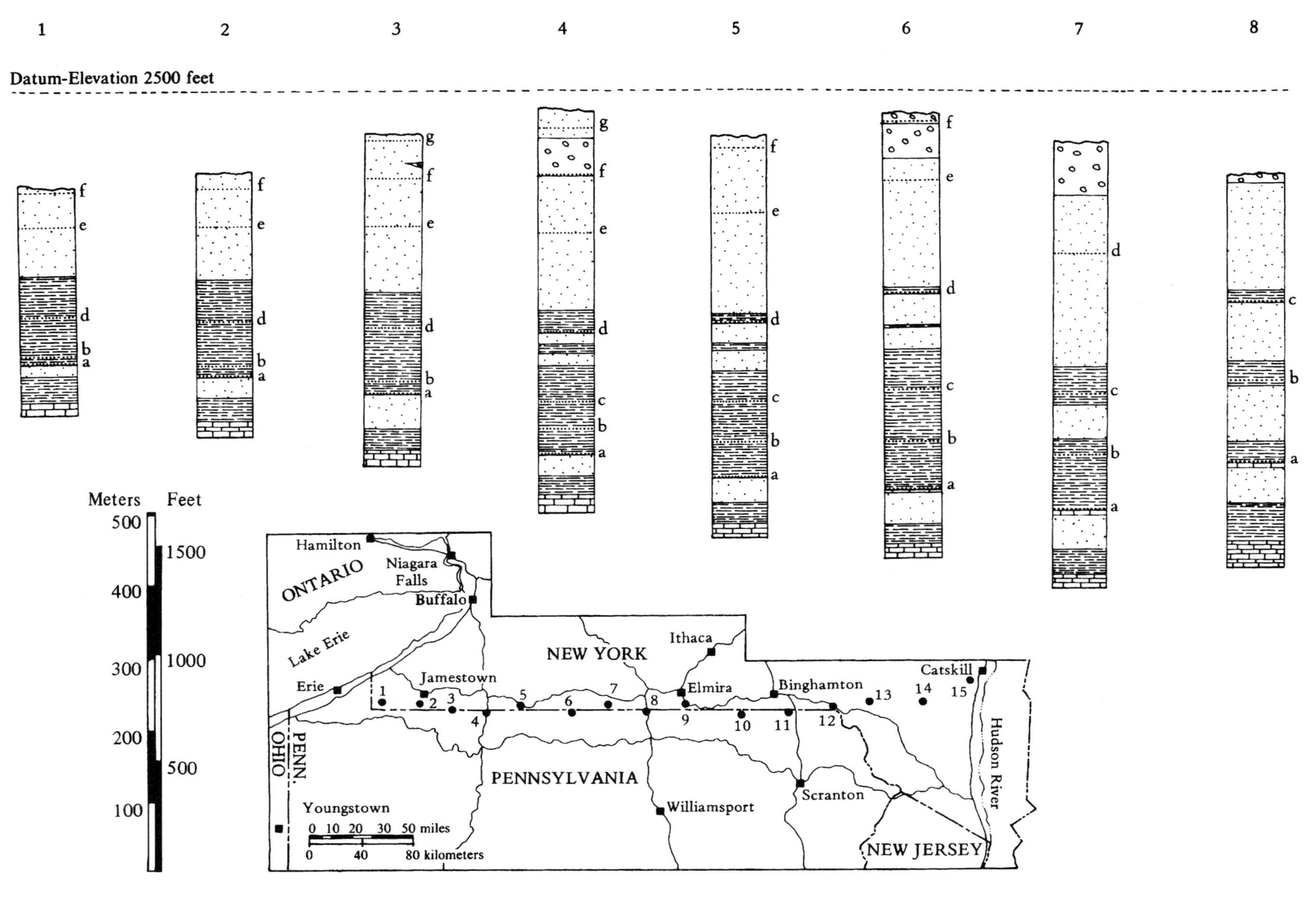

Figure 7.5 Left A series of 15 stratigraphic columns of the Devonian rocks in southern New York and northern Pennsylvania. This sequence is a classic example of integrating facies.

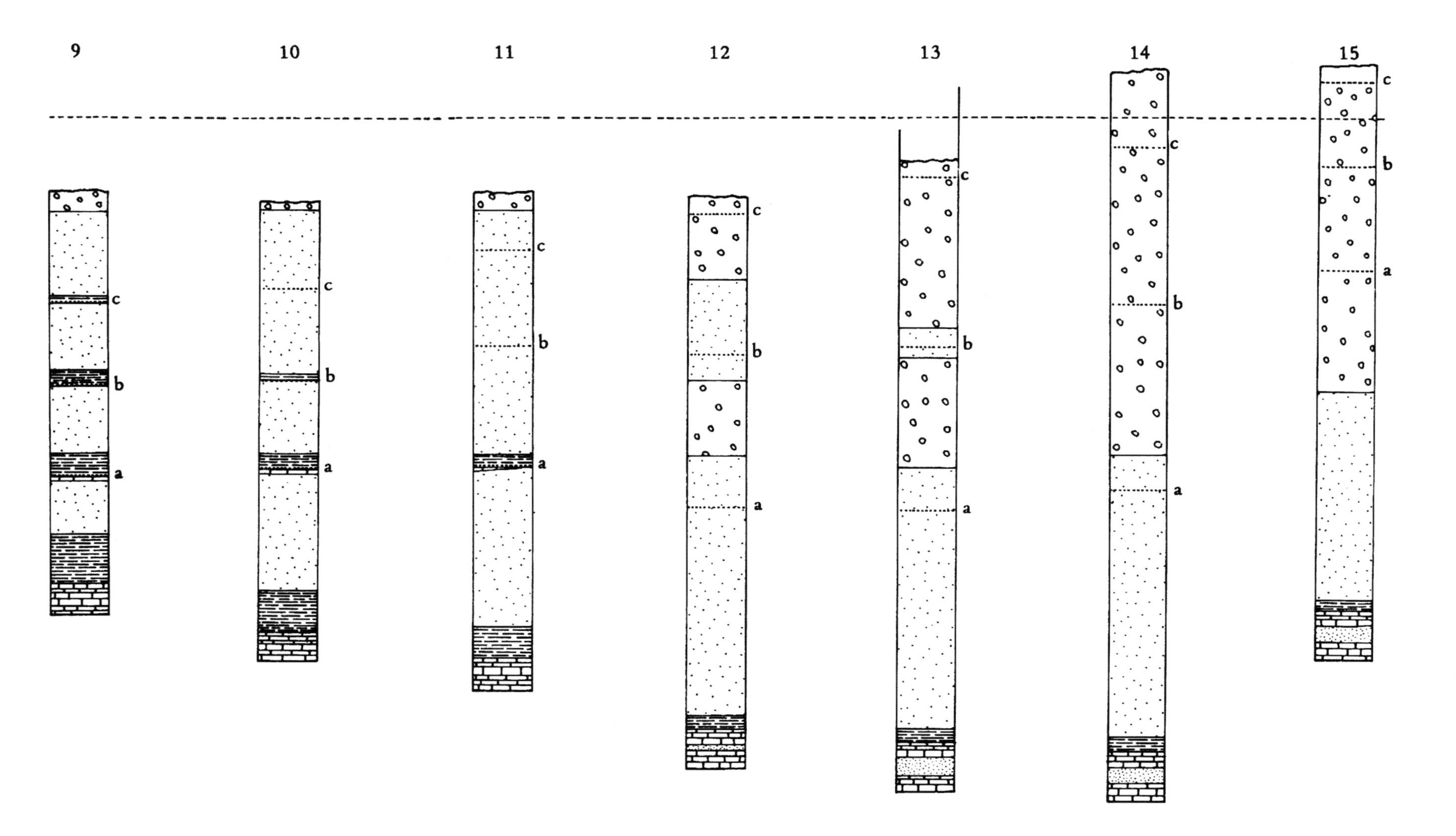

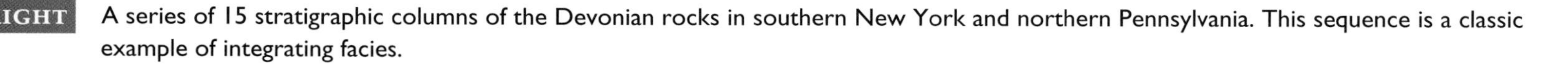

Figure 7.5 Right A series of 15 stratigraphic columns of the Devonian rocks in southern New York and northern Pennsylvania. This sequence is a classic example of integrating facies.

PART B

Correlation of Carbonate Rocks

Ten stratigraphic columns shown in figure 7.6 document the facies relationships through the classic Permian (Capitan) Reef in the Guadalupe Mountains of Texas and New Mexico, one of the major carbonate reef masses in North America. This series of stratigraphic sections is oriented approximately northwest to southeast (or from onshore to offshore). Bedded limestone is shown with the normal brick-like symbol, but massive limestone is shown with an open, discontinuous pattern. Evaporites (gypsum) are shown with a close-spaced diagonal grid in sections at both the eastern and western margins.

1. Construct a restored cross section across the traverse represented by the 10 sections. Detach both pages of figure 7.6 (left and right). Placing them side by side the long way, tape the two pages together and proceed. The lines that connect adjacent stratigraphic sections are time lines and are labeled with letters (a–e).

2. Are the massive reef limestones of section 3 the same age as the massive reef limestones of section 7? Are those of section 3 contemporaneous to those of the upper part of section 6?

3. What is the age of the massive dolomite in section 2 in relationship to the limestone beds of the Cherry Canyon Formation?

4. What is the age of the thin gypsum bed at the top of section 1 in relationship to the rocks in sections 8 and 9?

5. What is the direction of growth through time of the reef mass?

6. It is generally felt that the Cherry Canyon Formation contact with the top of the Brushy Canyon Formation remained essentially horizontal during deposition of these Permian reefs and that the top of the reef or the top of the massive dolomite was near sea level. In what depth of water was the gypsum in the top of section 1 deposited? What was the water depth during the deposition of the gypsum in sections 9 and 10?

7. From the diagram, which lithology would be termed *back reef* and which lithologies would represent *basin* or *fore-reef facies*?

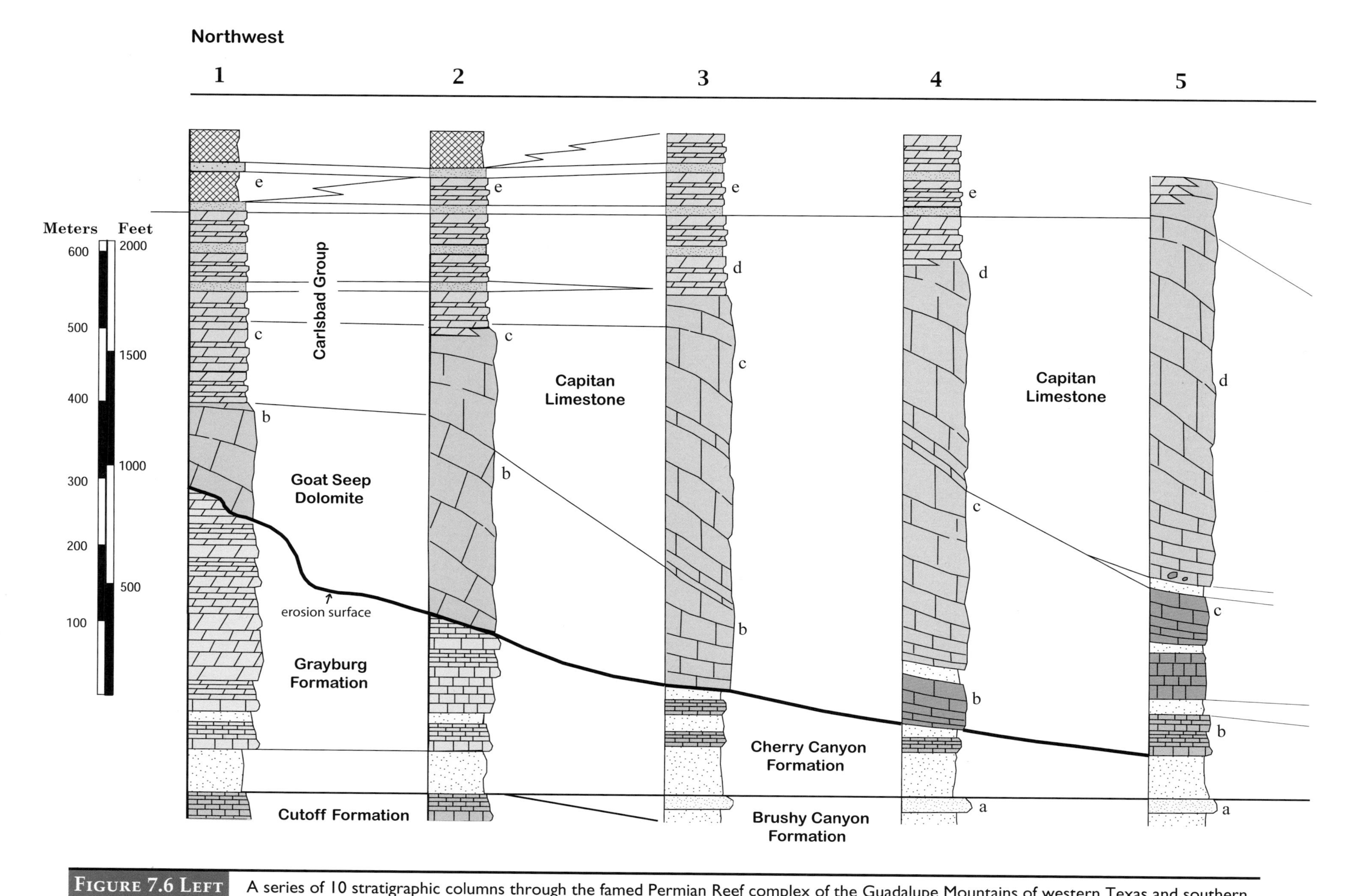

FIGURE 7.6 LEFT A series of 10 stratigraphic columns through the famed Permian Reef complex of the Guadalupe Mountains of western Texas and southern New Mexico. Horizontal distance from section 1 to 10 is approximately 8 km.

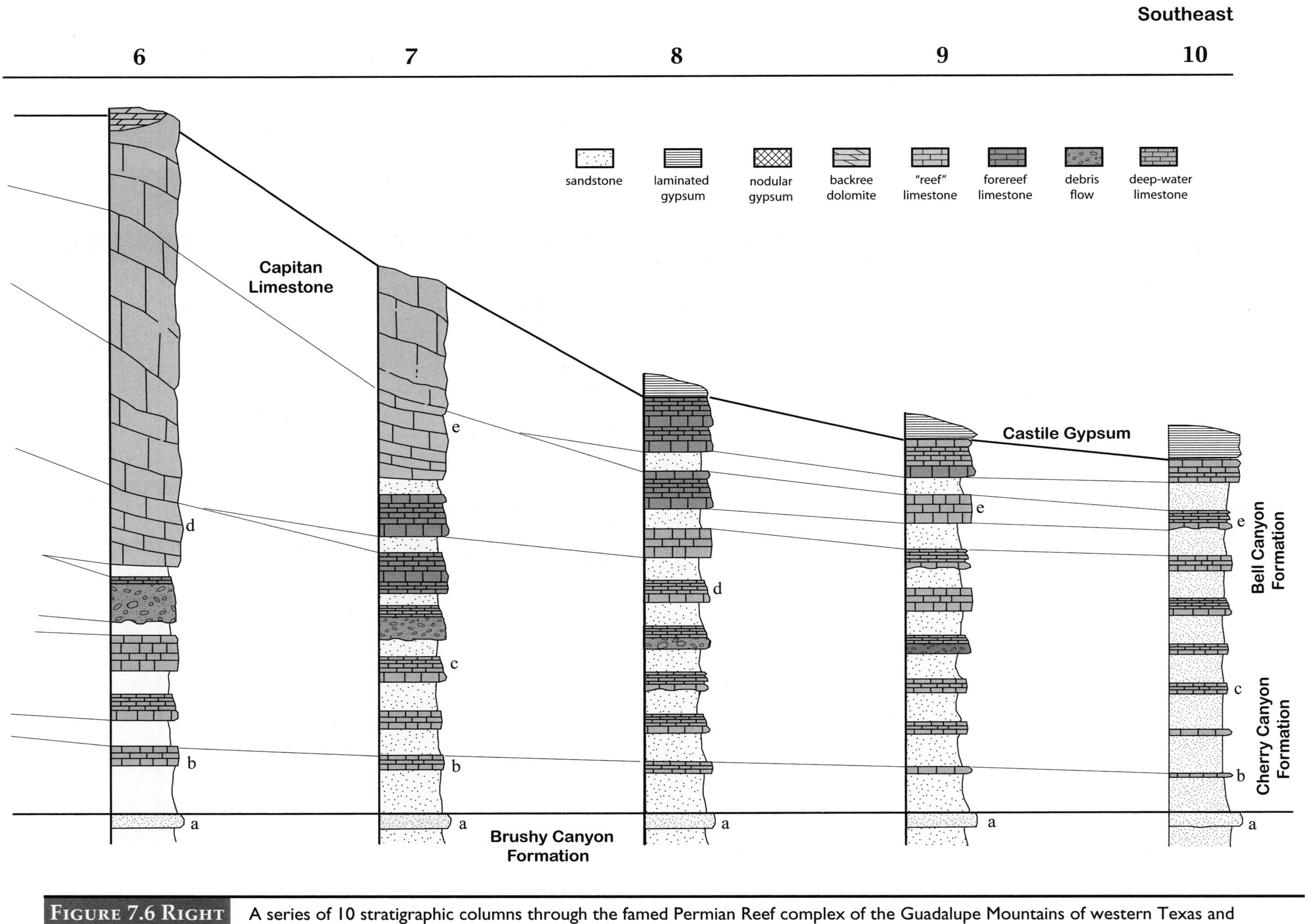

FIGURE 7.6 RIGHT A series of 10 stratigraphic columns through the famed Permian Reef complex of the Guadalupe Mountains of western Texas and southern New Mexico. Horizontal distance from section 1 to 10 is approximately 8 km.

PART C

Facies and Sea-Level Change

The Pleistocene Epoch began 2 million years ago and ended approximately 10,000 years ago. Climatic cooling during this time resulted in growth of huge ice sheets at both the North and South Poles. During glacial maxima, 30% of Earth's land surface was covered with ice, whereas only 10% is covered today during the current interglacial period. The shifting of water from the ocean reservoir to icecaps during glacial maxima resulted in a **eustatic** sea level fall of nearly 100 m (300 ft). If the polar icecaps and pack ice were to melt, sea level would rise 70 ft.

Although the Pleistocene is called the Great Ice Age, the climate was not cold everywhere, nor was it continually cold. Episodes of ice build-up were separated by interglacial episodes when ice would melt and sea level would rise. At least 18 Pleistocene glacial-interglacial cycles have been detected thus far. This means that sea level rose and fell dramatically at least 18 times during the last 2 million years.

1. How did these oscillations of sea level effect the face of south Florida?

2. How will continued melting affect Florida's future?

Figure 7.7 shows the subaerial and submarine topography of southern Florida today. Above current sea level (green area) the contour interval is 20 ft. Below current sea level (blue area) the contour interval is 10 ft down to a depth of 100 ft, after which the contours change to 100-ft intervals. Figure 7.8 shows the global sea-level history for the last 15,000 years with feet on the left and meters on the right. Note that sea level began to rise abruptly about 15,000 years ago, but that the rate of sea-level rise has tapered off over the last 6,000 years. Given these data:

3. Calculate the average annual rate of sea-level rise from 15,000 to 10,000 years ago.

4. Calculate the average rate of rise from 5,000 years ago to the present.

5. Using the graph below the map in figure 7.7, draw a topographic profile from A to A′. At this scale, the slope of the western Florida "ramp" is highly exaggerated. The actual slope is approximately 1 degree.

6. How far must you travel from the west coast of Florida to encounter water depths of 200 ft?

7. How far do you have to travel from the east coast of Florida to reach water depths of 200 ft?

8. Miami is currently situated at an elevation of 10 ft above sea level. What would Miami's approximate elevation have been 15,000 years ago?

9. Assume that seafloor sedimentation is controlled primarily by depth as follows:

- > 10 ft = soil genesis and small scale karstification of limestone bedrock
- 0 to 10 ft = coal swamp
- – 30 to 0 ft = quartz sand (quartz sandstone)
- – 60 to – 30 ft = fine clay and silt (shale)
- – 300 to – 60 ft = lime packstone with deep-water coral mounds

Using Maps A through D, draw the shoreline of Florida as it appeared 7,200 years ago (Map A), 5,000 years ago (Map B), and as it will appear at two points in the future when sea level is 20 ft (Map C) and 70 ft higher (Map D) than it is today. Map the facies boundaries using the depth-sediment relationships listed above. Neatly fill in the facies belts with the appropriate symbols (gray for coal, stippled pattern for sandstone, dashes for shale, brick pattern for limestone). On this and subsequent maps, fill in shorelines, facies belts, and symbols for only the area north of the line labeled A to A′.

10. Using the rate that you calculated in question (3), calculate how long it would take to flood Florida to a depth of 70 ft above present sea level.

11. What percentage of modern-day Florida would remain emergent?

12. On figure 7.9 draw a stratigraphic column showing the vertical succession of Holocene strata that would develop as a function of the marine transgression that you just mapped. To obtain the necessary information, stack the four maps on top of one another with Map A on the bottom and Map D on top. Leaf through the stack and record the facies pattern developed at the position indicated by the red dot on figure 7.7 beginning with Map A. Transfer this information to the stratigraphic column using the thicknesses provided below. Use the same symbols on the stratigraphic column that you did on the facies maps.

- coal = 10 ft thick
- sandstone = 30 ft thick
- shale = 15 ft thick
- limestone = 30 ft thick

What vertical pattern is produced by this transgression of the sea across south Florida? How does this compare with the horizontal distribution of facies belts at any given time?

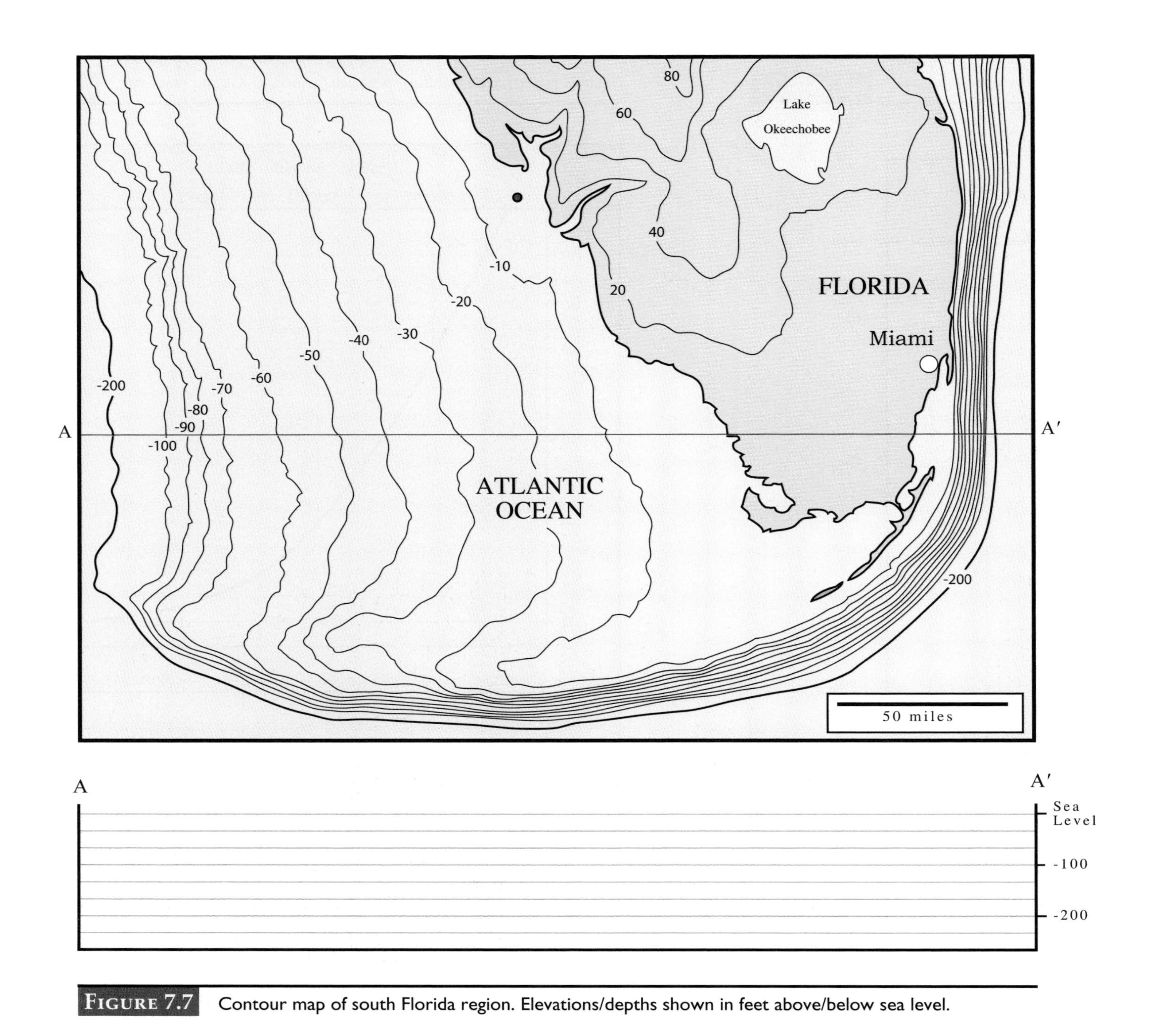

FIGURE 7.7 Contour map of south Florida region. Elevations/depths shown in feet above/below sea level.

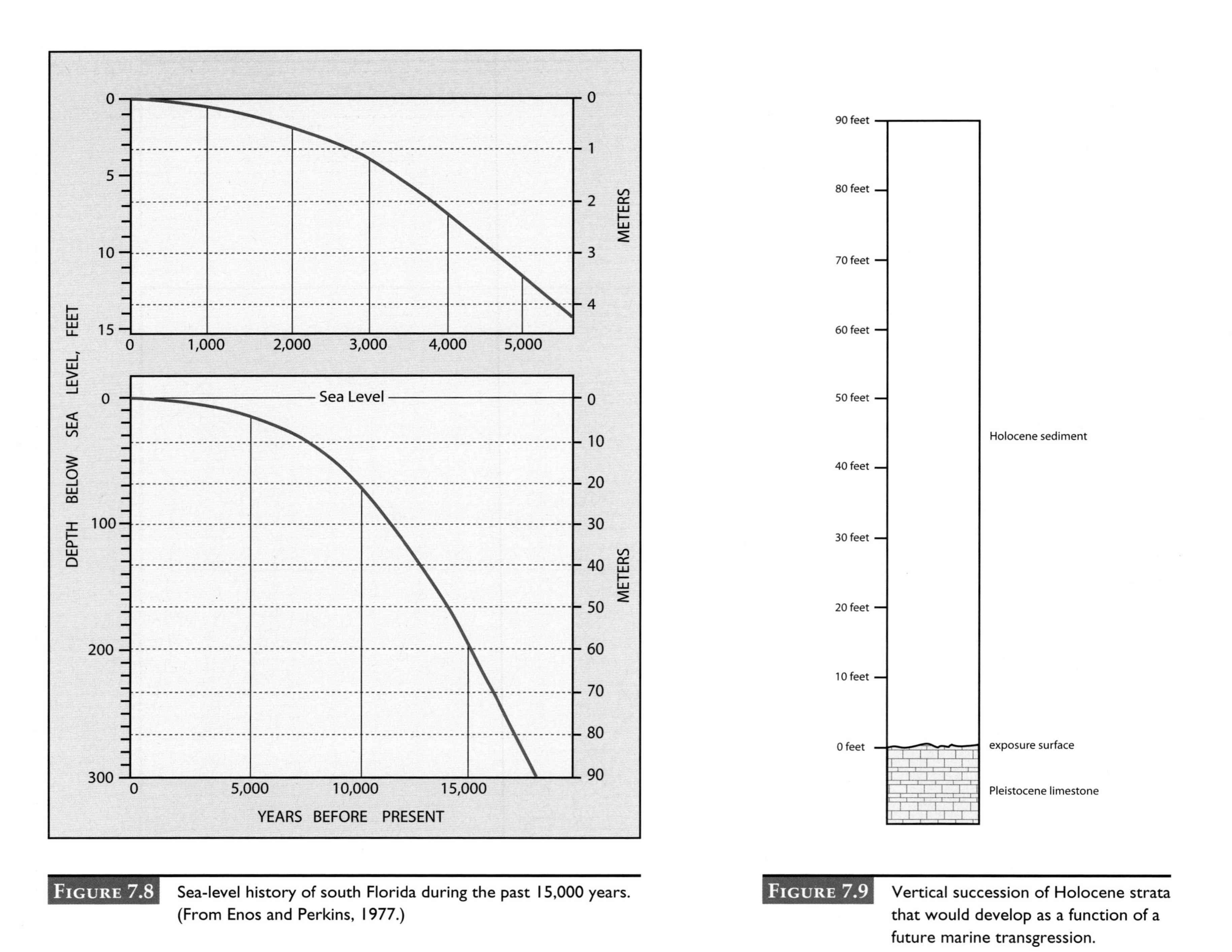

FIGURE 7.8 Sea-level history of south Florida during the past 15,000 years. (From Enos and Perkins, 1977.)

FIGURE 7.9 Vertical succession of Holocene strata that would develop as a function of a future marine transgression.

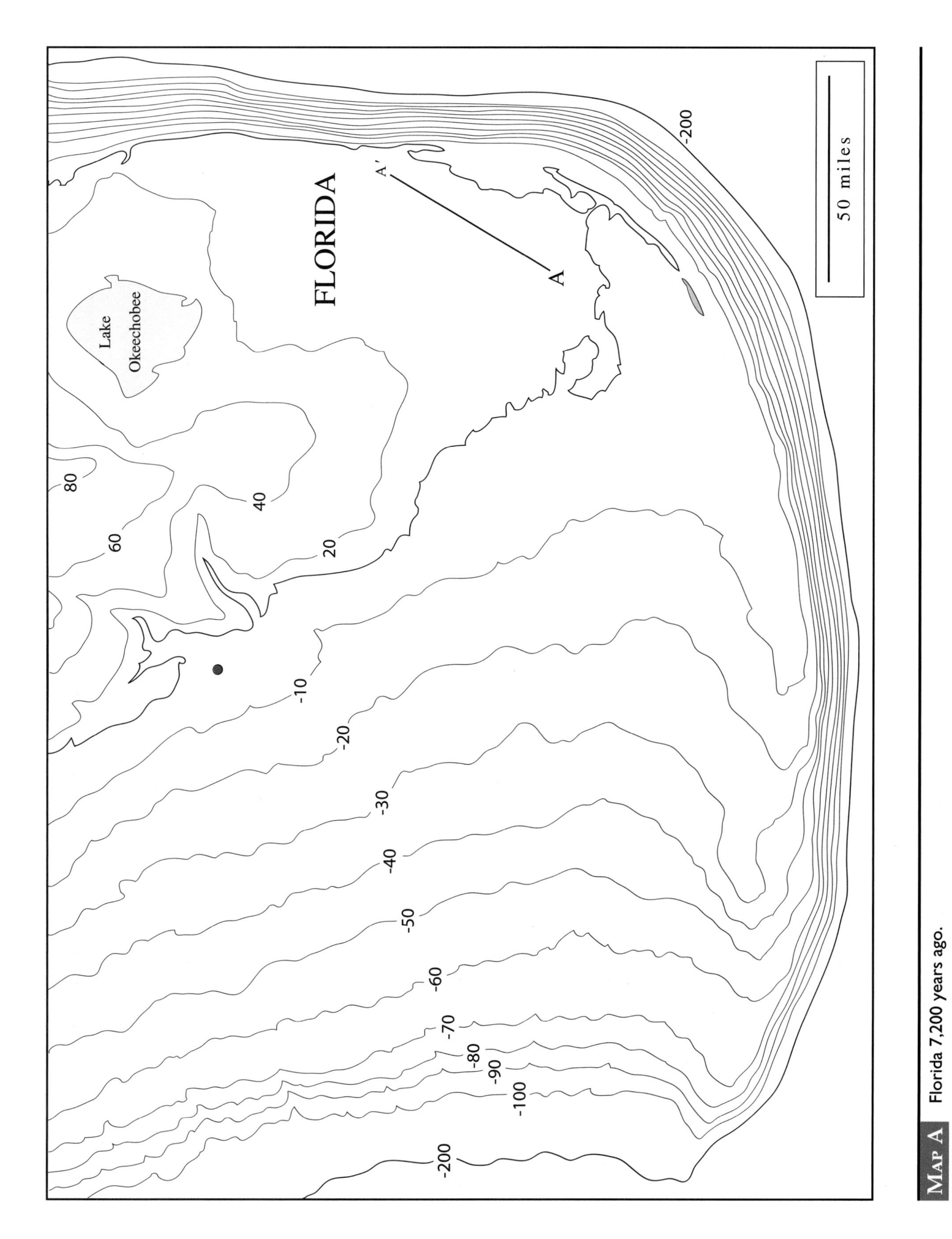

MAP A Florida 7,200 years ago.

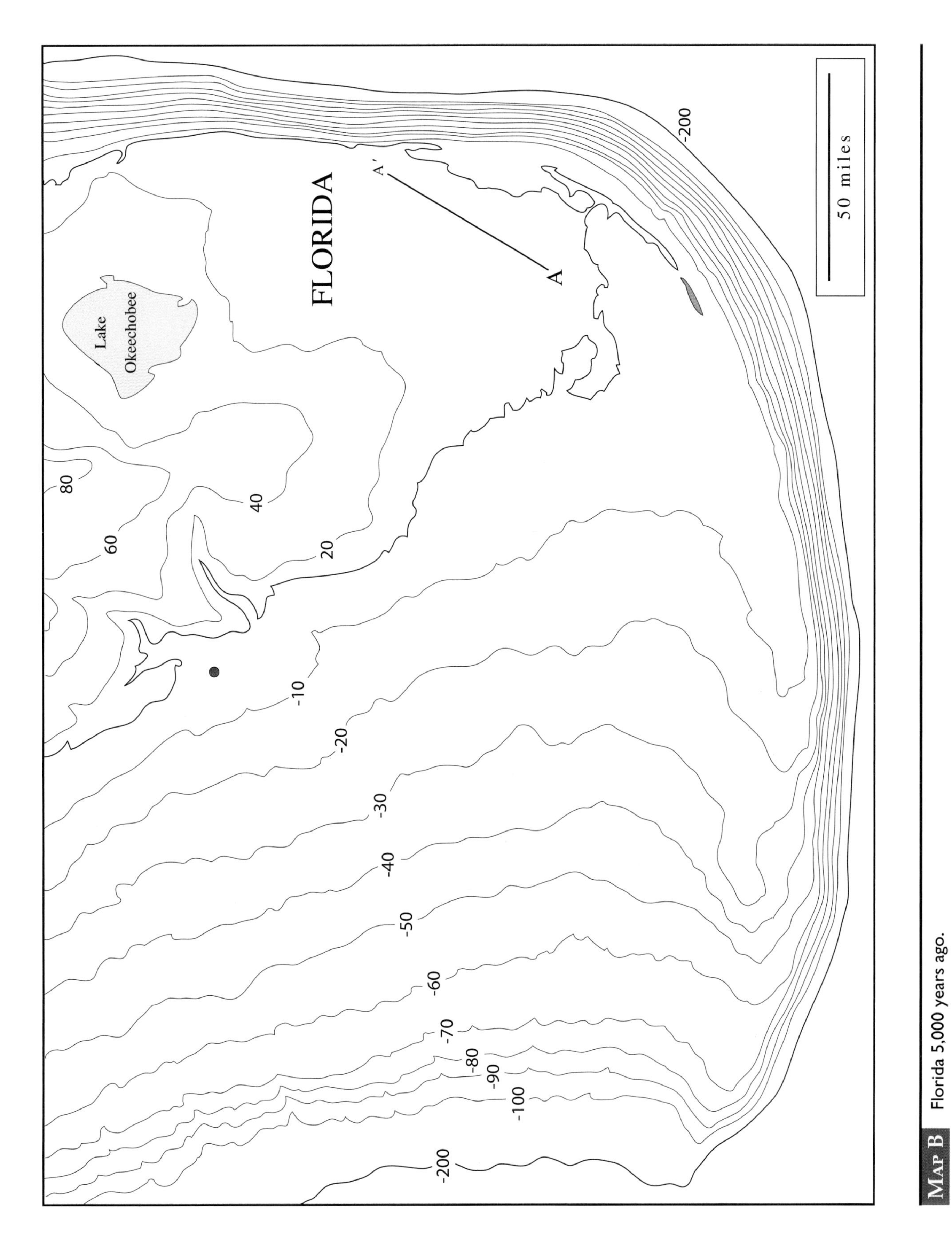

Map B Florida 5,000 years ago.

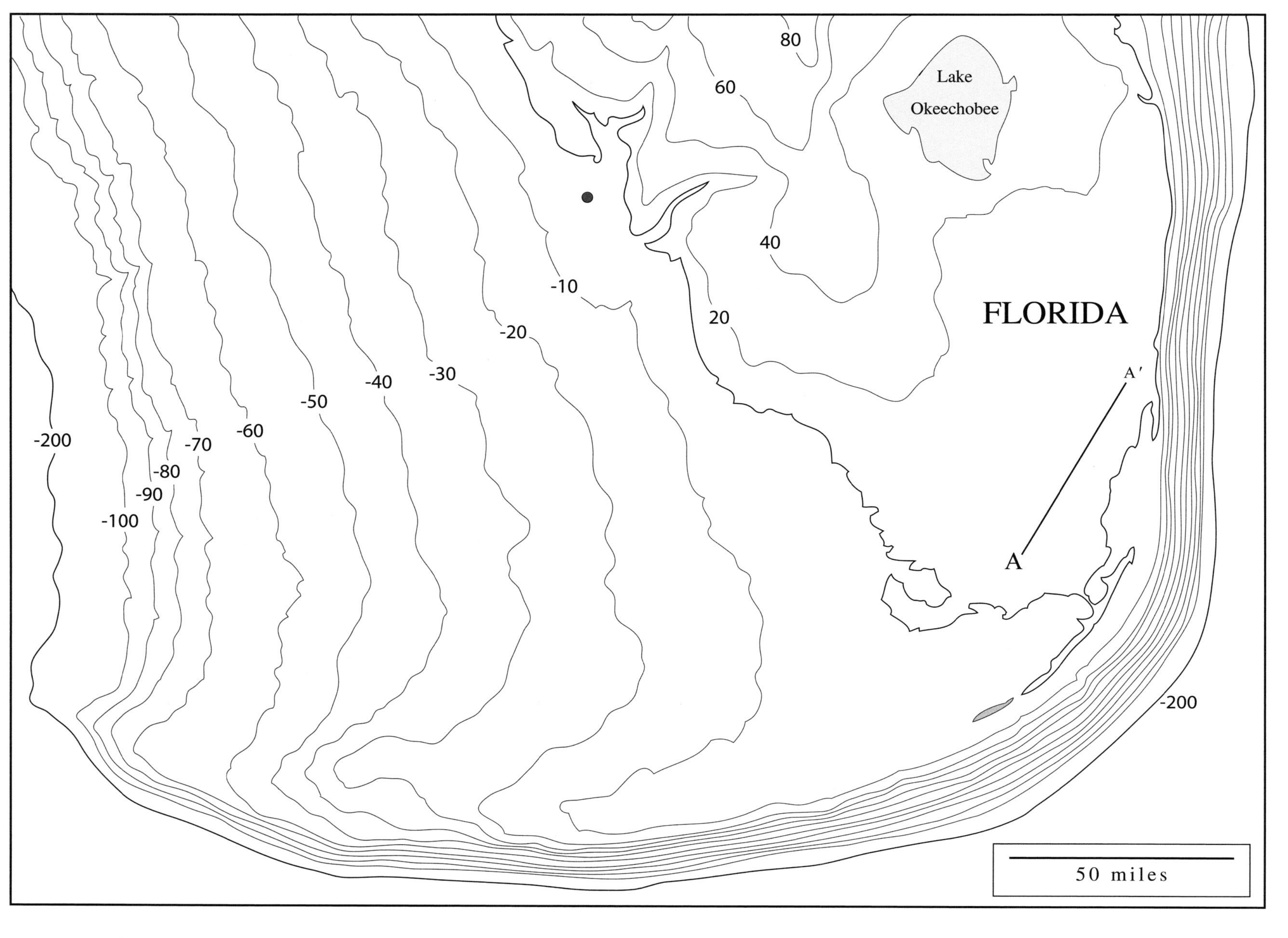

MAP C Florida at modern sea level plus 20 feet.

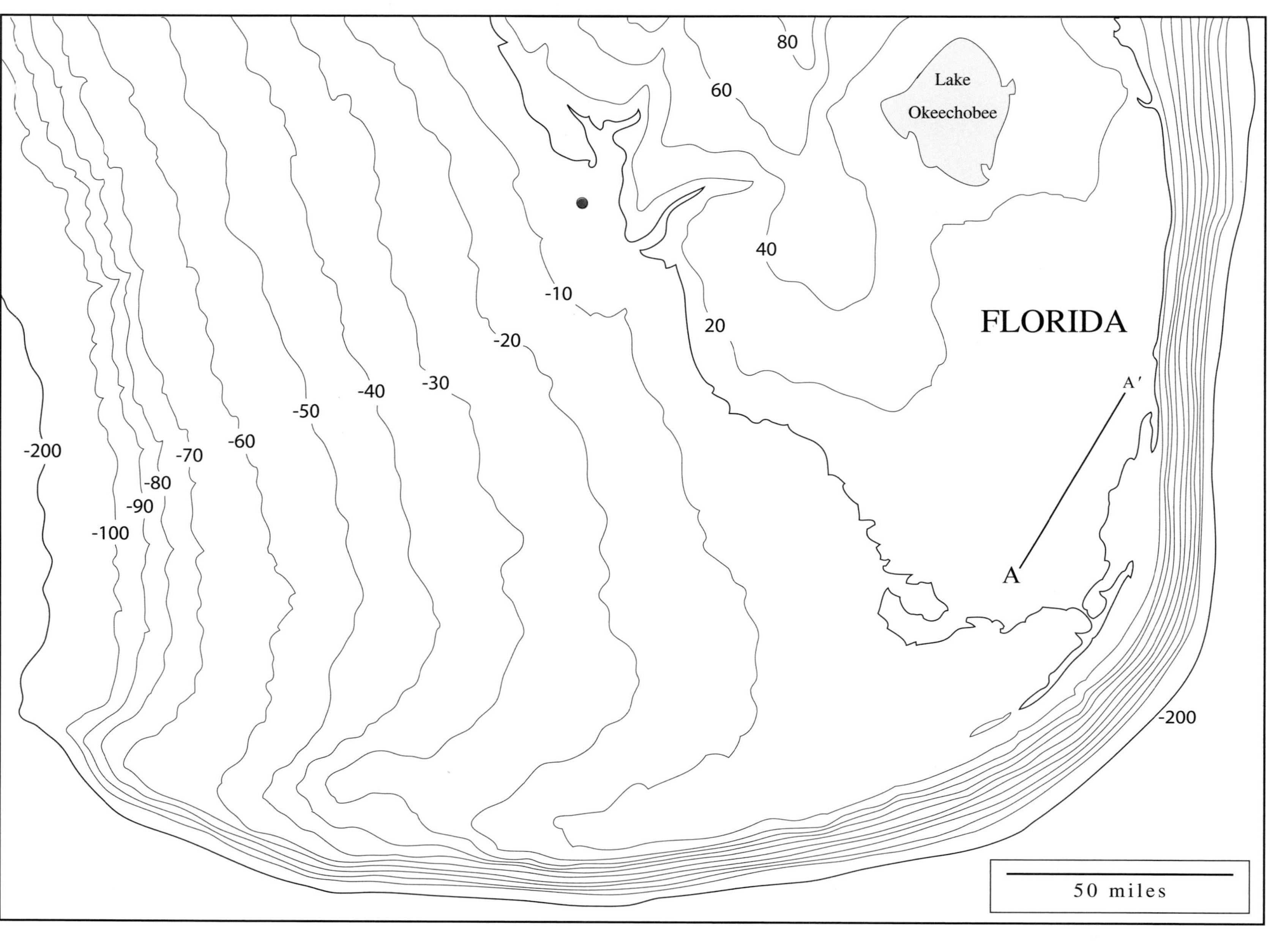

MAP D Florida at modern sea level plus 70 feet.

Fossils and Fossilization

Exercise 8

Learning Objectives

After completing this exercise, you will be able to:

1. define what a fossil is;
2. explain the different modes of fossilization;
3. indicate which type or types of preservation a particular specimen has undergone;
4. list the major phyla that comprise the majority of the fossil record; and
5. recognize representative specimens of each of the major invertebrate fossil groups.

Fossils, as remains of once-living organisms, are equal to rocks in terms of their importance as documents of Earth history. The term **fossil** has been variously used, but is here interpreted to be any evidence, direct or indirect, of the existence of organisms in prehistoric time. The term *fossil* comes from the Latin word *fodere,* which means to dig, or *fossilis,* which means dug out or dug up. An exact definition as to what constitutes a fossil, or an upper limit of prehistoric time, is difficult to establish. For example, are the hollow molds of a mule or a man buried in ash in the city of Pompeii fossils, or is the well-tanned and pickled Irish man, buried in a peat bog in the twelfth century, a fossil? Are the remains of a mastodon or wooly mammoth fossils when they are still well-enough preserved that the meat can be eaten rather palatably, even though the animals lived in the Pleistocene? The upper fringe of prehistoric time is a difficult and sometimes arbitrary area in the definition of the term *fossil.*

Evidence of fossils may range from exceedingly well-preserved forms like the mastodons frozen in the tundra of Alaska and Siberia, to less well-preserved forms such as clam shells that have been replaced and recrystallized. Direct evidences of fossils usually tell us much about the shape of the original animal or plant and commonly contain materials that were actually precipitated by the life processes of the organism. Indirect fossils, on the other hand, are normally only impressions, yet still suggestive of size and proportions. A track, a trail, a burrow-filling, a "stomach stone," or a coprolite are indirect evidences of organic activity and tell us something about the organisms. It has been suggested that perhaps only one organism in a million, living in even a preservable environment, is likely to be fossilized.

Several factors are normally requisite in order that fossilization may take place. First, a body to be fossilized ordinarily must contain **hard parts**, such as teeth, bones, or shells. Hard parts are much more easily preserved because of their relatively inert nature compared to soft tissues. Soft-bodied forms are occasionally preserved, as the impressions of jellyfish, leaves, the tentacles of octopus-like organisms, and the trails of gastropod feet have all been noted in the geologic record. The vast majority of soft-bodied forms, however, have left no record or have left a very incomplete fossil record. Second, the organism must be **buried rapidly** before the structures have a chance to decompose, erode, or disarticulate. Third, burial must effectively **seal the organism** from aerobic bacterial action and from decomposing chemical fluids.

Rapid burial, such as submergence in a peat bog, burial in sand or mud along a marine shore, trapping in pitch of conifer trees, or burial in ancient tar pools, would effectively seal organisms from bacterial action and preserve the hard parts. On the other hand, rapid burial in a broad alluvial fan or in a sandy area through which water can rapidly percolate is not likely to produce a fossil, because of chemical and biological decay, which will quickly set in and destroy any evidence of animal or plant.

Fossils are preserved and found in three forms: **unaltered** soft or hard parts, **altered** hard parts, and **trace** fossils. Essentially unaltered organisms are rare in the geologic column and each time one is described it causes some excitement in scientific circles. Those fossils that are essentially unaltered have undergone little chemical or physical change since the death of the organism. Wooly mammoths and rhinoceroses trapped in glacial sediments in Siberia, Alaska, and the Yukon Territory are examples of unaltered preservation. To be frozen intact is a temporary type of fossilization, because a constant frigid environment is necessary. Such preservation is only temporary in terms of Earth's history.

Insects, spiders, and various other arthropods have been preserved in amber in Tertiary deposits along the Baltic Sea and in Mexico. The hard parts of these animals preserved under such circumstances must have remained unaltered for a long time, although most of the soft parts were desiccated.

Unaltered hard parts have also been collected from ancient tar deposits, such as those at Rancho la Brea in Los Angeles, California, or in Argentina. At these sites, bones of Pleistocene birds, mammals, and some reptiles have been preserved and sealed from bacterial action by entombment in the tarry deposits. Because of the limited number of tar seeps, such fossilization is rare.

Much more common unaltered hard parts are the shells or tests of various invertebrate animals. Oysters and other mollusks have been preserved with the inner shell of "mother of pearl" layer still intact. Many conodonts, inarticulate brachiopods, and fragments of echinoderms have been preserved essentially unaltered because of the stable nature of their skeletal material. Often foraminifera and diatoms have their hard parts preserved unaltered in relatively young rocks. From the standpoint of total numbers of fossils, however, preservation of essentially unaltered materials is rare.

Another method of unaltered fossilization is **desiccation**, or driving off of water from tissues. This type of preservation is not common and results in rather temporary fossils in areas with an arid climate. For example, an extinct mummified ground sloth was still articulated, and ligaments held the bones firmly together. Pieces of skin were still attached to the carcass in places, but in other areas the skin and muscles had been consumed by carnivores trapped in the same volcanic pit. Desiccated mummies of southwestern Indians are examples of the same type of preservation, although in this instance the organisms are very young geologically.

A second major method of fossilization is **alteration of hard parts**. In general, three types of alteration of the hard parts can be recognized: recrystallization, permineralization, and replacement. Figure 8.1 illustrates a hypothetical clam shell undergoing a variety of changes on its way to becoming a fossil.

In **recrystallization**, the original skeletal materials have been reorganized into different minerals or larger crystals of the same mineral. Aragonite, a skeletal material precipitated by many mollusks and other organisms, is a form of calcium carbonate that is unstable over long periods of time. Aragonite recrystallizes easily into the more stable form of calcium carbonate, called calcite. In recrystallization, no new material is added or taken away, but simply a rearrangement of the crystalline substances occurs. Recrystallization generally does not change the external form of the hard part but obscures or destroys internal structures precipitated by the organism.

Permineralization is another type of alteration and is particularly common in porous substances such as bone or wood. Permineralization is accomplished by the addition of materials to fill the pores of the structure and generally produces a very heavy bone or piece of wood. The skeletal structure may be in its original condition or it may have been replaced or recrystallized. Many dinosaur and mammal bones in Mesozoic and Cenozoic deposits found in the western portion of the United States, and the much sought-after fossil wood of Mesozoic and Cenozoic deposits in many parts of the country, are a result of permineralization.

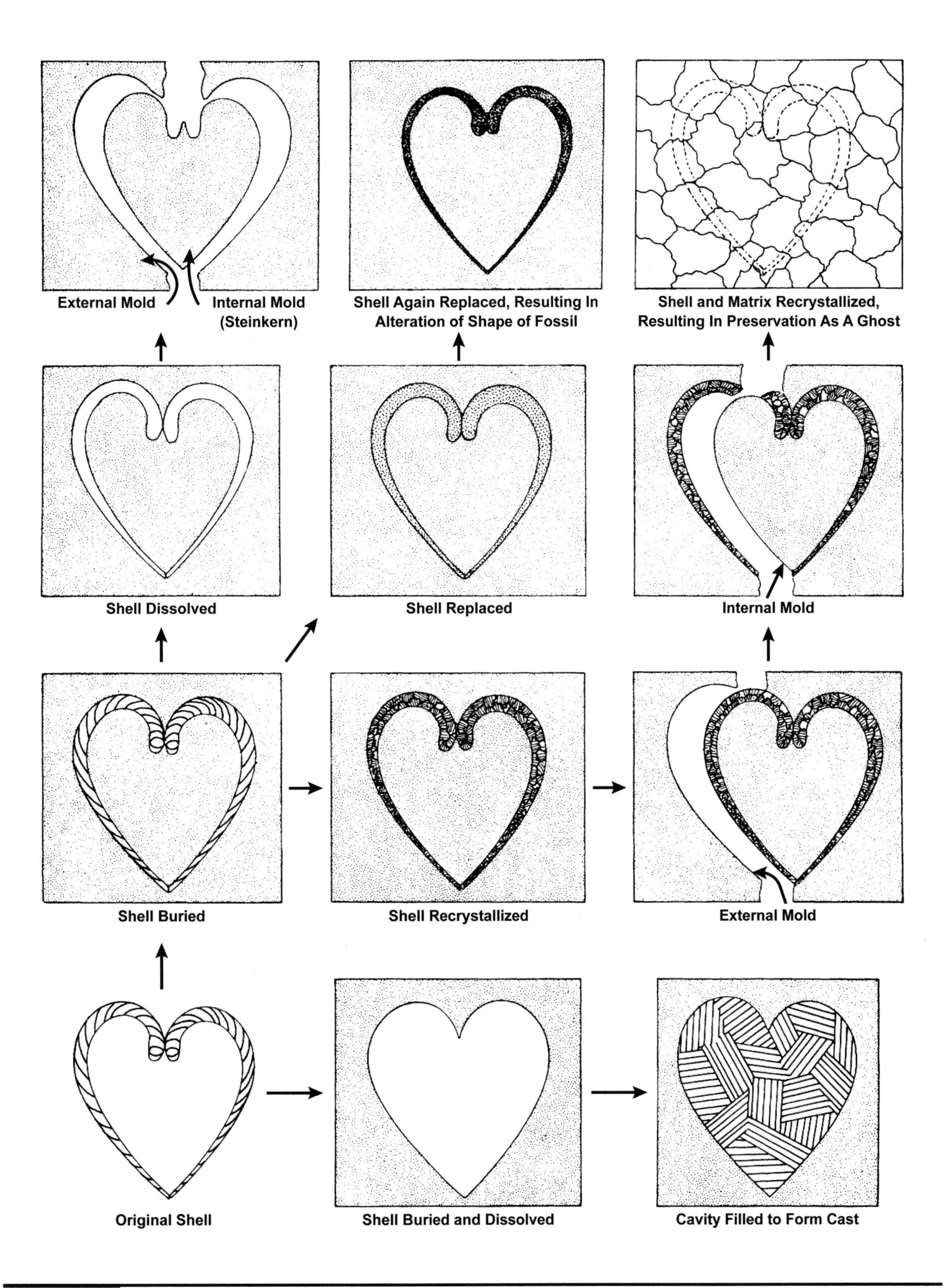

FIGURE 8.1 A diagram showing the various types of fossil preservation and their relationships. Uniform stipple pattern represents similar matrix. The arrows show possible sequences of preservation.

Replacement results in the removal of original skeletal material and subsequent replacement by a secondary compound. An example is original calcite or aragonite of invertebrate shells replaced by iron sulfide in the form of pyrite or marcasite. Silica is another common replacing material, and in instances where silica has replaced the shells in limestone, they are easily recovered from the surrounding rock by solution of the matrix in dilute hydrochloric acid. One of the most famous silicified faunas in North America is that of the Permian rocks of west Texas, where large numbers of organisms have been delicately silicified and preserved by replacement. Calcium carbonate may in turn replace structures that were originally siliceous. The original opaline spicules of fossil sponges are often replaced by coarse crystalline calcite if the matrix in which the sponges occur is calcareous. Limonite, gypsum, dolomite, native silver or copper, copper sulfate, and various phosphates have all been noted as substances that have replaced original material during fossilization. In replacement, permineralization, and recrystallization a solid, hard duplicate structure results from the chemical modification of the original hard parts. In only rare instances are soft parts replaced or permineralized as part of the fossilization process.

In many areas, particularly in humid regions, skeletal materials may be dissolved away, leaving a hollow impression in the rocks called a **mold**. This natural cavity may preserve, in varying detail, the shape of the original structure. A model of the outside of the shell is termed an **external mold**, whereas that of the inside is termed an **internal mold**. In some organisms, such as a clam, the inside area between the closed valves, or shells, may be filled with sediment, forming an internal mold of thc clam shell. This sediment-filling of the inside of the clam shell is termed a **steinkern** and is a common method of preserving various mollusks, such as clams and snails. If one fills the mold, either naturally or artificially, with a foreign substance, the foreign material then duplicates the shape of the original and is termed a cast. A **cast** is a replica of the original form. Natural casts occur where mineralizing solutions fill original molds by precipitation. Characteristic styles of fossilization for the most common organic materials are shown in figure 8.2.

In order to determine the preservational process that has affected a fossil, it is important to know the original mineralogy of the specimen under scrutiny. Table 8.1 shows the common minerals and organic compounds used by the major fossil groups in con-

FIGURE 8.2

A chart showing types of preservation characteristics of various organic materials. Dark blue circles represent common modes of fossilization, stippled circles denote less common modes, and light blue circles denote rare modes of preservation.

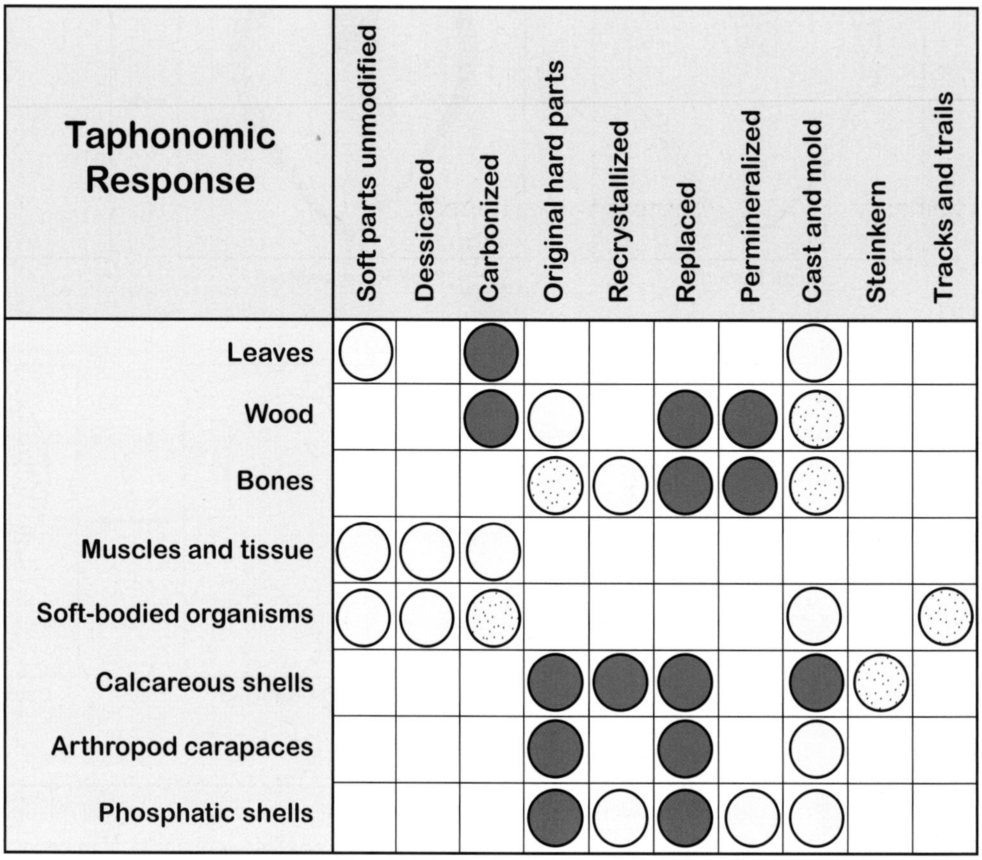

Taphonomic Response	Soft parts unmodified	Dessicated	Carbonized	Original hard parts	Recrystallized	Replaced	Permineralized	Cast and mold	Steinkern	Tracks and trails
Leaves	○		●					○		
Wood			●	○		●	●	◌		
Bones				◌	○	●	●	◌		
Muscles and tissue	○	○	○							
Soft-bodied organisms	○	○	◌					○		◌
Calcareous shells				●	●	●		●	◌	
Arthropod carapaces				●		●		○		
Phosphatic shells				●	○	●	○	○		

structing their shells and other hard parts. As a rule, most mollusk shells are comprised of unstable aragonite, making them readily susceptible to recrystallization and dissolution. By contrast, articulate brachiopods and bryozoans use low magnesium calcite in constructing their shells. As a result, even Paleozoic representatives of these groups may display only minor alteration to their hard parts. Conodonts, inarticulate brachiopods, and other vertebrate hard parts are constructed of phosphatic minerals.

Trace fossils are of increasing interest to paleontologists. **Tracks**, **trails**, or **burrows** are commonly the only vestige of large populations of land or marine animals. Most tracks and trails, however, show us little of the real configuration of the organism, although they do tell us something of its size, weight, and perhaps even its feeding pattern. By their very nature, interpretation of tracks and trails is more problematic than interpreting most other types of fossils because of the lack of direct information concerning the shape of the original living organism.

Animal borings may also tell us something of the life habits of the organism. Because of the recent interest shown in interpretation of sedimentary sequence where tracks, trails, and borings are the only fossils preserved, the new field of **ichnology**, a study of indirect fossils, has developed within historical geology.

Carbonization, as a method of fossilization, occurs when the organism remains are preserved as a film of carbon. Such carbon films may show structures of the organism in great detail, although they are now preserved only as a flat impression. Leaves, graptolites, crustaceans, and fish have been preserved in great numbers in limited areas by this method. One of the most famous carbonized faunas is that of the Middle Cambrian Burgess Shale, discovered near the turn of the twentieth century in British Columbia. Here carbonized films of even soft-bodied forms have been exquisitely preserved, so that details of the digestive tract, nervous system, and surficial hairlike bristles have been preserved as

TABLE 8.1 Original mineralogy of common invertebrate fossil groups.

Fossil Group			Calcite	Aragonite	Silica	Chitin	Phosphate
PHYLUM	CLASS	ORDER					
Sarcodina		Foraminifera	X				
		Radiolaria			X		
Porifera	Calcarea		X				
	Hexactinellida				X		
Cnidaria	Anthozoa	Rugosa	X				
		Tabulata	X				
		Scleractinia		X			
Bryozoa			X				
Brachiopoda	Inarticulata						X
	Articulata		X				
Mollusca	Bivalvia			X			
		"oysters"	X				
	Cephalopoda			X			
	Gastropoda			X			
Arthropoda	Trilobita		X			X	
	Eurypterida		X			X	
	Ostracoda		X				
Echinodermata	Crinoidea		X				
	Echinoidea		X				
	Blastoidea		X				
	Asteroidea		X				
Conodonta							X
Hemichordata	Graptolithina					X	

part of the fossil. Even though common under certain circumstances, carbonization is quantitatively a relatively minor method of fossilization.

Coprolites are solid excretory waste pellets of animals that are occasionally preserved. They may contain undigested material and thus give considerable information concerning the diet of the animal, its size, and where the animal may have lived. Coprolites may contain teeth, scales, plant remains, hard parts, or material that is not easily digested; as such they may be significant finds in terms of historical geology.

Gastroliths, or "stomach stones," have been described as associated with Mesozoic dinosaur remains throughout the world. These stones are highly polished and rounded, a result of grinding against one another as part of the fragmentation process in eating and digestion. Not all rounded, highly polished stones from even the most favorable horizons should be considered gastroliths, because many other processes can produce similar highly polished and rounded stones as well. Differentiation usually requires much experience, or an unusual association, to clearly identify polished gastroliths. One of the most diagnostic features on some of the least questioned occurrences is an acid-like pitting in depressions beneath the general rounded, polished surface of the stone.

Major Invertebrate Fossil Groups

Classification of fossil plants and animals is called **taxonomy**. The basis for taxonomy is similarity of morphology and shape, and phylogenetic relationships. In practice, all fossils are assigned two-part names, following what is called the **binomial system**. The first part of the name is the **generic** name; the second is the **specific** name. Occasionally, a third or even a fourth part is used to show combinations of subgenera or subspecies. Ordinarily, only genus and species names are used. The Roman alphabet is used for all names in texts of any language. Conventionally, names are derived from Latin or Greek roots. The basis for subdivision and classification of organisms is the species. **Species** are defined as groups of organisms that normally interbreed and produce fertile offspring. This definition, although usable for neontologists or taxonomists of modern species, is not usable for the study of fossils in paleontology where the basis of classification is in fact morphology or similarity in form. Thus, we only infer in paleontology that morphological species were in fact interbreeding groups of organisms during their lifetime.

	Dog	Human
Kingdom	Animalia	Animalia
Phylum	Vertebrata	Vertebrata
Class	Mammalia	Mammalia
Order	Carnivora	Primates
Family	Canidae	Hominidae
Genus	*Canis*	*Homo*
Species	*familiaris*	*sapiens*
Individual	Rover	John Brown

TABLE 8.2 Hierarchy of taxonomy.

In table 8.2, the taxonomic hierarchies of a man and a dog are used as examples in the classification from kingdom to individual. Generic, subgeneric, specific, and subspecific names are italicized. Many of the scientific names are used as common words or in general ways with English plurals and nonitalicized forms. Table 8.3 provides an abbreviated classification of animals and plants.

Protozoans

Protozoans are unicellular organisms that range in size from microscopic to several millimeters in diameter. Protozoans live in marine environments as plankton and benthos, in freshwater, and as parasites in many living organisms. These small single cells perform all the living functions necessary for complete life cycles.

In spite of their size and relative simplicity, protozoans display a great variety of shapes and forms, and species are differentiated on that basis. Of the great variety of protozoans, only two groups that possess hard parts are important as fossils (figure 8.3). The group called foraminifera (or forams) are extremely abundant and are important in stratigraphy as time indicators. The foraminifera have shells or tests composed of calcium carbonate or of fine grains of minerals or rocks cemented together. The tests of forams are composed of chambers assembled in a variety of ways and patterns. One group of Late Paleozoic foraminifera, called fusulinids, built shells that resemble grains of wheat or rice.

Another group of protozoans is the radiolarians, a group that is abundant in modern seas and

Animal Kingdom

Phylum Protozoa
Single cells, or groups of cells, generally microscopic foraminifers, radiolarians, fusulinids

Phylum Archeocyatha
Sponge-like animals with a cone-in-cone structure

Phylum Porifera
Sponges and stromatoporoids

Phylum Cnidaria
Corals—tetracorals, hexacorals, and tabulate corals

Phylum Bryozoa
Moss animals—small colonial animals

Phylum Brachiopoda
Bivalved invertebrates with unequal dorsal and ventral valves

Phylum Echinodermata
Animals generally with fivefold radial symmetry—starfish, sand dollars, echinoids, sea lilies or crinoids, blastoids, cystoids

Phylum Mollusca
Bivalves (Pelecypods—clam, oyster)
Gastropods—snail, slug
Cephalopods—squid, octopus, nautiloid, ammonoid

Phylum Annelida
Segmented worms
Scolecodonts

Phylum Arthropoda
Invertebrate animals with jointed legs, insects, lobsters, crabs, trilobites, eurypterids

Phylum Hemichordata
Graptolites

Phylum Conodonta

Phylum Vertebrata
Animals with notochords and articulated backbones
Pisces (fish)
Amphibians
Reptiles—dinosaurs, ichthyosaurs, plesiosaurs, mosasaurs
Aves (birds)
Mammals—warm-blooded animals, including humans

Plant Kingdom

Division Cyanophyta	blue-green algae
Division Chlorophyta	green algae
Division Phaeophyta	brown algae
Division Rhodophyta	red algae
Division Bryophyta	liverworts, hornworts, mosses
Division Psilophyta	psilophytes
Division Lycopodophyta	club mosses
Division Arthrophyta	horsetails
Division Pterophyta	ferns
Division Pteridospermophyta	seed ferns
Division Cycadophyta	cycads
Division Ginkophyta	ginkgos
Division Coniferphyta	confiers
Division Anthophyta	flowering plants
Class Dicotyledonae	dicots
Class Monocotyledonae	monocots

TABLE 8.3 Abbreviated classification of animals and plants (as used in paleontology).

that has tests composed of concentric spheres or helmet-shaped, spiny structures of silica. Protozoans are known from rocks as old as Cambrian, and even perhaps Precambrian, and they range to the Recent. Both foraminifera and radiolarians are abundant in the modern seas, where their tests accumulate to form radiolarian and foraminiferal oozes. Fusulinids are used as stratigraphic time indicators in Carboniferous and Permian rocks where they are abundant; other small foraminifera are used in the dating and correlation of Cretaceous and Tertiary rocks. Radiolaria are used as stratigraphic indicators for deep-marine sediments. Because of their small size, protozoans not only require identification and study with a microscope, but also can be recovered in cuttings from drill holes of oil wells and are one of the most common and valuable fossils for stratigraphic studies in late Mesozoic and Cenozoic rocks.

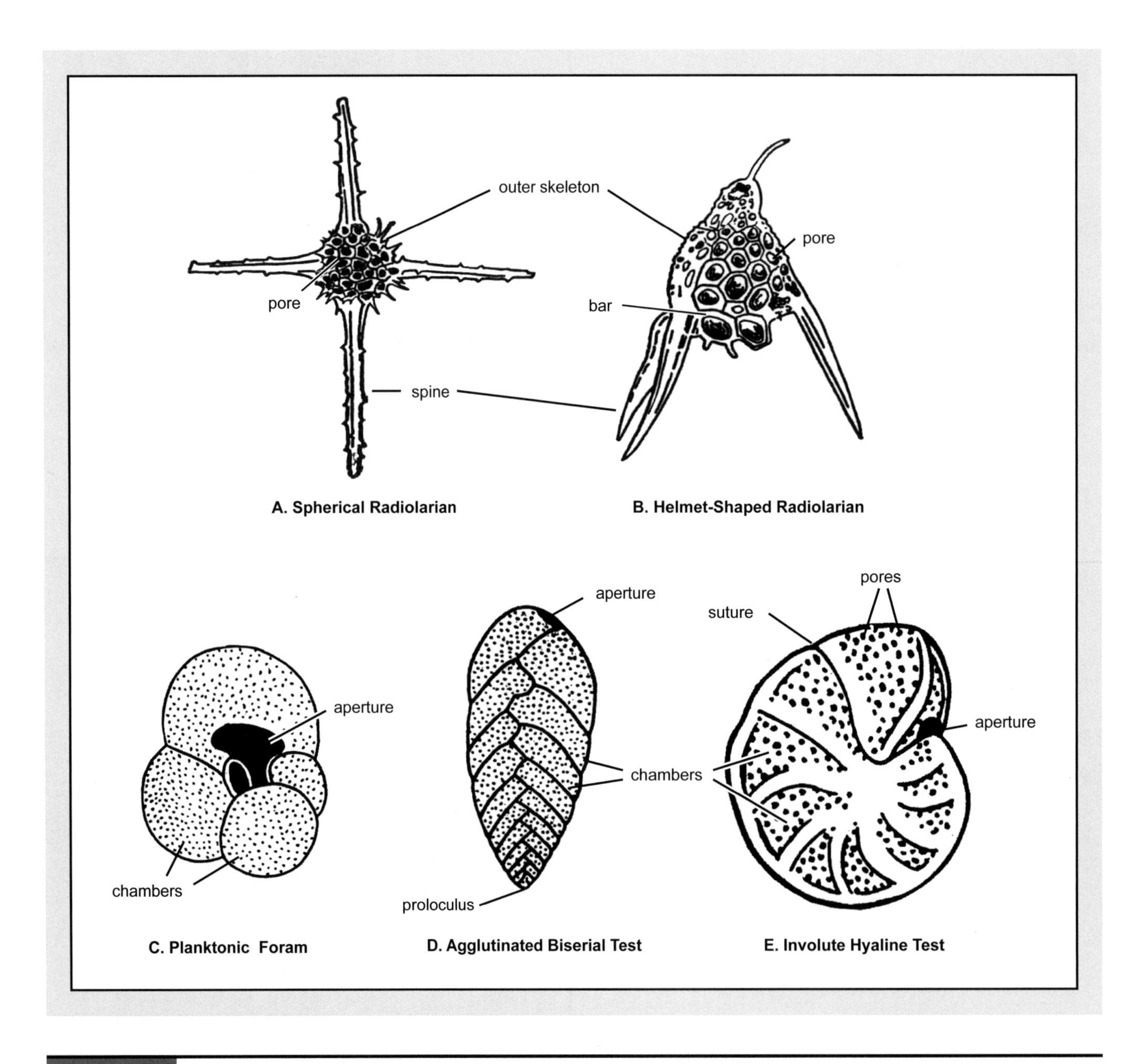

FIGURE 8.3 Morphology of the protozoans. Upper row indicates morphology of radiolarians. Lower row shows planktonic and benthonic foraminifera. (Modified from Moore, Lalicker, and Fischer, 1952.)

Archeocyatha

Archeocyathids (ancient horns or cups) are extinct invertebrate animals characterized by a calcareous (calcite) cone-in-cone skeleton (figure 8.4). The inner cone is called the inner wall. It is separated from the outer wall by open spaces called intervalla (pl.) and radial partitions called paries. All structural elements are perforated by small pores, which presumably permitted nutrient-rich, oxygenated water to pass through the archeocyathid's body, thereby allowing it to eat and breathe. This phylum appeared early in the Cambrian Period, thrived until Middle Cambrian time, and then began to decline in both diversity and geographic distribution, until its demise in the Late Cambrian. This is the only invertebrate phylum whose fossil remains are restricted to a single geological system. Although short-lived and relatively rare, archeocyathids were important elements of Cambrian marine ecosystems and were the first metazoan reef builders of the Paleozoic Era.

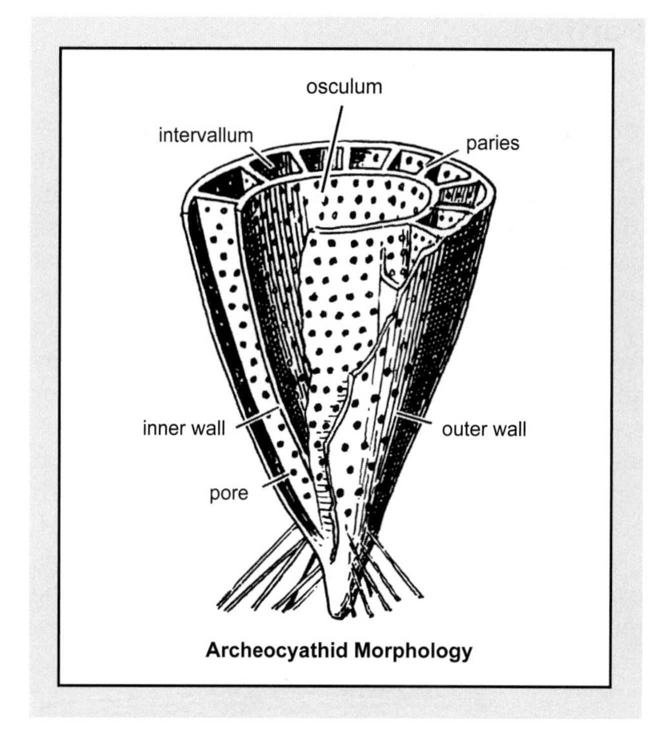

FIGURE 8.4 Skeletal morphology of archeocyathids. (After Moore, Lalicker, and Fischer, 1952.)

Porifera

Porifera, or sponges, are simple multicellular animals that live attached to the substrate (figure 8.5). A sponge may be thought of as a vase-shaped animal. Its body walls are penetrated by a series of canals of varying degrees of complexity. Microscopic food particles are removed from currents of water that pass through the canal system of the sponge. Most sponges are marine organisms, but some occur in freshwater. The sponge is held more or less rigid by an internal stiffening skeleton of fibers of spongin or spicules made of calcium carbonate or silicon dioxide. The commercial sponge is the flexible skeleton (spongin) of a modern marine sponge from which all of the once-living protoplasm has been removed.

Because of the nature of the body of the sponge, complete animals are rarely preserved as fossils. The most common fossil remains are the individual minute spicules of silica or calcium carbonate that provided support for the soft parts. The most complete fossil sponges are of groups where the skeleton was solidly fused during life. These sponges appear as conical, spherical, and plate-like fossils that are perforated by numerous small canals.

A group of sponges important as fossils are the stromatoporoids (figure 8.5). They were a colonial group of organisms that secreted a calcareous laminated skeleton. These laminated colonial skeletal structures resemble heads of cabbage or flat-lying sheets. Other more unusual stromatoporoids formed twig-like structures with a dendritic or branching mode of growth. Stromatoporoids are found in rocks ranging in age from Cambrian to Cretaceous and are quantitatively important in rocks of Ordovician and Devonian ages where they comprise large reef masses.

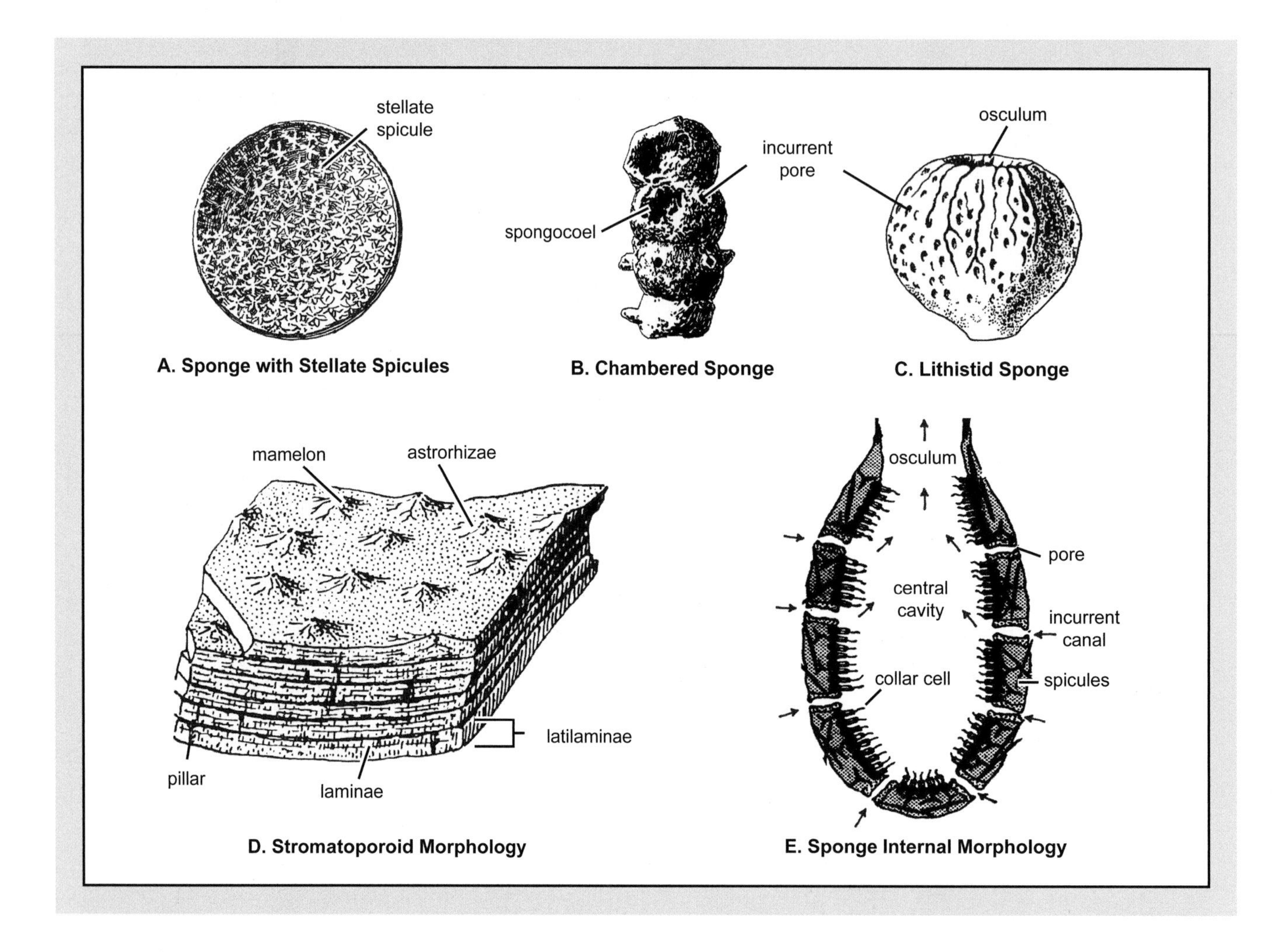

FIGURE 8.5 Morphology of the sponges. (After Moore, Lalicker, and Fischer, 1952.)

Phylum Cnidaria

Within the phylum Cnidaria, two groups are important as fossils. The first group, the corals, are among the most abundant fossils in sedimentary rocks, and are typically found as either "horn corals," the skeletal remains of a single organism (figure 8.6A), or as a group of individual skeletons cemented together to form a colony (figure 8.6B and C). Solitary corals have a pit in which the animal was attached to the top, broad part of the calcareous "horn." The "horn" functioned as an external skeleton and was built in daily increments by the organism. Corals first appeared in Ordovician rocks; continued as a major group through the Paleozoic, Mesozoic, and Cenozoic; and are common in modern seas. Corals form reefs in modern seas and inhabit the shallow areas of the ocean near the equator, and it is presumed that they did so in the past. Paleozoic corals (rugose and tabulate corals) constructed their skeletons using low magnesium calcite. Mesozoic and Cenozoic corals, known as scleractinians, used aragonite in constructing their skeletons.

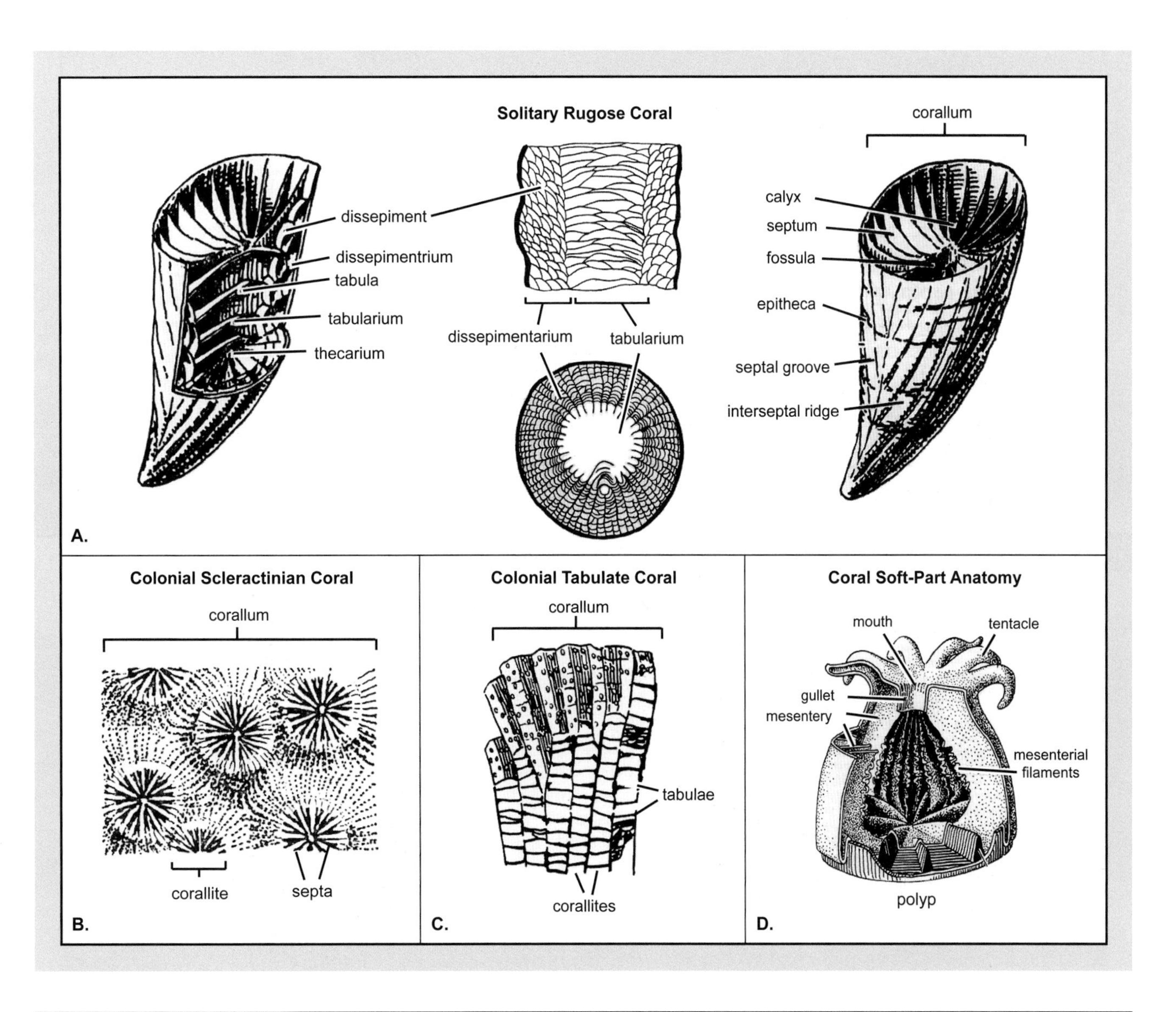

FIGURE 8.6 Morphology of the solitary and colonial corals. (After Moore, Lalicker, and Fischer, 1952.)

Phylum Bryozoa

Bryozoans, the moss animals, are small aquatic organisms that secrete colonial calcareous external skeletons (figure 8.7). Most bryozoans were marine but a few freshwater forms are also known. They are more advanced than corals and have a nervous system and a U-shaped digestive tract. The colonial organisms lived in the minute pores or tubes that perforate the stony skeletal structure. Fossil bryozoans resemble bits of lace or small twig-like structures, often encrusting other organisms or fossils. They are usually preserved lying parallel to the bedding planes of the enclosing layers. Bryozoa are found in rocks that range in age from the Cambrian to the Recent and are particularly abundant in rocks of Mississippian to Permian ages. Because of their small size, relatively slow evolution, and difficulty of identification, fossil bryozoans have not been used extensively in stratigraphic determinations.

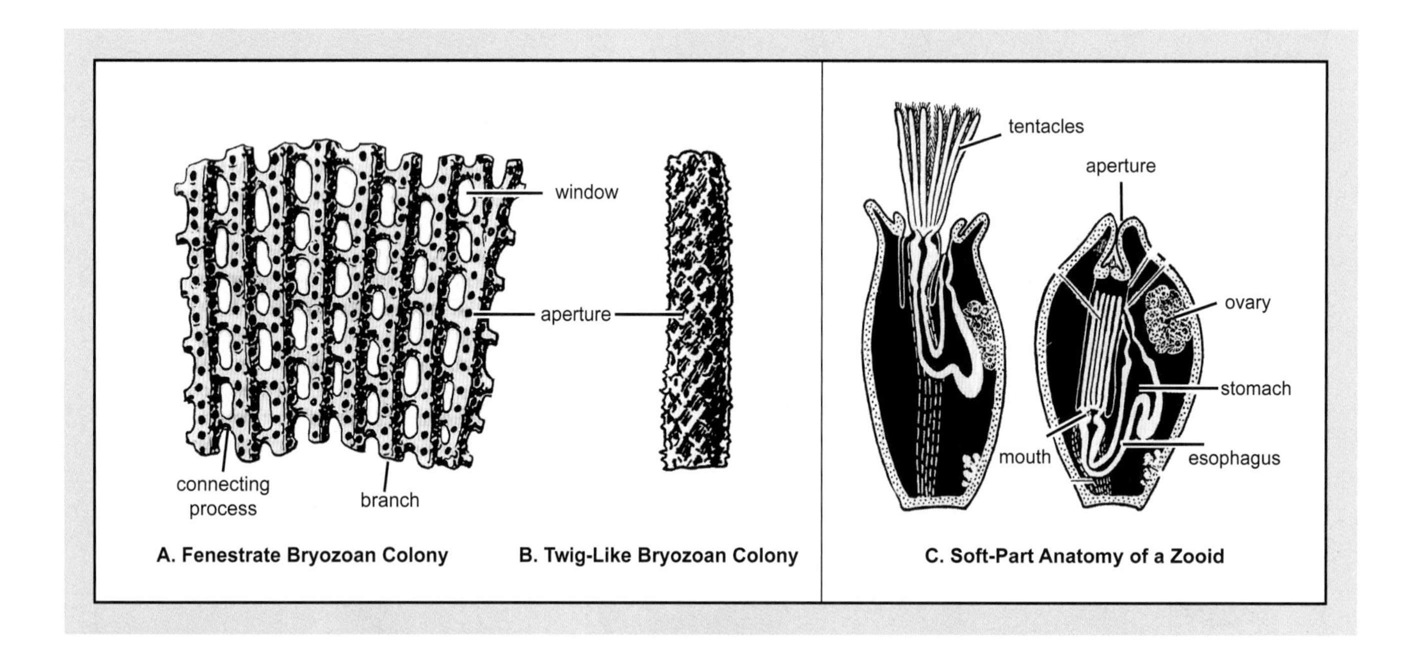

FIGURE 8.7 Morphology of bryozoans. (After Moore, Lalicker, and Fischer, 1952.)

Brachiopods

Brachiopods are marine invertebrates that were much more abundant in the seas of the Paleozoic Era than they are today (figure 8.8). They range in size from less than an inch to approximately 6 in at their broadest point. Their calcareous or chitino-phosphatic shells consist of two unequal valves that are symmetrical when divided into lateral halves. This shell shape distinguishes the brachiopods from the bivalves (clams, etc.), which are equivalved with right and left valves that are essentially mirror images of each other (figure 8.9B). They are among the most abundant fossil types found in rocks of Paleozoic age. Their shells are preserved in nearly every type of sedimentary rock. Because of their abundance, their great variety in shell form, and ease of identification, brachiopods are extremely useful as time and ecologic indicators in stratigraphic studies.

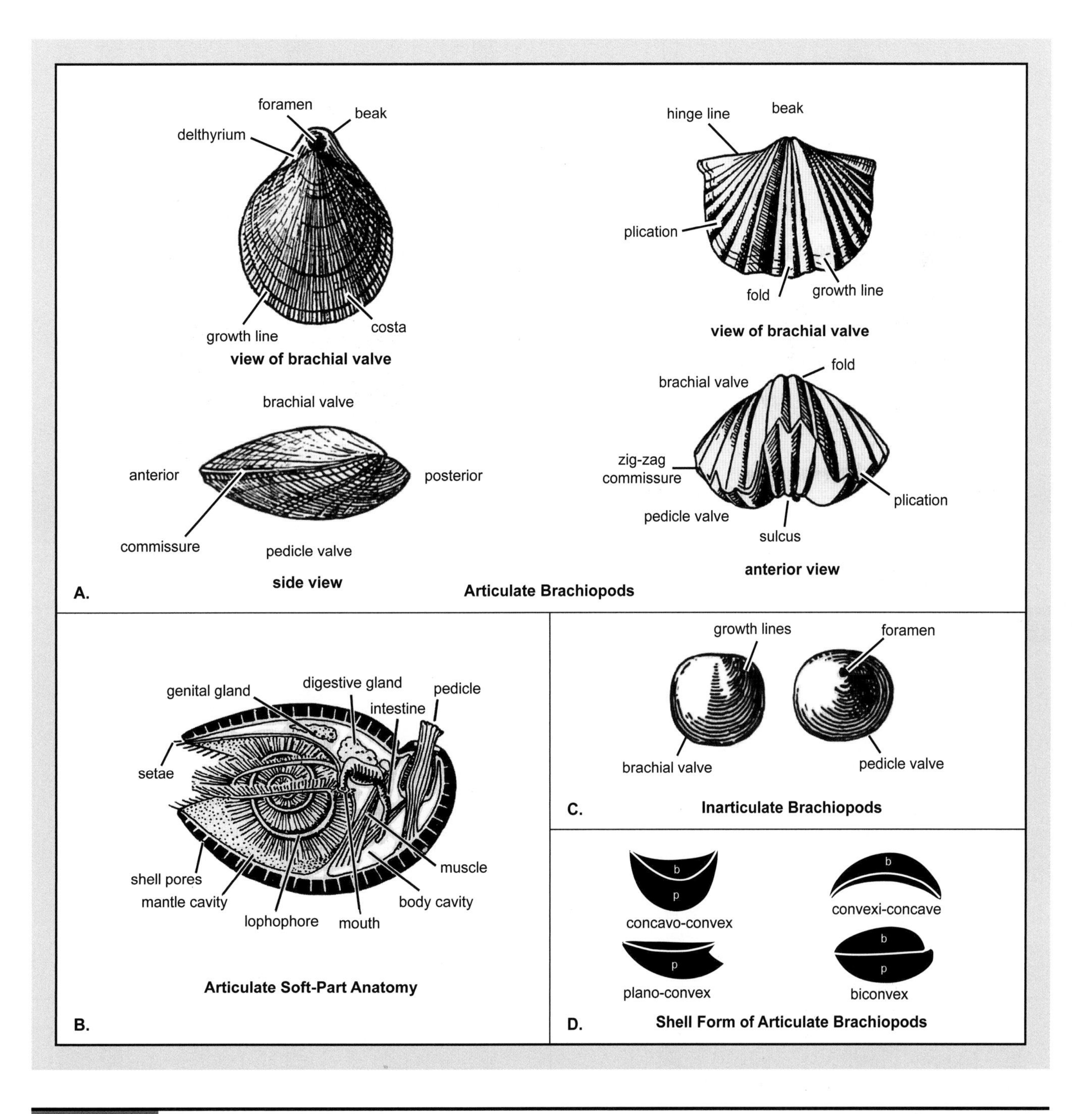

FIGURE 8.8 Morphology of articulate and inarticulate brachiopods. (After Moore, Lalicker, and Fischer, 1952.)

Phylum Mollusca

The phylum Mollusca includes animals that are zoologically similar, but superficially different because of the great diversity in the shapes of their shells. Included in the phylum are the gastropods (snails, slugs, pteropods) (figure 8.9A), bivalves (clams, scallops, and oysters) (figure 8.9B), cephalopods (*Nautilus,* squid, octopus, and cuttle fish) (figure 8.10), and other less common forms such as the chitons and scaphopods.

As fossils, mollusks are common in marine and nonmarine rocks from the Cambrian to the Recent. Their calcareous shells are valuable indicators of time and ecology. One group of cephalopods, the ammonoids (figure 8.10B), have complexly chambered shells and are used as a worldwide standard of reference in biostratigraphy from the Devonian through the Cretaceous. Mollusks have adapted at present to many ecologic niches from deep ocean benthonic environments to air-breathing existence on mountain peaks at 18,000 ft above sea level.

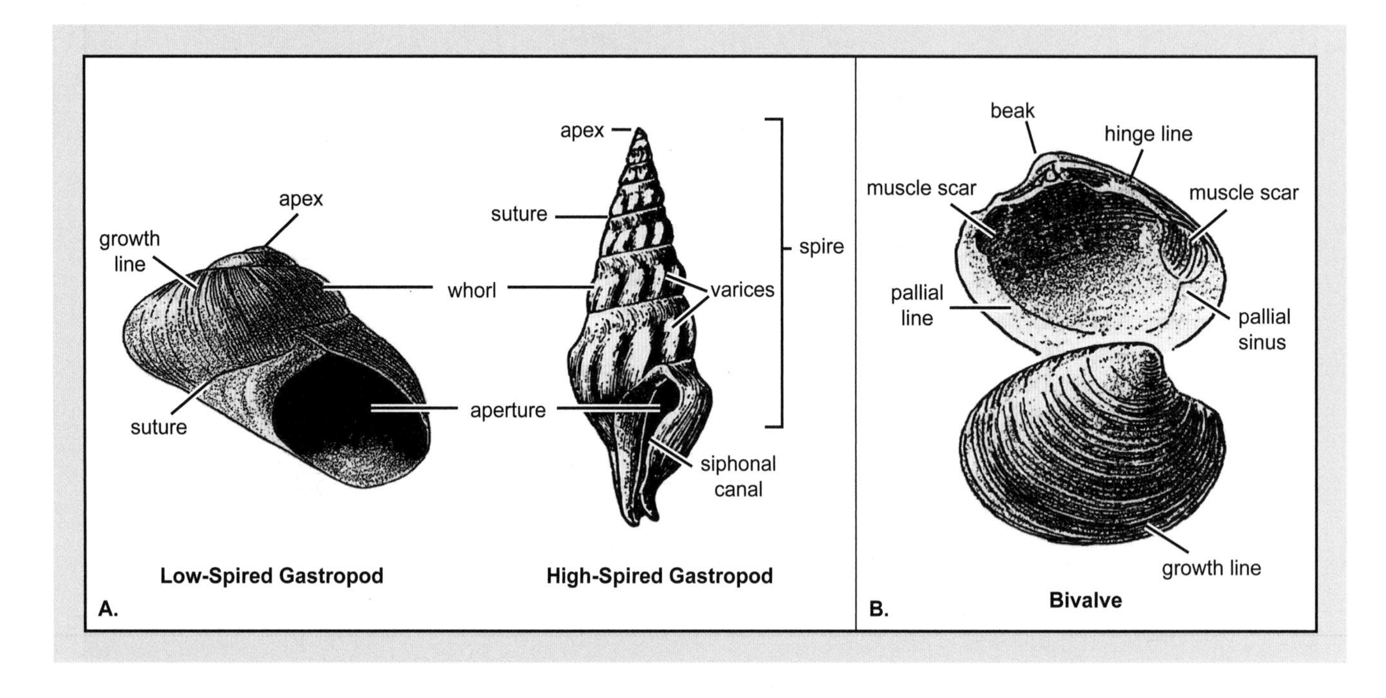

FIGURE 8.9 Morphology of bivalves and gastropods. (After Moore, Lalicker, and Fischer, 1952.)

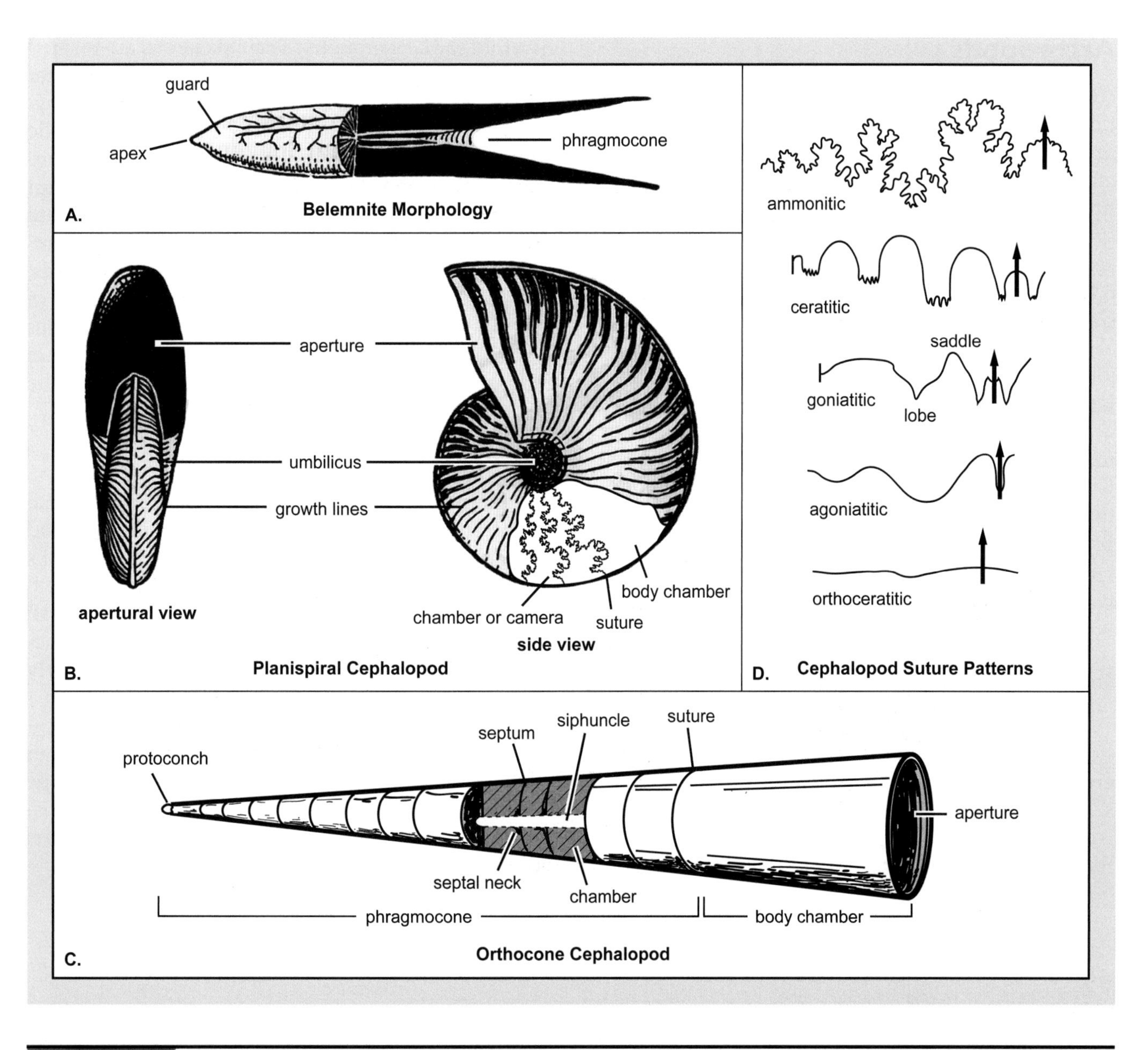

FIGURE 8.10 Morphology of cephalopods. A. Belemnite, B. ammonoid, C. nautiloid, and D. cephalopod suture patterns. (Figures A, B, and D after Moore, Lalicker, and Fischer, 1952.)

Arthropods

The arthropods are characterized by their chitinoid, segmented exoskeleton, or carapace. This group includes insects, crustaceans (crabs, lobsters, etc.), chelicerates (spiders, etc.), myriapods (centipedes and millipedes), and trilobites. Of these groups, only the trilobites and a type of crustacean, the ostracodes, are common as fossils (figure 8.11). Trilobites especially are abundant fossils in Cambrian, Ordovician, and Silurian rocks. Because arthropods shed their chitinous exoskeletons along joints, their fossil remains often consist of disarticulated heads, tails, or body segments.

Both trilobites and ostracodes are important fossils as time indicators. Trilobites lived from the Cambrian through the Permian, and the ostracodes lived from the Cambrian to the Recent. Trilobites are particularly useful guide fossils for Cambrian and Ordovician rocks, and ostracodes have been most used in rocks from the Ordovician, Silurian, Devonian, and Cenozoic.

Echinoderms

Echinoderms, or "spiny-skinned animals," are exclusively marine organisms and include the modern starfish, brittle stars, sand dollars, sea urchins, sea cucumbers, and sea lilies. This group of organisms was much more common in the geologic past than it is at present. Echinoderms typically display fivefold radial symmetry. Their skeletons are formed of calcite plates, which are secreted inside an outer tissue layer, and form an external skeletal covering. Echinoderms that are important as fossils include crinoids, blastoids, cystoids, and echinoids (figure 8.12). Echinoids (figure 8.13) are found throughout the entire geologic column from Lower Cambrian to Recent; however, they are more typical of Carboniferous, Permian, and Tertiary deposits. Blastoids, cystoids, and crinoids all reached high points of evolution during the Paleozoic. Many limestones of Mississippian age contain abundant crinoid stem fragments, sometimes making up the bulk of the rock mass.

Graptolites

Graptolites are a group of extinct colonial organisms whose proteinacious, sawblade-shaped remains appear superficially similar to pencil marks on the surface of rocks, thus their name, which means "writing on rocks." They are typically small fossils; colonies ordinarily measure a few centimeters across. The individuals of a colony are nearly microscopic in size (figure 8.14).

Graptolites are most important as time indicators in rocks of Ordovician, Silurian, and Devonian age. The Ordovician Period is often called the Age of Graptolites. Some graptolites lived into the Mississippian, but the group is not useful for time determination above the lower part of the Devonian.

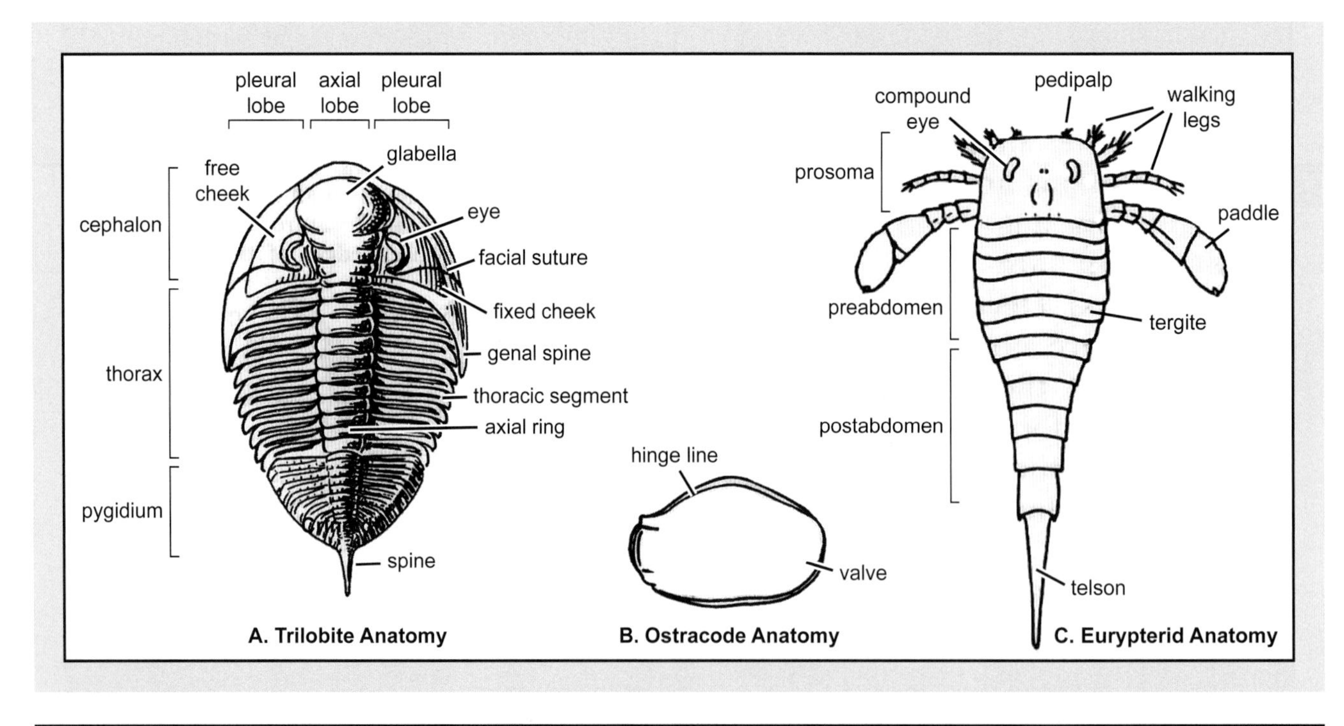

FIGURE 8.11 Morphology of three common fossil arthropods. (After Moore, Lalicker, and Fischer, 1952.)

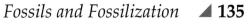

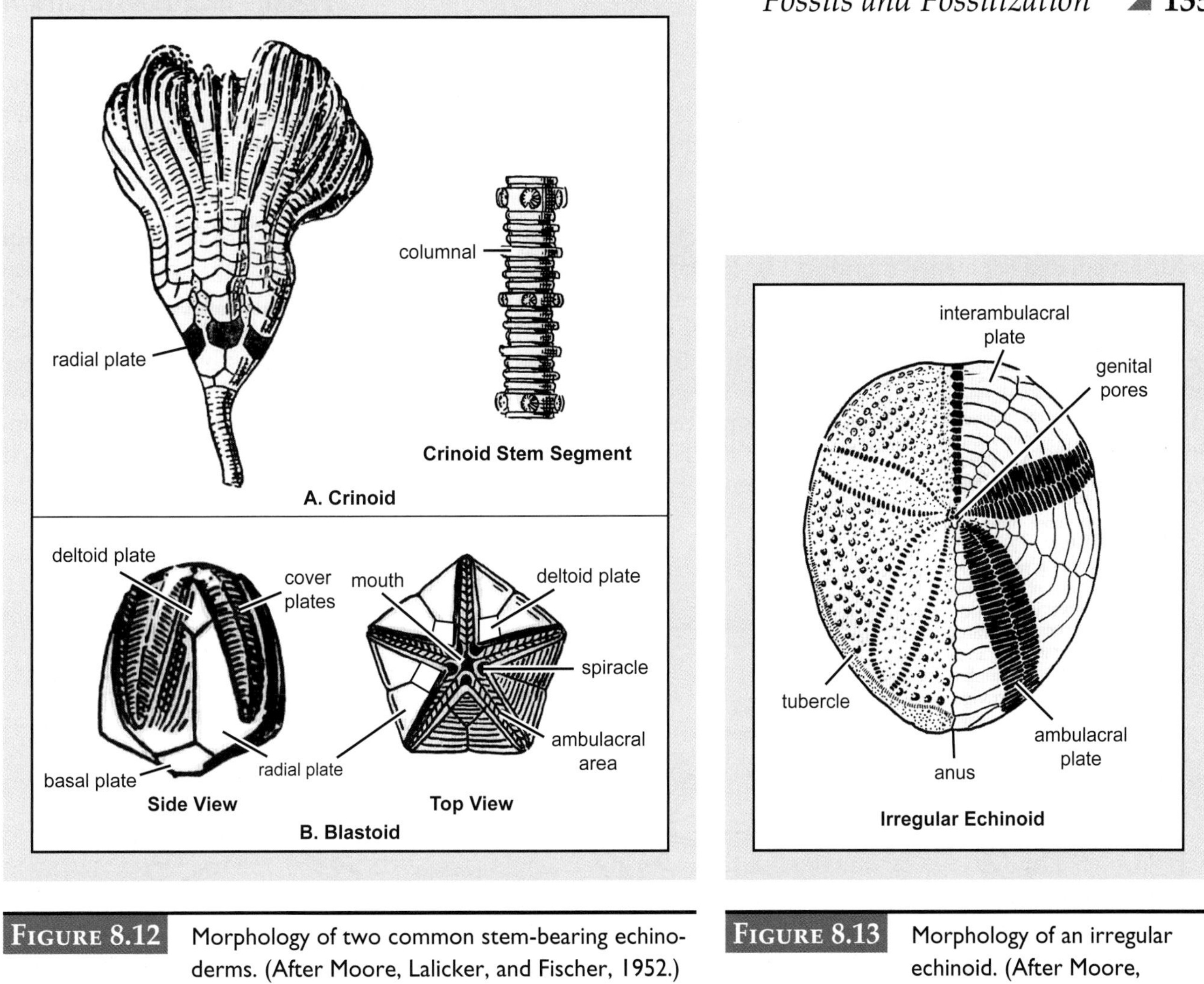

FIGURE 8.12 Morphology of two common stem-bearing echinoderms. (After Moore, Lalicker, and Fischer, 1952.)

FIGURE 8.13 Morphology of an irregular echinoid. (After Moore, Lalicker, and Fischer, 1952.

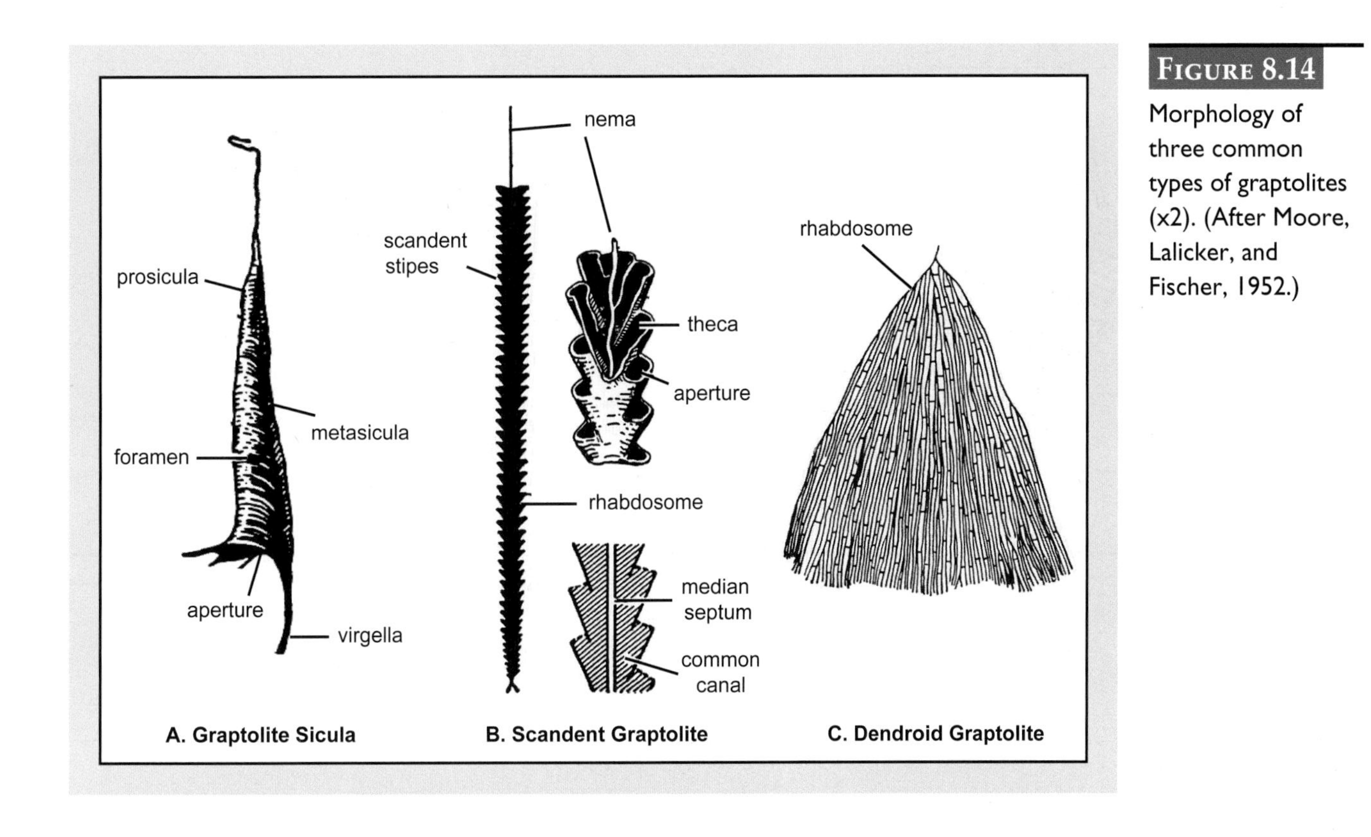

FIGURE 8.14 Morphology of three common types of graptolites (x2). (After Moore, Lalicker, and Fischer, 1952.)

Conodonts

Conodonts are an extinct group of phosphatic microfossils whose zoological affinities were unknown until recently. They are now placed in a separate phylum, Conodonta. Conodonts are found as disarticulated elements (figure 8.15). Elements were parts of an apparatus within the head region of the conodont animal. Conodont elements are generally designated P, M, and S according to their presumed positions within the apparatus. The contained elements as well as the resulting apparatus are unique for any given taxon. Generic and specific names of conodonts refer to the makeup of the apparatus, whereas the separate elements represent only a disarticulated portion of the remains of the conodont organism. Conodont elements have either right- or lefthand symmetry. Each symmetrical type is positioned on the appropriate side of the medial line of symmetry.

Most conodont elements are less than 1 mm in size and are medium to dark brown in color. They can be found in almost all kinds of sedimentary rock from the Late Cambrian to the Late Triassic. In the Ordovician, Devonian, Mississippian, and Triassic periods, they are the most useful fossils available for intercontinental correlation because of their widespread occurrence, abundance, and highly predictable evolutionary patterns.

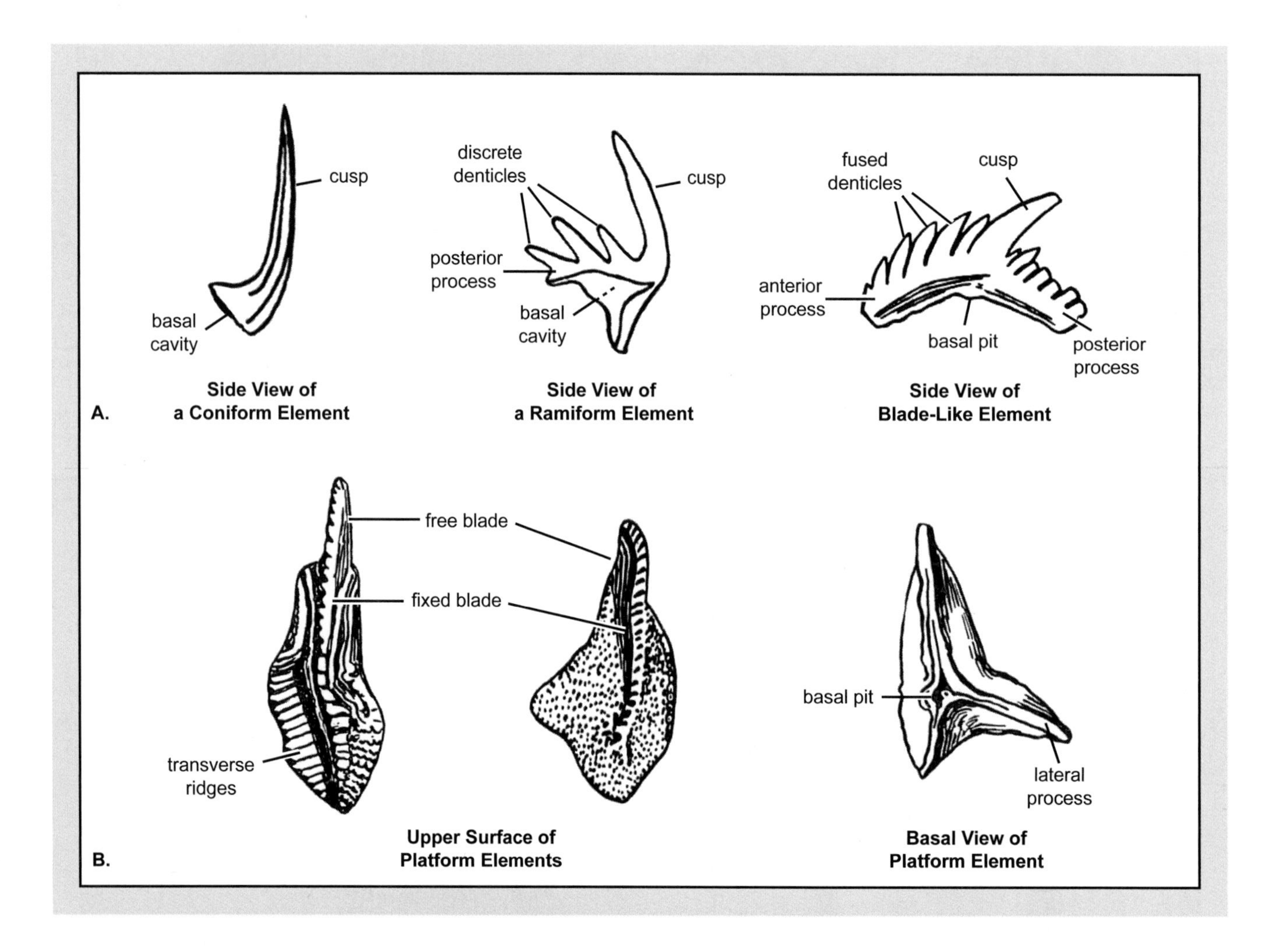

FIGURE 8.15 Morphology of conodont elements, showing parts of six common types (x50). (After Moore, Lalicker, and Fischer, 1952.)

PROCEDURE

The information presented in this exercise is provided to acquaint the student with the important morphologic features of the major fossil groups and to see examples of the different types of fossilization. While studying actual specimens provided by your instructor, identify the morphological features that characterize each group of fossils. It is common for a single specimen to display some morphologic elements, while others are not preserved. Observe several specimens of each type of fossil until you have personally observed all of the important morphological elements.

1. Examine the various fossils prepared for the exercise and determine the taxon to which it belongs. Record the type of fossil, original mineralogy, current mineralogy, mode of fossilization, and other preservational features in the appropriate boxes in table 8.4.

Specimen #	Taxon (trilobite, sponge, etc.)	Original Mineral Composition	Mineral Composition Now	Mode of Preservation	Biostratinomic Processes (disarticulated, broken, etc.)
GC-53	bivalve	aragonite	silica	replacement (silicification)	disarticulated

TABLE 8.4 Morphological elements of observed fossil specimens.

Exercise 9 Evidence of Evolution

Learning Objectives

After completing this exercise, you will be able to:

1. define the concept of organic evolution; and
2. indicate how vestigial and homologous structures support the theory of evolution.

In only a few centuries, by selective breeding, humankind has produced an amazing array of variation in several species, for example, dogs, cats, horses, fruits, and cereal grains. This human-made change through time is much more rapid but otherwise very similar to nature's change through time—**evolution**.

The results of evolution are evident upon close examination of living as well as fossilized organisms. Through the study of fossils, at least three lines of evidence have contributed to our understanding of evolution: vestigial structures, ontogenetic change, and homologous bone structure.

Vestigial structures are organs or features that at one time were functional but have now lost their usefulness. For example, the pineal body in humans is a remnant of a third eye that is still well developed in some reptiles. A small bony tail and ear-moving muscles are present in humans although their functions have been lost, and the small toe shown in figure 9.5G was a vestigial structure in three-toed horses. These organs are vestiges of once-functional features.

Ontogeny describes the growth or development of individuals throughout their life spans. Some organisms retain all growth stages in their development, from immature to mature growth forms. One such group is the ammonoids, whose growing shell encompasses the previously deposited part of the shell to form an expanding planispiral structure. By examining the growth stages of ammonoids (figures 9.1 and 9.2), as well as some other forms, it was seen that the development of an individual resembles the evolutionary stages in the history of the individual's lineage, or **phylogeny**.

PROCEDURE

PART A

In figure 9.1, the ontogeny of an ammonoid genus, *Perrinites*, of Permian age, is illustrated. The sutures A through F represent growth stages of the individual, A being the earliest, and F representing the mature stage.

In figure 9.2, only the mature sutures of six ammonoid genera from Pennsylvanian and Permian rocks are illustrated. These genera are not contemporaneous, but are found in rock units of various ages, although within these two periods.

1. The stratigraphic positions of the ammonoids, as represented by their sutures, have intentionally been rearranged. By studying the ontogeny of *Perrinites* (figure 9.1), rearrange the sutures in figure 9.2 into proper stratigraphic order, based on the biogenetic law, and compare with the order given by your instructor.

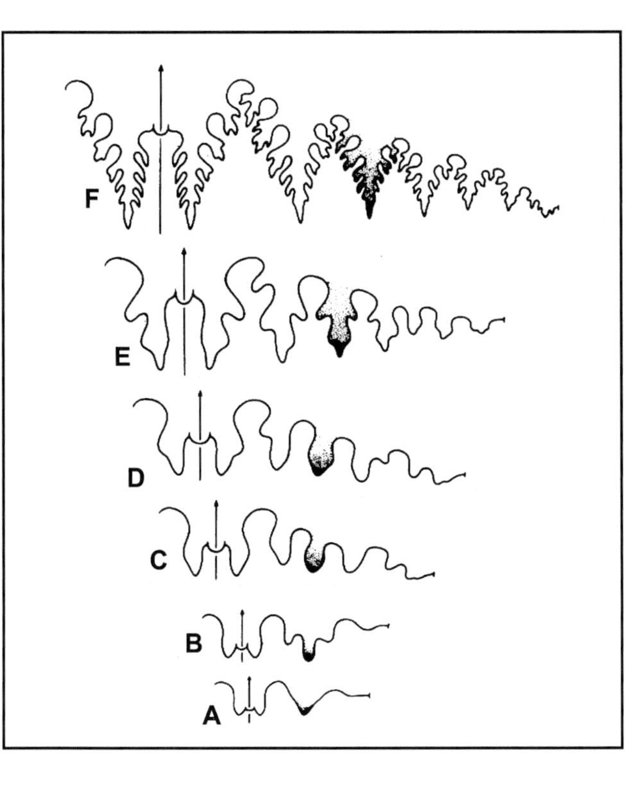

FIGURE 9.1 Drawings of sutures showing the ontogenetic development of the Permian ammonoid genus *Perrinites*. (Paleontological Institute, n.d.)

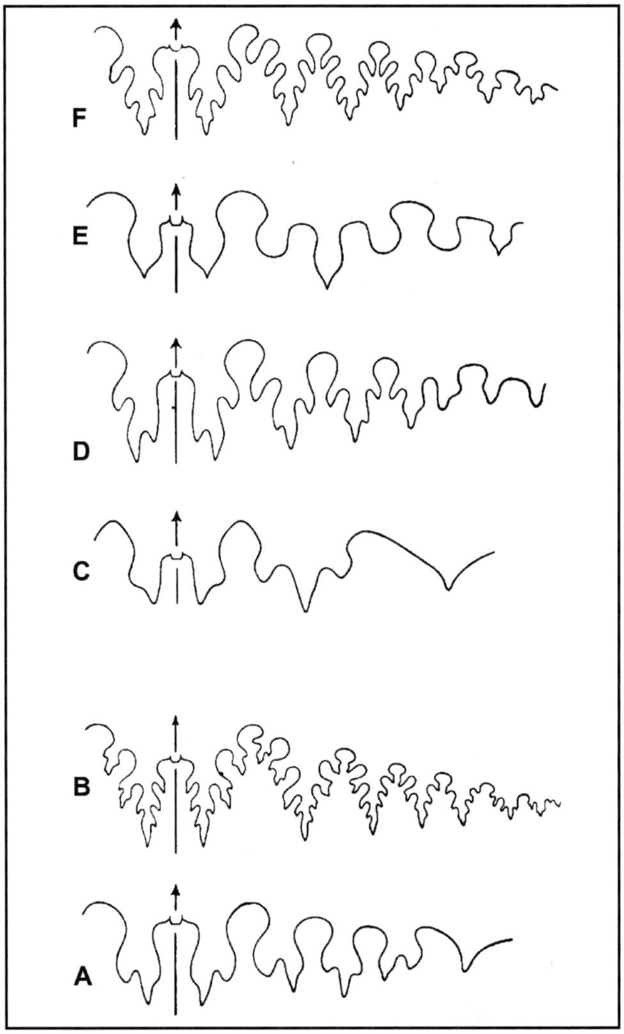

FIGURE 9.2 Drawings of adult sutures of six ammonoids, in a random arrangement of ages.

A. *Shumardites*.

B. *Perrinites*.

C. *Aktubites*.

D. *Properrinites*.

E. *Parashumardites*.

F. *Metaperrinites*.

Part B

Skeletal elements of structures that appear similar in different groups of organisms are termed **homologous structures**, even though their functions and overall shapes change slightly from group to group. For example, the arm and hand of a human are similar to, or have homologous bones with, the front limb of a horse, the wing of a bat or bird, or the front limb of a marine reptile. The relative position and structural relationships of the respective bones remain the same, but various types of limbs can be produced by changing the shape of individual bones. This similarity is the result of modification through time to accommodate a particular environment or organic function.

1. Various bones are identified on the skeleton of a human in figure 9.4. Homologize as many of these bones as possible on the skeleton of *Dimetrodon* (figure 9.3). *Dimetrodon* is one of the fin-backed, mammal-like reptiles of the Late Paleozoic.

2. Color each of the homologous bones of the front limb of the mammals, reptiles, and birds illustrated in figure 9.5. Color the scapula green, the humerus red, the radius blue, the ulna pink, the metacarpals yellow, and the phalanges purple.

3. Summarize the modifications that are evident in the various limbs and explain, if possible, the reason for nature's selection of the modification, in view of the environment where the various organisms lived.

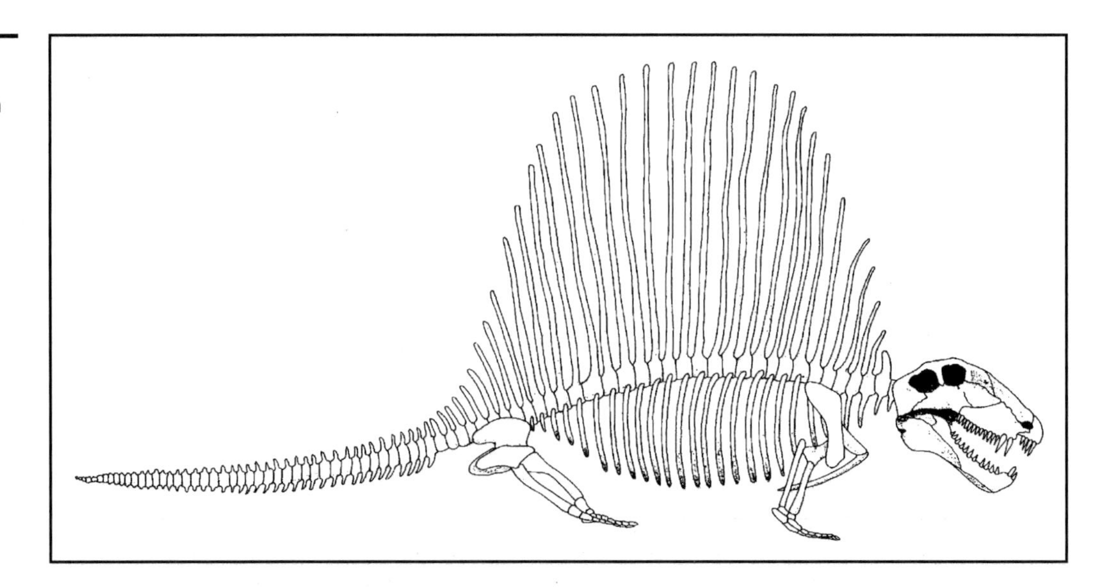

Figure 9.3

Diagram of a skeleton of the fin-backed Permian reptile, *Dimetrodon*.

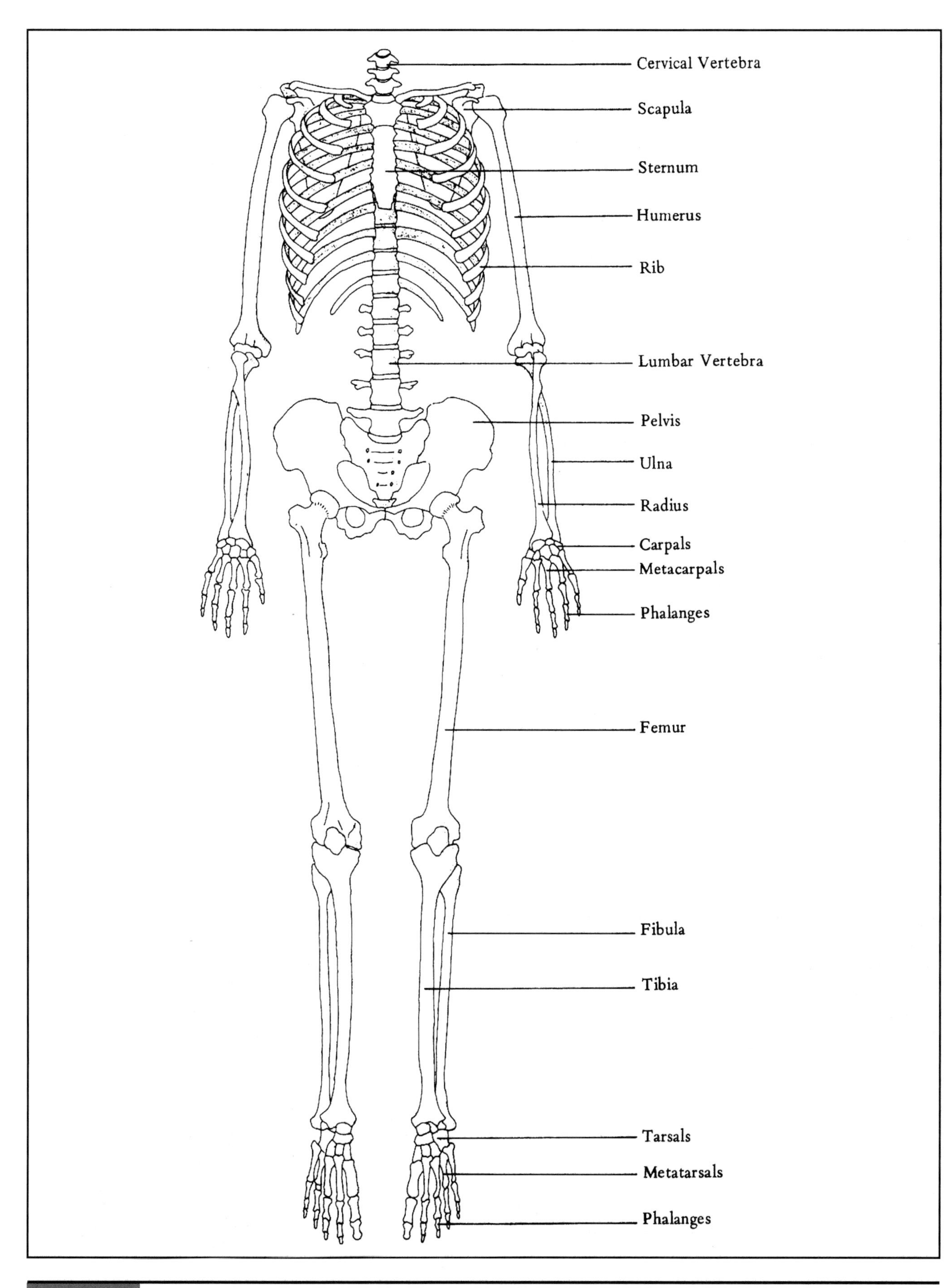

FIGURE 9.4 Diagram of the skeleton of a human, with various bones labeled.

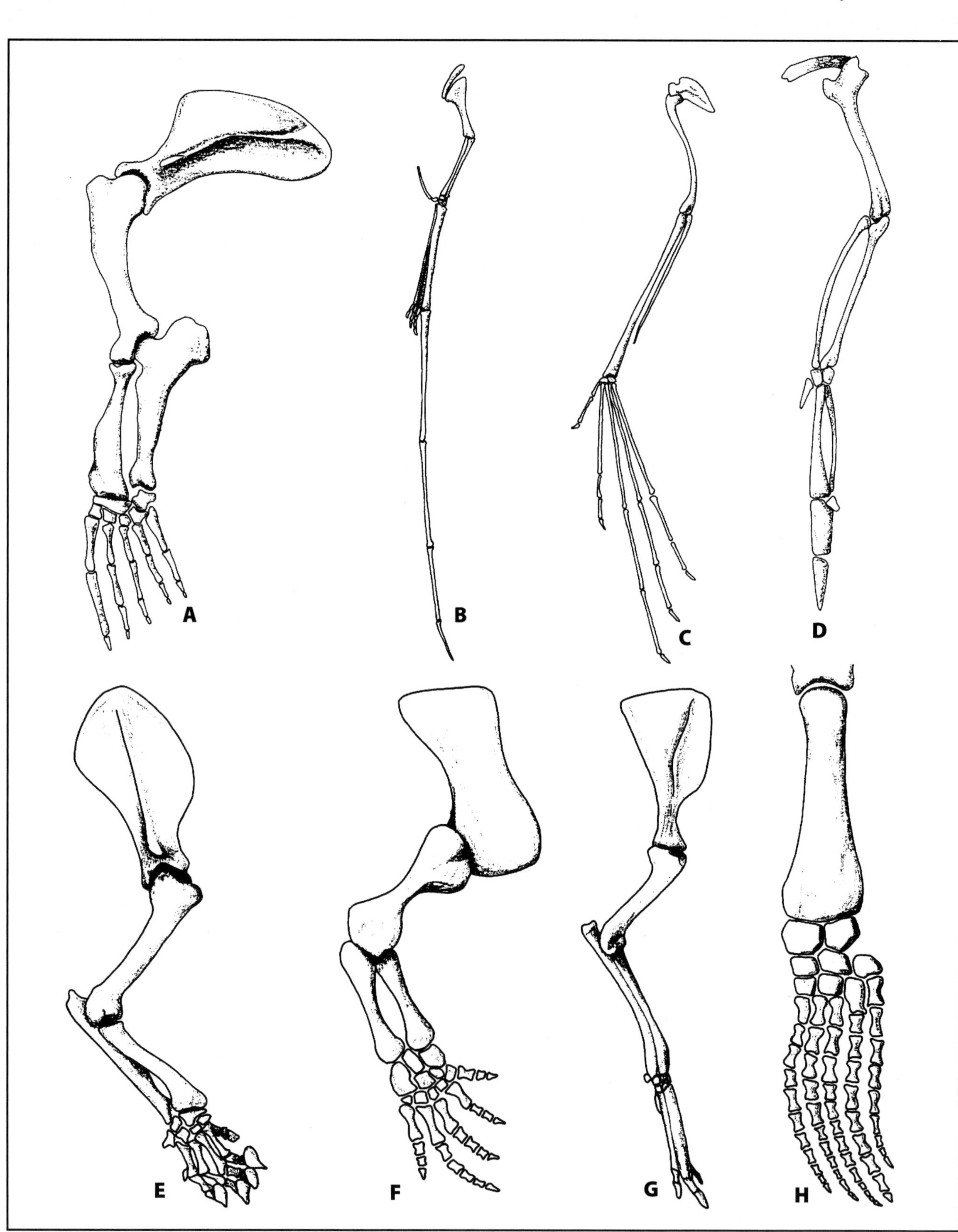

FIGURE 9.5 Diagrams of the forelimbs of eight vertebrate animals.

A. Modern seal.
B. Cretaceous pterodactyl.
C. Modern bat.
D. Modern bird.
E. Quaternary sabre-toothed cat.
F. Permian mammal-like reptile, edaphosaur.
G. Oligocene three-toed horse.
H. Jurassic plesiosaur.

Patterns of Evolution

Exercise 10

Learning Objectives

After completing this exercise you will be able to:

1. explain the concept of adaptive radiation, convergence, and divergence; and
2. understand the differences between phyletic gradualism and punctuated equilibrium.

The fossil record is one of the most conclusive pieces of evidence of change in organisms through geologic time. Since the first appearance of life on Earth (sometime during the Precambrian), organisms have undergone continual change, or evolution. Lines of descent of organisms through time are termed **phylogenies**. The study of enzymes and proteins of living things (molecular biology) is the principal means of reconstructing the universal tree of life. The fossil record, however, is better suited for studying long-term evolutionary patterns of both organisms and communities.

Generally speaking, the soft parts of fossil organisms are not available for study, because only hard parts are preserved. This preservation naturally limits the interpretive value of fossils in reconstruction of lineages because many of the most characteristic features of organisms are lost in the process of fossilization. The aspect of the organism that remains is usually only the form or shape of the original hard parts. A phylogeny of fossil material is, therefore, an interpretation based on comparison of similar forms, their stratigraphic occurrence, and apparent changes in morphology. Such studies are called **comparative morphology**.

Various patterns of evolution, or inferred phylogenies, are illustrated by fossil study. Some of these patterns are illustrated on figure 10.1. Convergence is one pattern commonly followed by evolving organisms. **Convergence** is a pattern of evolution that is the result of dissimilar or unrelated organisms becoming more similar-appearing through time, probably in response to life in similar environments. If the convergence is essentially contemporaneous in both lineages, the pattern is termed **isochronous** convergence. An example is shown in figure 10.1D. If the convergence, however, occurs at different times, such as illustrated in figure 10.1E, the pattern is termed **heterochronous** convergence. Several remarkable examples of convergence are found in fossil records (for example, the development of the marsupial Tasmanian wolf of the Australian region to that of the placental doglike forms in the Northern Hemisphere). The convergence in gross shape of mammalian dolphins and whales to that of rapidly swimming fish is immediately apparent. Several fossil reptiles also show the same general trend toward streamlining. The latter instance is an example of heterochronous convergence since the swimming reptiles were at their peak of development in the Mesozoic Era and the mammals at their peak in the later Cenozoic. The spectacular development of flight in reptiles (pterodactyls), mammals, and birds is another example of convergence because of the adaptation of each group to a similar mode of life. Examples can also be drawn from among invertebrate organisms. Various genera of brachiopods and bivalves converge at different times to form structures similar to solitary horn corals. Mimicry in insects might also be considered an example of convergent evolution where similarity to the successful group is protective.

Parallel evolution—where two lineages show the same general trend in morphology—is another pattern that is demonstrated in the fossil record. In dinosaurs, for example, there is a general tendency in several lineages for an increase in size. Fusulinid foraminifers show other examples of parallelism where several lineages show a gradual increase in wall thickness and a corresponding increase in complexity of wall folding and other structural elements. These fossils occur in the same beds and were undoubtedly contemporaneous. In ammonoid mollusks, the same patterns may be seen where separate lineages appear to independently develop increasingly complex shell chamber walls.

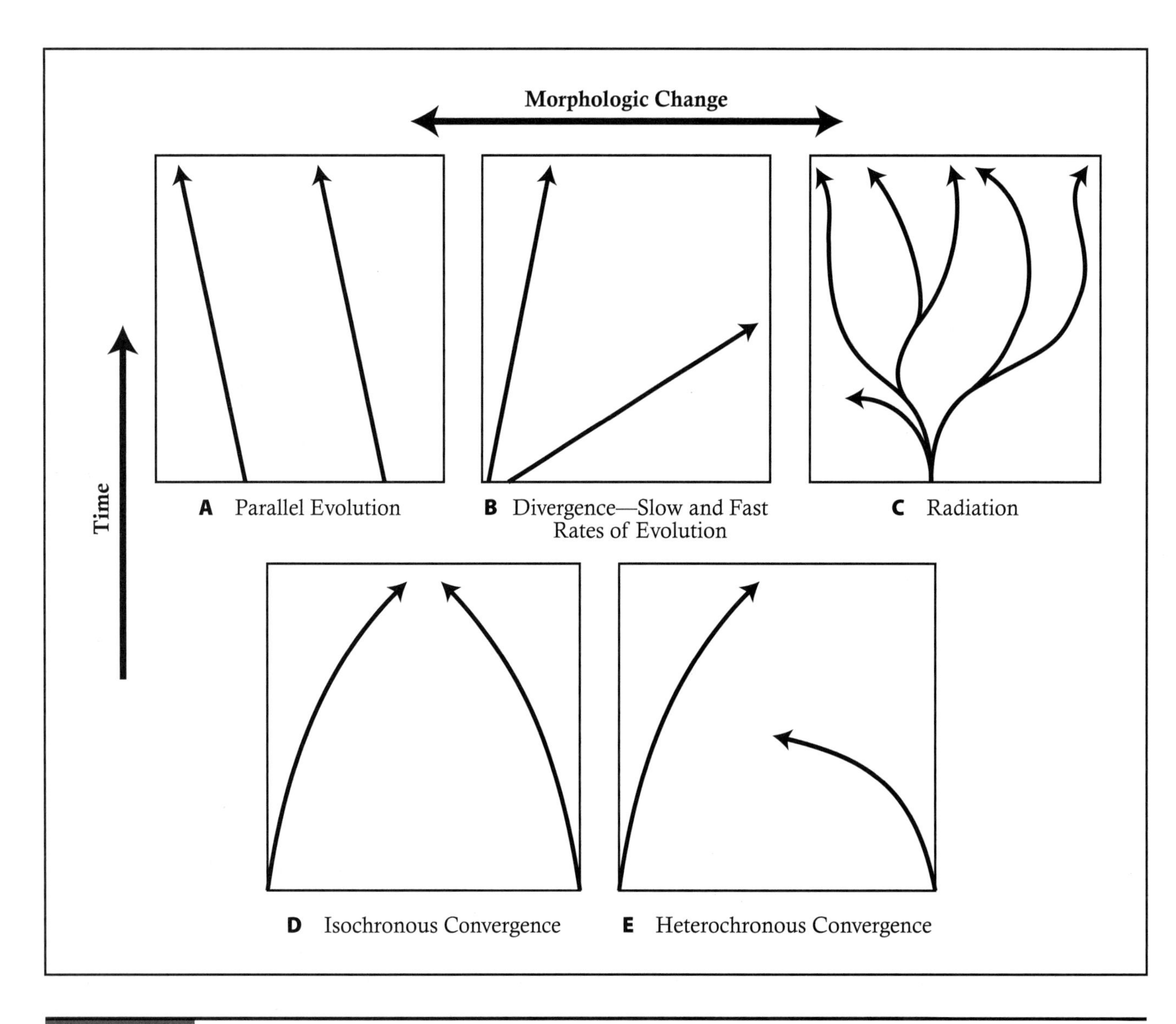

FIGURE 10.1 Graphs of patterns of evolution. Time is plotted vertically and morphologic change is plotted horizontally.

Divergence is one of the most common evolutionary patterns in the geologic record. Divergence occurs where two closely related lineages in the geologic record become less and less similar as they are traced in younger and younger rocks. Morphologic differences are interpreted to suggest change in environment, with one lineage specializing in one environment and the other in a distinctly different one. An example of divergence can be seen in horses, where browsing and grazing horses diverge from one another in the middle part of the Tertiary Period. The illustration of divergence, figure 10.1B, shows relative rates of evolution with a slow-rate and a fast-rate line. The slow-rate line illustrates the least change over the longest period of time and is the steeper of the two orthogenetic trends shown in figure 10.1B. The fast-rate line has the more gentle gradient and the greatest amount of structural change in a short time. Examples of very rapid evolution can be seen in many groups of invertebrate fossils and in some of the more specialized vertebrate fossils as well, where considerable morphologic change was accomplished in a very short time. Ammonoids, some mammals, and graptolites all have high-rate evolutionary lineages. Slow evolutionary rates are exemplified by some inarticulate brachiopods, which have changed little since the group appeared in the Cambrian over 500 million years ago. Other classic examples are forms like the opossum, which has changed little since the early Tertiary.

Radiation is a common pattern and is seen where several lineages diverge from one another. An example of radiation is seen in the Mesozoic history of reptiles, which radiated to become the largest herbivores and carnivores on land. Some reptiles began to fly, and others returned secondarily to the sea. Several lineages of reptiles and mammals have returned to the sea; however, they have not developed gills but have retained lungs like their terrestrial ancestors.

Not since the years directly following Darwin's monumental publication has paleontology more strongly influenced evolutionary thought than in the last 25 years. Happily, the fossil record, now much better understood than in the late nineteenth century, has become a fertile field of data to test modern ideas. For the past several years, two ideas have been popular points of debate among paleontologists and neontologists. The ideas, called **phyletic gradualism** and **punctuated equilibria**, are models for the mode and tempo of evolution. They do not challenge the idea of evolution, only the mode and tempo of the theory. The basic question debated is, does evolution proceed gradually in a continuous pattern, or is it episodic?

If gradualistic, then the fossil record should display intermediate forms as progenitors, responding to a slowly changing environment, giving rise to descendants. The two species would show such a complete transition in their morphology that separating the two would be arbitrary. On the other hand, if episodic, the fossils should show abrupt change between species followed by a relatively long period of stasis (figure 10.2A). Proponents of punctuated equilibria have discovered numerous examples from the fossil record to support their interpretation. Supporters of the gradualistic model also found support from some fossil groups, especially mammals and protozoans (figure 10.2B). Most students today see evidence for both (figures 10.2C, D).

PROCEDURE

PART A

The drawings in figure 10.3 illustrate geometric forms from eight successive stratigraphic horizons. The oldest (T_1) represents the lowermost series of beds in the sequence and contains the most primitive shapes. Younger beds successively yield newer or more specialized kinds of fossils, until the youngest fossils are encountered in the topmost bed of the sequence.

Because many students have trouble manipulating complex shapes, such as those characteristic of many fossil lineages, the shapes here have been simplified to give only a sequence of geometric forms. These forms were constructed to show various kinds of evolutionary patterns, such as convergence, parallel evolution, divergence, and radiation. In several instances, there is no single correct or right answer. There are some forms that—by convergence—may be produced as a result of either of two or more converging lineages. These are termed **polyphyletic** forms. Other lineages seem best interpreted as deriving from a single root stock, which is called **monophyletic**.

1. Begin with the oldest series of shapes (layer A) and work upward, connecting by pencil lines or identifying by letter and number the likely phylogenetic series in each successive layer.

2. Tear out (or copy) the page and cut the fossils apart, rearranging them into patterns somewhat like that of the graph of radiation in figure 10.1C. You may identify the fossils by letter and number and report your work.

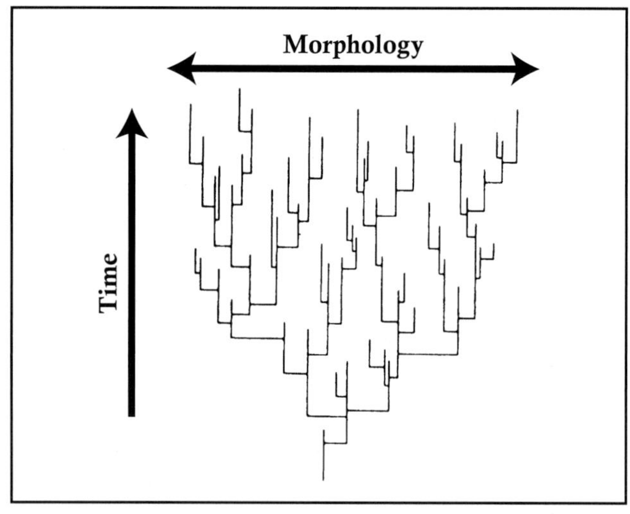

A. Extreme view of punctuational model. Sudden change in morphology (evolution) followed by static condition.

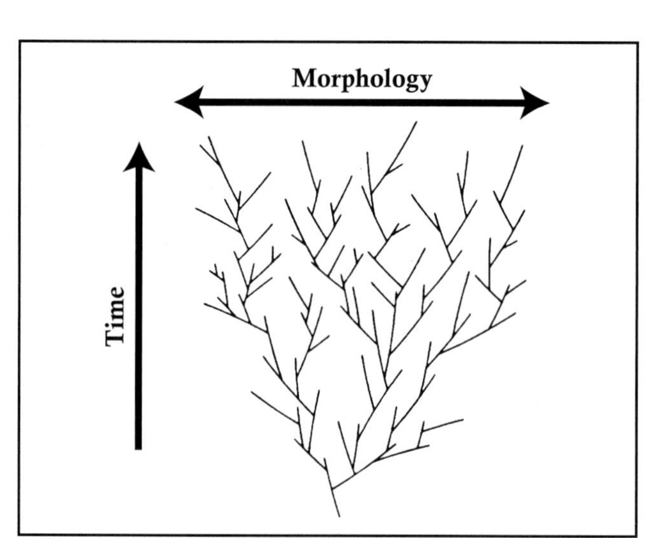

B. Extreme view of gradualistic model. Evolution is gradual, continuous, and nearly uniform as shown by slope of lines representing species.

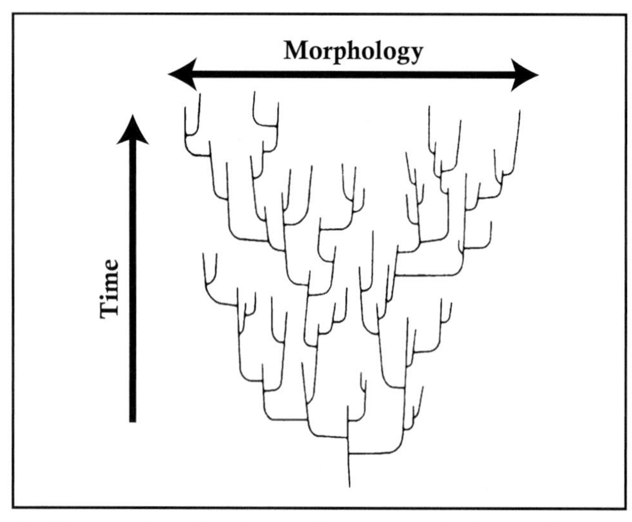

C. Punctuationalist view with some gradualistic influence.

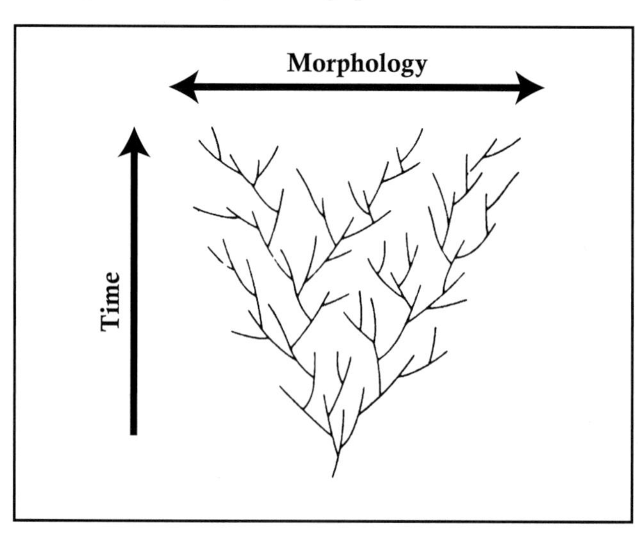

D. Gradualistic model with some accommodation for punctuationalism.

FIGURE 10.2 Hypothetical phylogenies. (After Stanley, 1979.)

3. Identify by number and letter an example of each of the following: a convergent, parallel, divergent, and radiating trend in the evolutionary pattern. Cite examples of isochronous as well as heterochronous convergence, if they occur. Can you see monophyletic and polyphyletic lines?

4. Connect the various series with arrows to show trends. Indicate with asterisks those lineages that became extinct based on the horizon or level of occurrence of the last form in the series.

5. Is the dominant mode and tempo of evolution in the relationships shown in figure 10.3 gradualistic or punctuational?

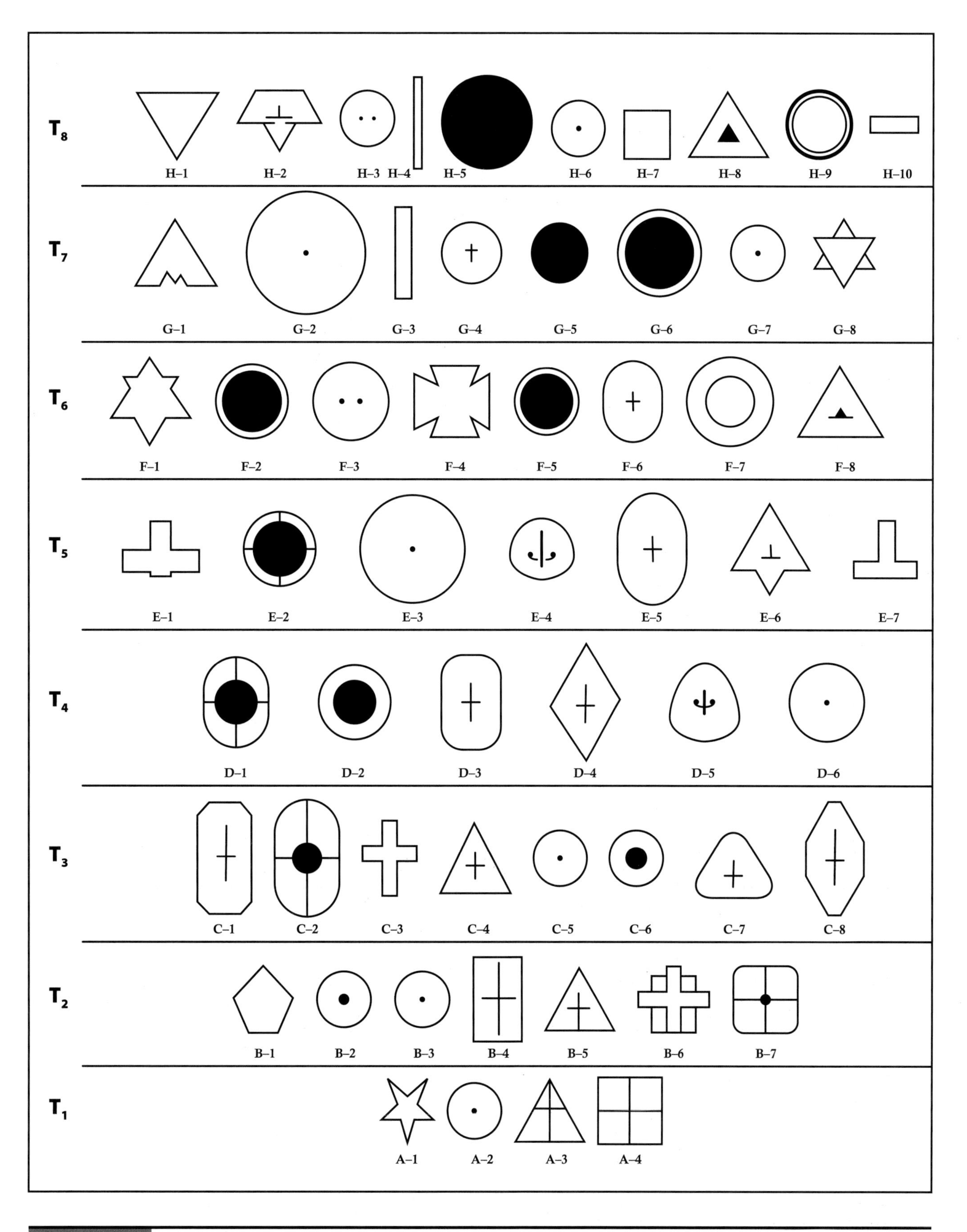

FIGURE 10.3 A diagram of various geometric shapes to simulate kinds of fossils in eight stratigraphic intervals, to be used in a problem showing evolutionary patterns and phylogenetic trends. Time is represented by the letter "T"; the oldest horizon is labeled T_1.

PART B

Figure 10.4 illustrates 10 stratigraphic horizons of graptolite faunas, with the oldest at the bottom and the youngest at the top. The entire sequence of organisms in figure 10.4 represents fossils that first appeared in Upper Cambrian or Lower Ordovician rocks and continued through the Silurian into the Devonian. Each of the collections is distinctive of a separate faunal zone of early Paleozoic graptolite evolution. Graptolites are an extinct group whose characteristics evolved rapidly through geologic time. Because of their rapid change, they are used as time stratigraphic indicators in Ordovician and Silurian deposits.

1. Establish possible lineages or phylogenies for the various graptolites shown, as was done on figure 10.3, with a series of pencil lines showing phylogenies on the page. A more clearly defined interpretation can be obtained by copying the page and cutting the various forms apart, rearranging them in likely series. Be careful to maintain the fossils at their correct stratigraphic horizon. Each of the graptolite zones represents several million years of time and therefore it is possible to have one form derived from another within the same stratigraphic interval. In many instances, the entire suite of fossils illustrated may occur on a single bedding plane and may record an instant in terms of organic evolution.

2. Locate with numbers and letters possible examples of parallel evolution, convergence, divergence, and a possible polyphyletic origin.

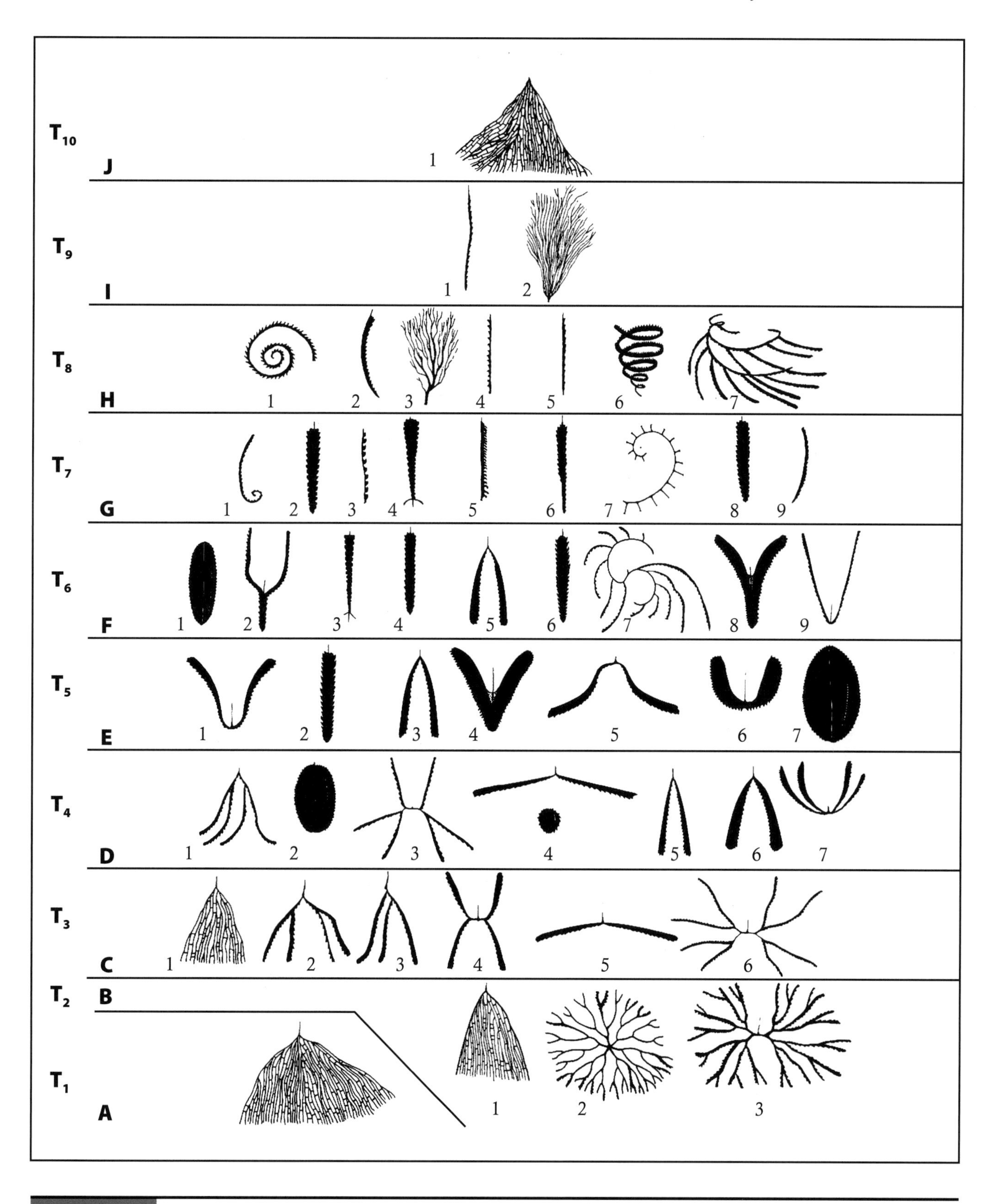

FIGURE 10.4 A diagram of graptolite occurrences from 10 successive stratigraphic horizons, a problem in reconstruction of phylogenic trends. Time is represented by the letter "T"; the oldest horizon is T_1.

Part C

Figure 10.5 illustrates mammal teeth in a stratigraphic succession like the graptolites in figure 10.4. By comparing the kinds of teeth shown in the illustration, reconstruct possible evolutionary trends in this group of vertebrates. The evolutionary trends can best be seen by studying the patterns of the tooth surfaces and the general outline of the individual teeth. The dark areas on the teeth are areas of exposed dentine due to wear. The basic pattern of this dentine is a distinctive characteristic of every mammal, and an excellent means of identification. Teeth serve well for such an exercise, because the tooth structures of each mammal are highly specialized and are a major basis of classification in this group of organisms.

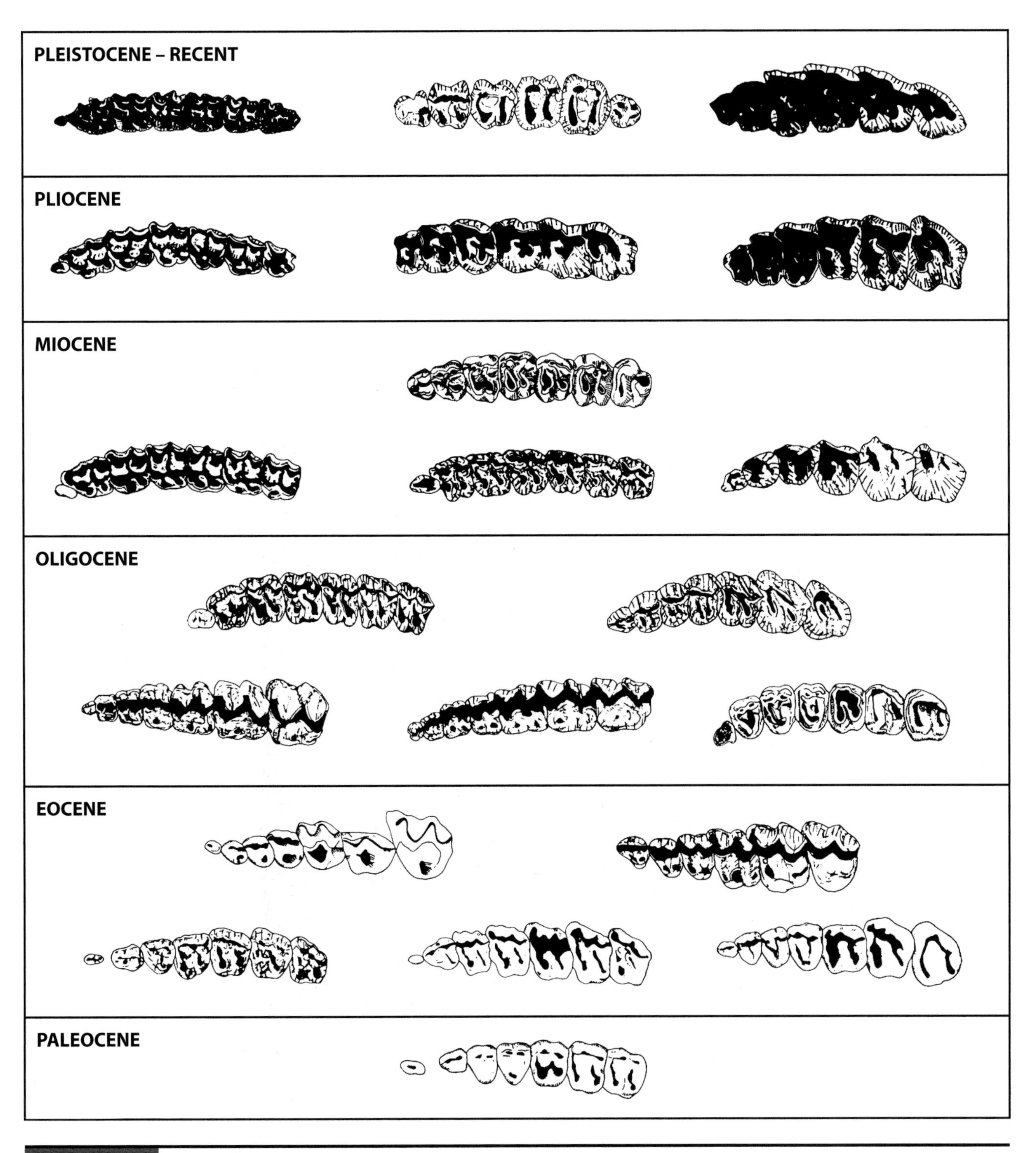

Figure 10.5 A sampling of mammal cheek teeth (upper) as they occur in Cenozoic strata.

Continental Drift and Plate Tectonics

Exercise 11

Learning Objectives

After completing this exercise you will be able to:

1. better understand the concept of continental drift as espoused by Alfred Wegener;
2. know how geologists use the distribution of earthquakes to define plate boundaries; and
3. explain how magnetic reversals are used to date seafloor rocks.

The modern theory of plate tectonics was developed from Alfred Wegener's (1912) earlier concept of continental drift. Geologists now explain the mechanism of geologic change in relation to at least 10 major plates making up the Earth's lithosphere. These plates are approximately 100 km thick and move over the Earth's surface at an average velocity of 2 to 4 cm per year.

Plate interactions are responsible for most of the Earth's major tectonic events and features. Plate margins are classified according to the relative motion of the plate across the margin, such as:

- divergent—plates moving apart;
- convergent—plates moving together; or
- transform faults—horizontal side-by-side movement.

Convergent plate margins are further classified on the basis of the types of crust involved in abutting plate margins, such as:

- continent–continent, where two continental masses collide along opposite sides of plate boundaries;
- continent–ocean, where continental and ocean crust collide; and
- ocean–ocean, where two slabs of oceanic lithosphere collide.

The combination of plate motion and physiographic character is a reasonable means of analyzing Earth dynamics.

PROCEDURE

PART A
Evidence of Continental Drift

Outline maps of South America, Africa, Europe, and North America are given in figure 11.1. The stippled area around each of the continents represents the area between the present shoreline and the 500-fathom (900-m) line (continental shelf). The latter is approximately the edge of the granitic continental crustal margin.

A fitting of the continents is best done on a small globe, but various map projections have been used by others. Because of the limitations of representing a curved surface on a plane, most map projections distort either the shape or size of the continents and alter their margins. Because of these difficulties, a modified Mercator projection has been used in the present exercise. This projection minimizes distortion of the coastline to be fitted. Various sections of Central America and Europe have been shifted slightly out of their present-day positions, but these are areas of relatively recent orogenic activity and may have had different positions in the past.

1. Trace the continental outlines following their true margin at the outer edge of the continental shelf from figure 11.1. Adjust their relative positions until the best fit is achieved, not only of continental outlines but also of geologic data, which are summarized on some of the maps.

PART B
Seafloor Spreading and the Magnetic Timescale

Available evidence suggests that until the Late Paleozoic, the continents of Africa and South America were essentially side-by-side, and that drifting and opening of the Atlantic Ocean began early in the Mesozoic Era. If the continents drifted and the Atlantic Ocean is still enlarging, there must be some evidence on the present ocean floor. Various workers studied the topography, sediments, and igneous rocks of the mid-Atlantic ridge to see if there were clues or evidence to document movement. It soon became apparent that a major linear topographic depression, or rift zone, is superimposed along the crest of the mid-Atlantic ridge along most of its length. Not only was the rift recognized in the Atlantic, but the linear topographic low appears to be part of a major series of rifts and ridges associated with mid-oceanic areas of basaltic volcanic activity and seismic activity. The distribution of this worldwide system is shown in figure 11.2.

The topographic and structural features were documented at about the same time that parallel belts of alternating reversed and normal polarity of the magnetic intensity, or magnetic field strength, were recognized in rocks along the ridge trend. These alternating bands of normal and reversed polarity in the basalts occur symmetrically along the rift zone in the Atlantic Ocean in the fashion illustrated in the generalized diagram in figure 11.3. The location of the segment of the seafloor represented in figure 11.3 corresponds with line A–A′ on figure 11.2. The sequence and ages of polarity reversals for the past 5.5 million years are shown in figure 11.4.

1. Why are the belts of polarity contrast situated symmetrically on either side of the rift zone on the mid-Atlantic ridge?

2. Using figures 11.3 and 11.4, calculate as closely as possible the rate, in centimeters per year, at which seafloor spreading is occurring in the Atlantic Ocean, based upon the paleomagnetic data. Your calculation is the rate at which a single plate is moving.

3. Enter the rate of plate motion determined in question 2 above in the blue box labeled "b." on figure 11.2.

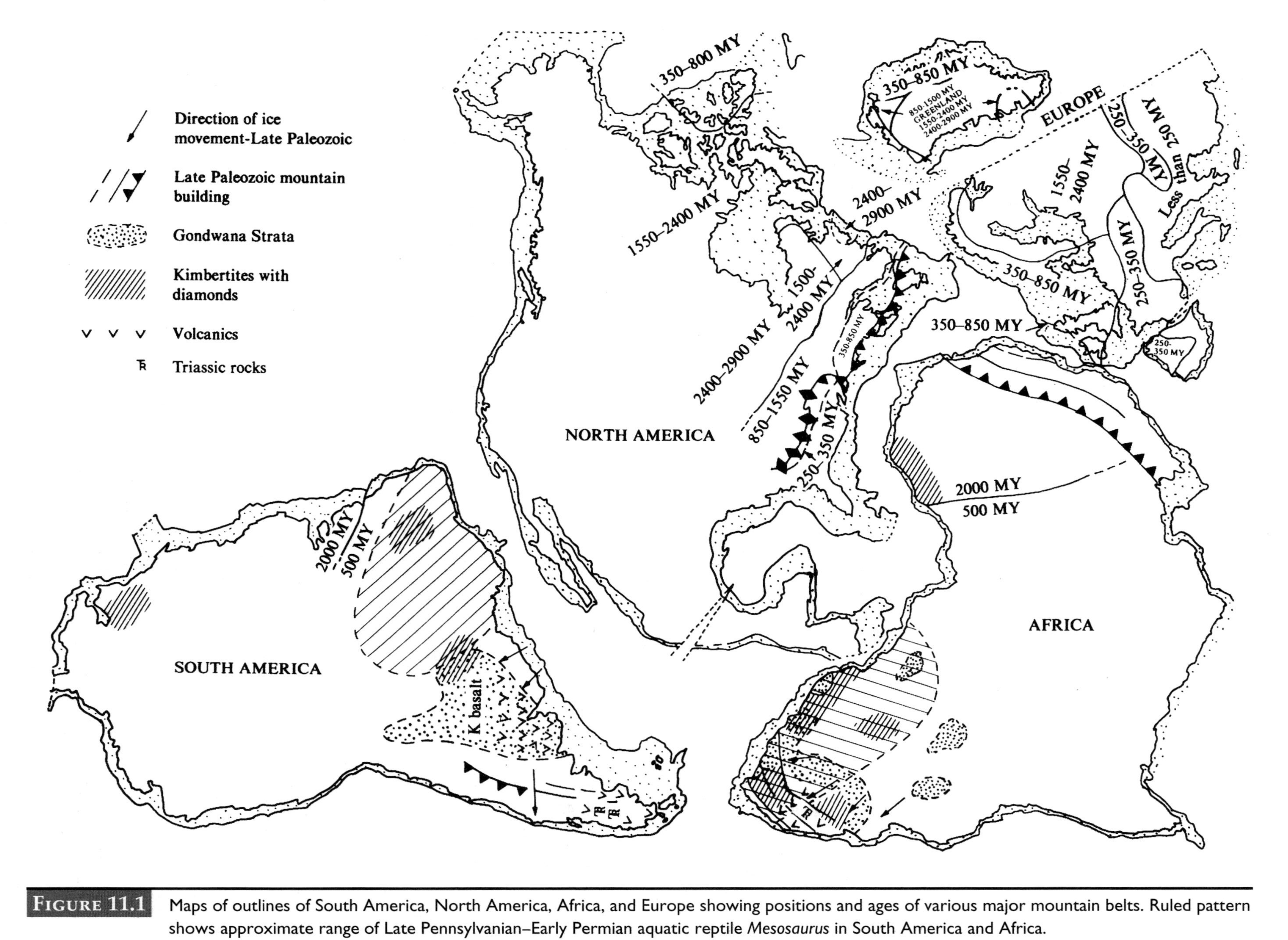

FIGURE 11.1 Maps of outlines of South America, North America, Africa, and Europe showing positions and ages of various major mountain belts. Ruled pattern shows approximate range of Late Pennsylvanian–Early Permian aquatic reptile *Mesosaurus* in South America and Africa.

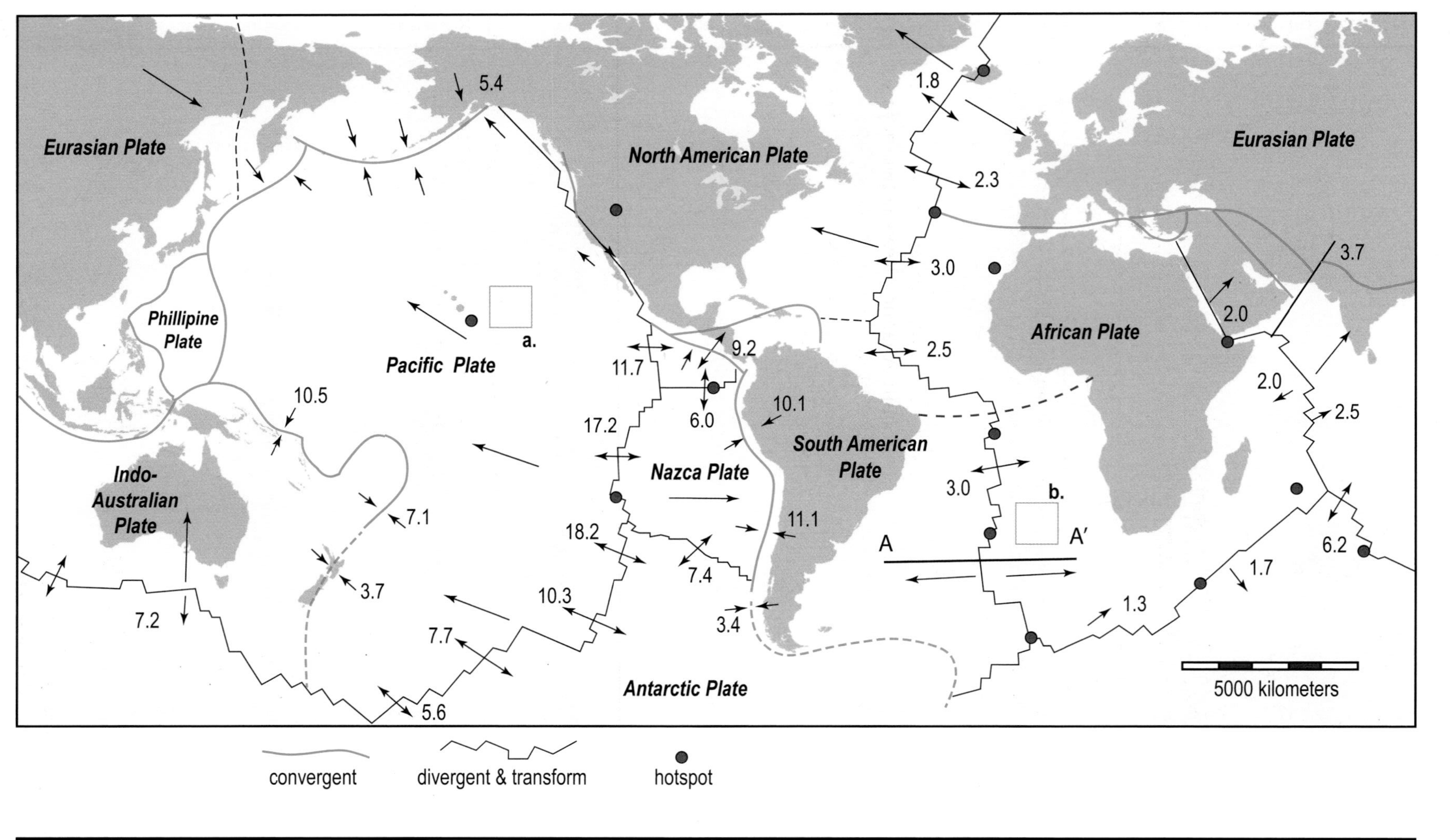

FIGURE 11.2 Map of the world showing the major plates, their boundaries, and direction of motion. Subduction zones are indicated by green lines. Red circles are presently active hot spots. Numbers indicate rate of motion in cm/yr. (Rates after McKenzie and Richter, 1976.)

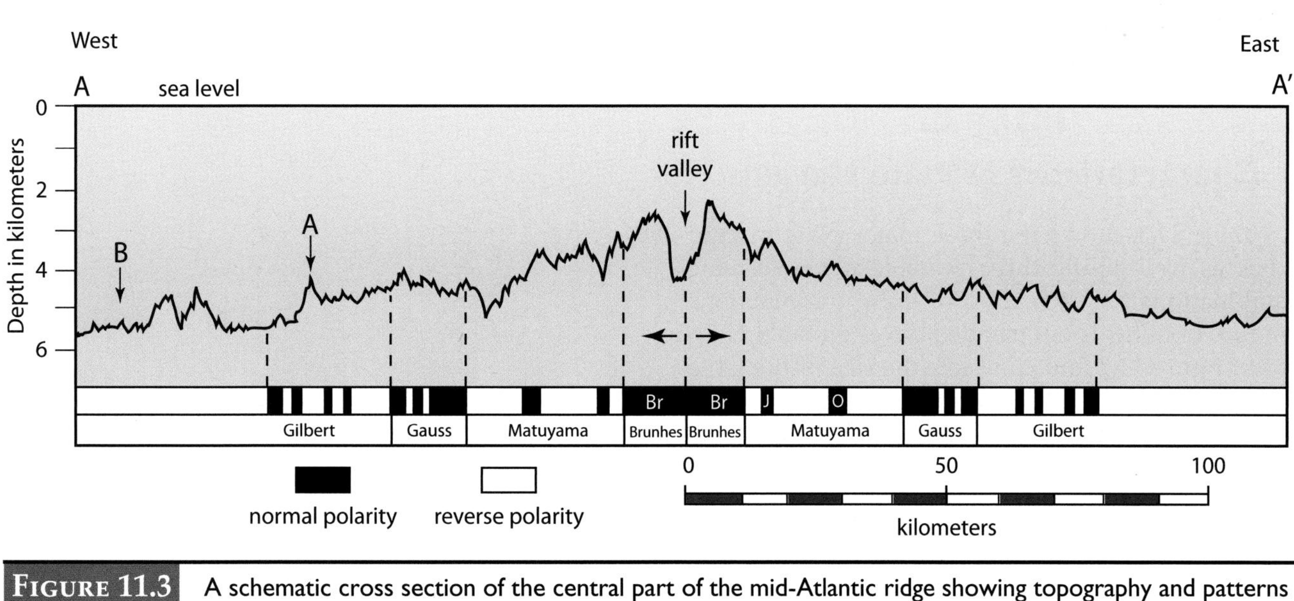

FIGURE 11.3 A schematic cross section of the central part of the mid-Atlantic ridge showing topography and patterns of paleomagnetic data. The dark and light areas correspond to the pattern shown on figure 11.4.

FIGURE 11.4 Magnetic polarity scale for the last 5.5 million years of Earth history. Purple bands indicate periods of normal polarity. White band indicate times when magnetic polarity was reversed compared to modern magnetic properties.

4. Estimate the time necessary for the spread of the South Atlantic Ocean between Africa and South America by dividing the present separation along the trace of the red dashed line on figure 11.2 by the rate of spreading as calculated in question 2.

5. What ages of basalt might be expected at points A and B on figure 11.3, assuming the rate of spreading to have been constant?

Part C

Characteristics of Plate Margins

Table 11.1 shows the three major plate margin types as well as the three possible physiographic combinations for each. Two of these combinations are rare or nonexistent (shaded boxes on table 11.1). Study figure 11.2 and complete the rest of the table by entering names of modern geographic locations that serve as examples for each of the seven characteristic plate margins.

Plate Margin	Modern Examples by Geographic Name		
	Continent-Continent	Continent-Ocean	Ocean-Ocean
Divergent • Tensional stresses • Crustal lengthening • New ocean crust • Normal faults • Shallow earthquakes • Basaltic volcanism			
Convergent • Compressional stress • Crustal shortening • Ocean crust destroyed • Reverse and thrust faults • Folds • New continental crust • Shallow and deep quakes			
Transform Fault • Shear stress • Shallow quakes • Crust neither created nor destroyed • Strike-slip faults			

Table 11.1 Table showing types and characteristics of plate margins.

Part D
Locating Plate Margins

Seismic or earthquake data provide us with direct information concerning plate interactions or tectonics. The US Geological Survey in Washington, DC maintains and publishes a monthly record of seismic events around the world. These records, distributed to the public by the National Earthquake Information Service, are entitled Preliminary Determination of Epicenters. They include date; time; location of epicenter; depth of focus, or hypocenter; and magnitude of every earthquake of magnitude 3 or greater occurring in the United States, and of those greater than magnitude 4 in other places in the world. Further information is included for some earthquakes. Occasional comments include the intensity as measured on the Modified Mercalli earthquake intensity scale. This kind of information is extremely useful for understanding the dynamic nature of our planet. By studying this information, a student can gain firsthand knowledge of the frequency, distribution, and energy of the Earth's seismic activity.

The epicenters of the selected earthquakes occurring during the month of January 1982 have been plotted on a world map and are illustrated as figure 11.5. In addition, the seismic activity for the week 24–31 December 1981, is presented in table 11.2. Abbreviations used in the headings are: UTC = universal time; GS = geological survey data; MB = magnitude based on "P" seismic waves; and MSZ = magnitude based on vertical surface waves. An asterisk (*) following the time indicates a less reliable solution in the calculation because of incomplete or less reliable data.

1. Using a colored pen or pencil, plot every other epicenter from table 11.2 on figure 11.5. When you are finished, your map will display essentially five weeks of seismic activity. The only information you need to complete this task is the longitude and latitude of the earthquakes.

2. Compare your finished map (figure 11.5) with figure 11.2. Is there a clear correlation between the distribution of recent earthquakes and the boundaries of plates? Is such a correlation expected?

3. The region near the Bonin Islands is a common site for deep earthquakes. Why is this?

4. What is the probable cause of the Kazakh seismic event of 27 December at 03:43 hours? What two things form the basis of your answer?

	Origin Time, UTC			Geographic Coordinates						Magnitudes, GS		
Day, UTC 12/81	**Hr.**	**Min.**	**Sec.**	**Lat.**		**Long.**		**Region and Comments**	**Depth**	**MB**	**MSZ**	**No. Sta. Used**
24	02	22	07.7	34.017	N	116.767	W	Southern California	20			10
24	04	24	53.6*	14.641	N	119.919	E	Luzon, Philippine Islands	33	4.6		8
24	05	33	21.5	29.956	S	177.701	W	Kermadec Islands	33	6.1	6.8	151
24	09	43	51.5*	30.041	S	177.563	W	Kermadec Islands	33	5.0	5.1	15
24	11	11	17.1*	22.061	S	175.916	W	Tonga Islands Region	63	4.9		14
24	13	02	40.4	29.925	S	177.374	W	Kermadec Islands	33	5.3	5.4	29
24	14	07	39.3*	39.952	N	77.366	E	Southern Sinkiang Prov., China	33	4.9		7
24	14	44	07.4*	14.282	S	74.321	W	Peru	108			8
24	19	44	53.1*	30.228	S	177.378	W	Kermadec Islands	33	4.9		17
24	22	00	51.6*	33.153	N	49.705	E	Western Iran	33	4.6	4.0	28
24	22	36	211.5	30.076	S	177.387	W	Kermadec Islands	36	5.2	5.4	40
25	00	06	09.5*	12.555	N	88.303	W	Off Coast of Central America	33	4.7		26
25	00	28	16.8	4.738	N	118.458	E	Kalimantan	52	5.4	5.2	47
25	04	55	52.7*	17.406	N	61.768	W	Leeward Islands	49	4.2		12
25	06	03	07.9*	77.013	N	6.601	E	Svalbard Region	10	4.4		14
25	08	28	48.0	6.561	N	73.039	W	Northern Colombia	195	4.5		13
25	09	12	06.4	30.313	S	177.489	W	Kermadec Islands	33	5.4	5.4	57
25	10	37	10.7*	30.200	S	177.411	W	Kermadec Islands	33	4.7		12
25	10	48	45.1*	30.324	S	177.301	W	Kermadec Islands	33	5.2		20
25	12	35	49.6*	11.172	N	62.474	W	Windward Islands	102	4.8		55
25	15	27	18.7*	13.613	S	76.425	W	Near Coast of Peru	33			5
25	15	50	33.3*	23.197	N	121.642	E	Taiwan	33	4.5		11
25	17	02	35.5	53.884	N	160.800	E	Near Coast of Kamchatka	33	4.6		19
25	22	26	41.0*	59.871	N	152.716	W	Southern Alaska	113	4.3		14
26	03	42	19.5*	37.952	N	22.702	E	Southern Greece	33	4.0		25
26	10	17	16.6*	2.209	S	139.847	E	Near N. Coast of West Irian	33	4.8	4.5	18
26	11	16	05.8	23.942	S	66.512	W	Jujuy Province, Argentina	207	4.9		49
26	14	29	11.1*	38.950	N	25.310	E	Aegean Sea	10	4.2	3.6	39
26	17	05	32.8	29.812	S	177.854	W	Kermadec Islands	33	6.3	7.1	149
26	17	53	30.6*	29.983	S	177.830	W	Kermadec Islands	33	5.2		17
26	17	53	38.4*	40.213	N	28.778	E	Turkey Felt in the Istanbul area	27	4.3		28
26	19	38	07.7*	9.773	S	119.218	E	Sumba Island Region	100	4.4		10
26	21	50	43.9	7.260	S	129.183	E	Banda Sea	150	5.3		21
26	22	02	18.5*	22.765	S	68.332	W	Northern Chile	119	4.8		11
27	03	43	14.1	49.923	N	78.876	E	Eastern Kazakh SSR	0	6.1	4.3	167
27	06	22	50.3*	30.063	S	177.622	W	Kermadec Islands	33	5.4	4.9	18
27	10	30	44.4	2.160	S	139.825	E	Near N. Coast of West Irian	33	5.6	5.9	60
27	13	25	33.3*	46.331	N	16.832	E	Yugoslavia	10			13
27	16	36	48.3*	5.090	S	139.311	E	West Irian	33	3.7		6
27	17	39	16.7	39.004	N	24.799	E	Aegean Sea Ten houses damaged on Evvoia. Felt strongly in eastern Greece. Also felt in the Izmir, Turkey area	33	5.3	6.5	112
27	20	21	05.9*	7.107	N	73.093	W	Northern Colombia	138	4.6		10
27	20	24	15.5*	34.250	N	117.617	W	Southern California	9			9
27	21	23	13.6	8.278	S	79.875	W	Near Coast of Northern Peru	33	5.2	4.4	41
28	01	53	02.2	6.889	S	130.004	E	Banda Sea	131	4.6		13
28	10	28	15.9*	54.642	N	160.380	W	Alaska Peninsula	33			9
28	12	40	18.4	14.974	S	168.121	E	Vanuatu Islands	33	5.7	5.2	89

TABLE 11.2 Preliminary determination of epicenters.

	Origin Time, UTC			Geographic Coordinates						Magnitudes, GS		
Day, UTC 12/81	**Hr.**	**Min.**	**Sec.**	**Lat.**		**Long.**		**Region and Comments**	**Depth**	**MB**	**MSZ**	**No. Sta. Used**
28	13	08	26.2	21.644	N	143.470	E	Mariana Islands Region	33	5.3	5.0	57
28	14	18	15.5*	21.495	N	143.583	E	Mariana Islands Region	33	5.0		30
28	14	38	20.9*	21.893	N	144.109	E	Mariana Islands Region	33	4.5		22
28	14	49	40.6	35.016	N	45.934	E	Iran–Iraq Border Region	33	5.0	4.0	69
28	15	20	11.4*	21.353	N	143.818	E	Mariana Islands Region	33	4.6		13
28	15	40	02.1*	21.642	N	143.718	E	Mariana Islands Region	59	5.1		25
28	16	11	00.4*	14.213	N	92.200	W	Near Coast of Chiapas, Mexico	68	4.7		7
28	16	37	35.8*	21.360	N	143.664	E	Mariana Islands Region	33	4.7		15
28	17	23	52.2	21.511	N	143.789	E	Mariana Islands Region	27	4.9		35
28	17	40	49.8	21.593	N	143.518	E	Mariana Islands Region	33	5.3		36
28	18	10	57.7	13.805	N	95.915	E	Andaman Islands Region	33	5.0	5.1	48
28	18	13	25.5*	21.487	N	143.500	E	Mariana Islands Region	33	5.0		28
28	20	56	01.0*	21.429	N	143.523	E	Mariana Islands Region	33	5.0		29
28	22	01	50.1	63.111	N	150.819	W	Central Alaska	151			13
28	22	45	42.1	37.211	N	114.980	W	Southern Nevada Felt (IV) at Las Vegas Felt in Clark and Lincoln Counties, Nev. Also felt at Toquerville, Utah, and Temple Bar, Arizona.	5			16
28	23	17	50.8*	21.489	N	143.621	E	Mariana Islands Region	33	4.7		11
28	23	35	59.6*	21.645	N	142.958	E	Mariana Islands Region	33	4.7		13
29	02	22	59.7	21.352	N	143.152	E	Mariana Islands Region	33	4.8		26
29	05	07	22.3*	21.337	N	143.680	E	Mariana Islands Region	33	4.6		14
29	06	39	29.8*	21.474	N	143.694	E	Mariana Islands Region	33	4.7		16
29	08	00	44.9	38.796	N	24.720	E	Aegean Sea Felt in the Khalkis–Thessaloniki area	10	4.7	5.3	83
29	09	26	36.0*	21.425	N	143.841	E	Mariana Islands Region	33	4.8		13
29	15	37	03.9*	21.858	N	143.581	E	Mariana Islands Region	33	4.7		12
29	15	38	26.1*	6.066	S	155.274	E	Solomon Islands	150	5.0		12
29	16	11	17.0*	21.459	N	143.643	E	Mariana Islands Region	33	4.6		9
29	16	37	17.9*	19.209	S	68.168	W	Chile–Bolivia Border Region	205	5.0		9
29	19	06	31.3	30.231	S	177.924	W	Kermadec Islands	60	5.5		36
29	22	49	00.4	21.508	N	143.379	E	Mariana Islands Region	33	5.3	4.8	56
29	23	37	53.3*	21.539	N	143.648	E	Mariana Islands Region	33	4.9		9
30	11	26	36.0*	43.756	N	147.678	E	Kuril Islands	33	5.2	4.1	56
30	13	47	27.3	64.589	N	148.080	W	Central Alaska	33	3.9		11
30	14	00	33.8	64.577	N	148.158	W	Central Alaska Felt (V) at Ester and (IV) at Fairbanks	27	4.8		37
30	15	00	53.8*	22.028	N	143.536	E	Volcano Islands Region	79	4.7		13
30	16	46	34.4*	38.812	N	20.803	E	Greece	33	4.4		11
30	17	44	09.6	6.734	N	126.957	E	Mindanao, Philippine Islands	77	5.0		31
30	20	32	36.1*	13.570	N	90.620	W	Near Coast of Guatemala	33	4.5		12
30	21	09	54.0*	4.349	N	126.008	E	Talaud Islands	104	4.7		12
31	05	08	13.5	27.694	N	139.680	E	Bonin Islands Region	494	4.7		38
31	06	54	51.1*	33.935	S	179.331	W	South of Kermadec Islands	33	5.2		28
31	09	20	05.9*	0.796	N	123.883	E	Minahassa Peninsula	294	5.1		20
31	12	15	54.4	61.910	N	151.758	W	Southern Alaska Felt at Wasilla and Houston	128	4.1		14
31	13	48	43.0*	2.150	N	126.451	E	Molucca Passage	125	4.9		14
31	21	38	13.9	21.448	N	143.716	E	Mariana Islands Region	33	5.0		12

TABLE 11.2 Preliminary determination of epicenters. *(continued)*

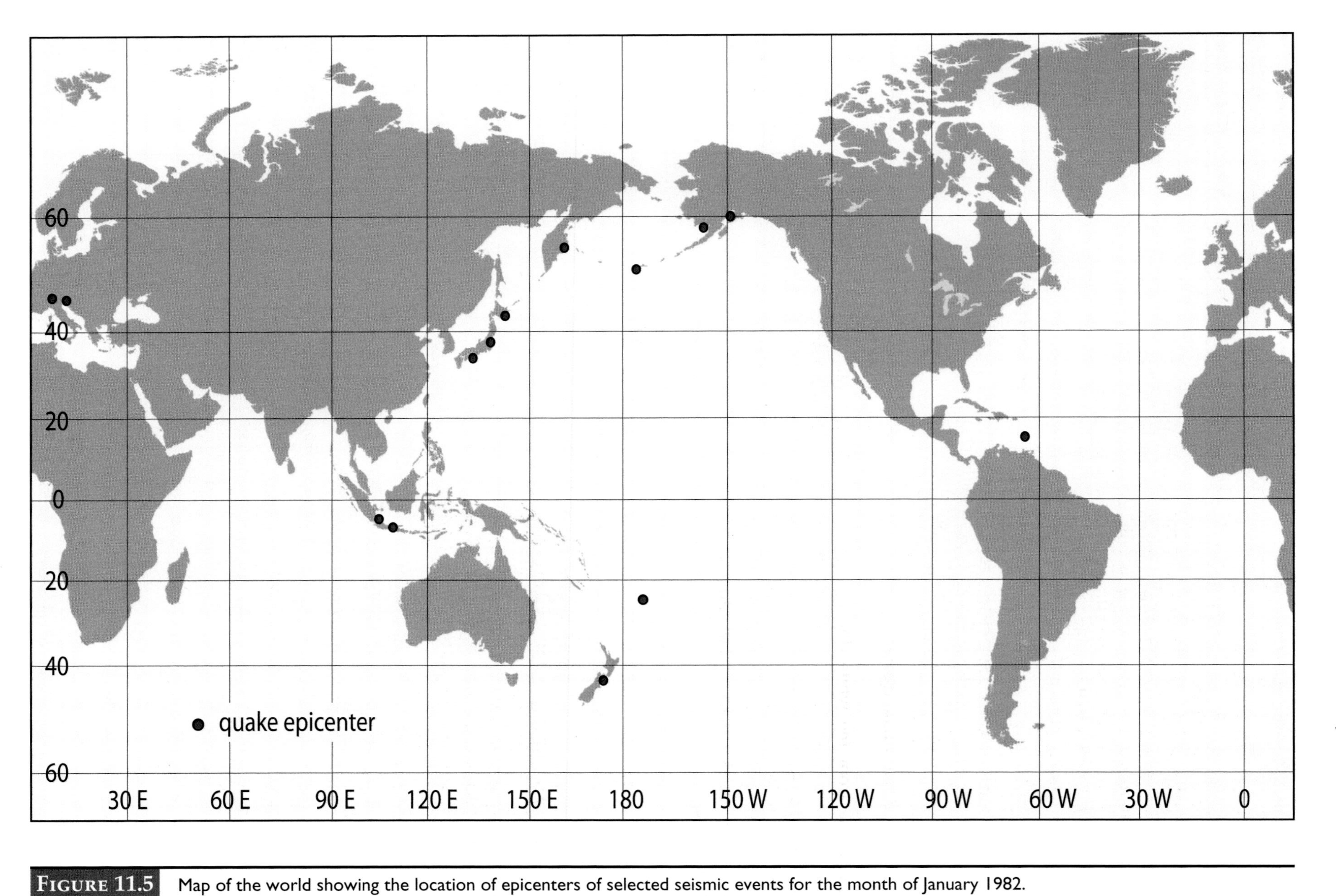

FIGURE 11.5 Map of the world showing the location of epicenters of selected seismic events for the month of January 1982.

PART E
Hot Spots and Plate Motion

Figure 11.6 is a map of the emergent portion of the Hawaiian Islands chain in the central Pacific Ocean. These islands have a volcanic origin resulting from the outpouring of vast amounts of basaltic lava from a hot spot located southeast of the "Big Island" of Hawaii. The hot spot remains fixed in the mantle while the Pacific plate moves across it in a northwesterly direction. The masses of volcanic rocks that comprise the emergent and submergent portions of the Hawaiian chain are all younger than the seafloor basalts (Pacific plate) onto which they were extruded. The average radiometric ages in millions of years (Ma) of the basalts at eight selected sample stations are shown on figure 11.6.

1. Draw a line from the white dot adjacent to recent basalt flows on the island of Hawaii and the white dot on the north side of Kauai. Using a protractor and the north arrow provided on the map, determine the average direction that the Pacific plate has been moving over the past 5.1 million years.

2. Using figure 11.6, fill in the data pertaining to distance and age differences between the sample station(s) located on each island in table 11.3. Next calculate the rate of plate motion that occurred between the extrusion of lavas at each sample area (right column of table 11.3). Finally, calculate the average rate of plate motion between Hawaii and Kauai (bottom row of table 11.3).

3. Has the rate of plate motion been constant for the past 5.1 million years? What is the range of values of plate motion?

4. Enter the average rate of motion into the blue box labeled "a." on figure 11.2.

STATION	DISTANCE (in km)	DISTANCE (in cm)	AGE DIFFERENCE (in years)	RATE (cm/yr)
Hawaii to SE Maui	125	12,500,000	700,000	17.9
SE Maui to NW Maui				
NW Maui to E Molokai				
E Molokai to W Molokai				
W Molokai to SE Oahu				
SE Oahu to NW Oahu				
NW Oahu to Kauai				
HAWAII TO KAUAI				

TABLE 11.3 Table provided for recording information on the distances of sample localities, ages, and rates of motion of the Hawaiian Islands.

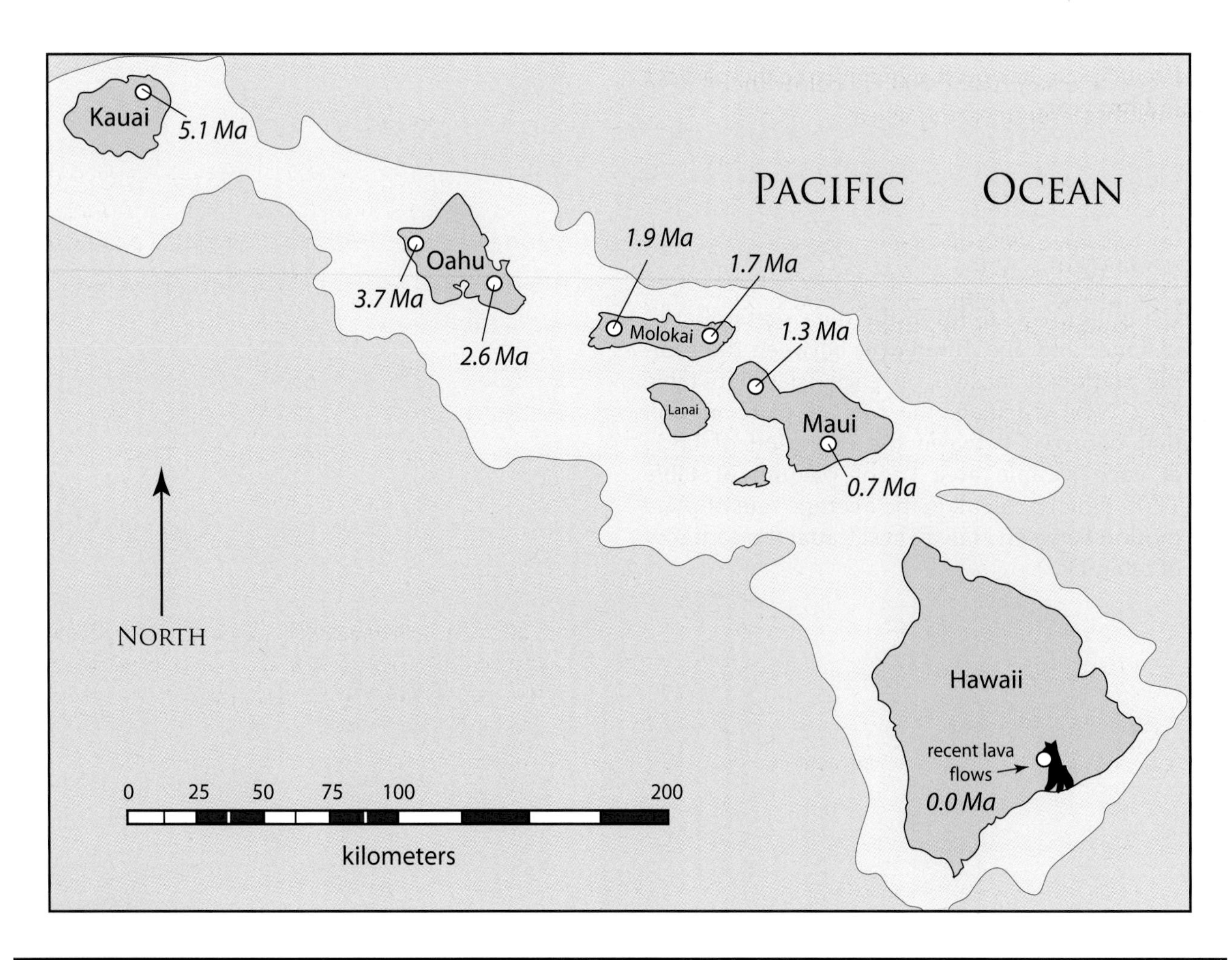

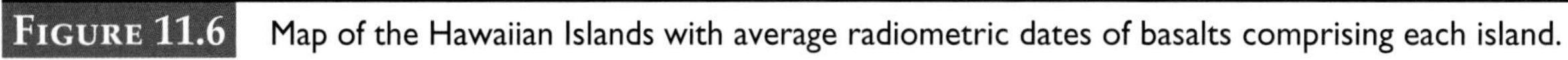
FIGURE 11.6 Map of the Hawaiian Islands with average radiometric dates of basalts comprising each island.

PART F
Plate Tectonics and the Aleutian Trench

Figure 11.7 is a map showing the boundary between the North American and Pacific lithospheric plates. Volcanic activity associated with this margin has produced a 3,000 km-long chain of volcanoes and volcanic islands (island arc) known as the Aleutian Islands. Figure 11.7A shows the location of the volcanic chain between Alaska and Russia. Figure 11.7B shows details of the central portion of the Aleutian chain. The dark line with triangular "teeth" on map B shows the position of the Aleutian Trench, a physiographic feature that marks the line of interaction between the Pacific plate and the North American plate. Red and green dots indicate the epicenters of earthquakes that have been recorded in the area from 1900 until 2010 (Benz et al., 2011). Red dots indicate the locations of shallow-focus earthquakes (those that originate between 0 and 70 km below the surface). Green dots represent the epicenters of earthquakes that originated at depths between 70 and 300 km below the surface (intermediate-depth earthquakes).

1. Plot the point of origination (focus) of each earthquake listed in table 11.4 on figure 11.8. To do this, first determine the position of the epicenter relative to the axis of the Aleutian Trench (red arrow). Positive values indicate quakes occurring in the Pacific plate south of the trench axis. Negative values indicate distance north of the trench axis. Once you have located the epicenter along the surface, use the depth data to locate the focus of the quake. Plot shallow- and intermediate-focus earthquakes using red dots and green dots, respectively.

2. What type of plate boundary is represented (convergent, divergent, transform)?

3. What is the range of earthquake depths in the vicinity of the Aleutian Trench?

4. Why are earthquakes more numerous north of the trench axis?

5. What is the dip of the earthquake zone? Since the horizontal and vertical scales are equal, you can make this determination with a protractor.

6. Label the North American and Pacific plates on the cross section. Draw arrows to illustrate relative motion of the Pacific and North American plates along the zone.

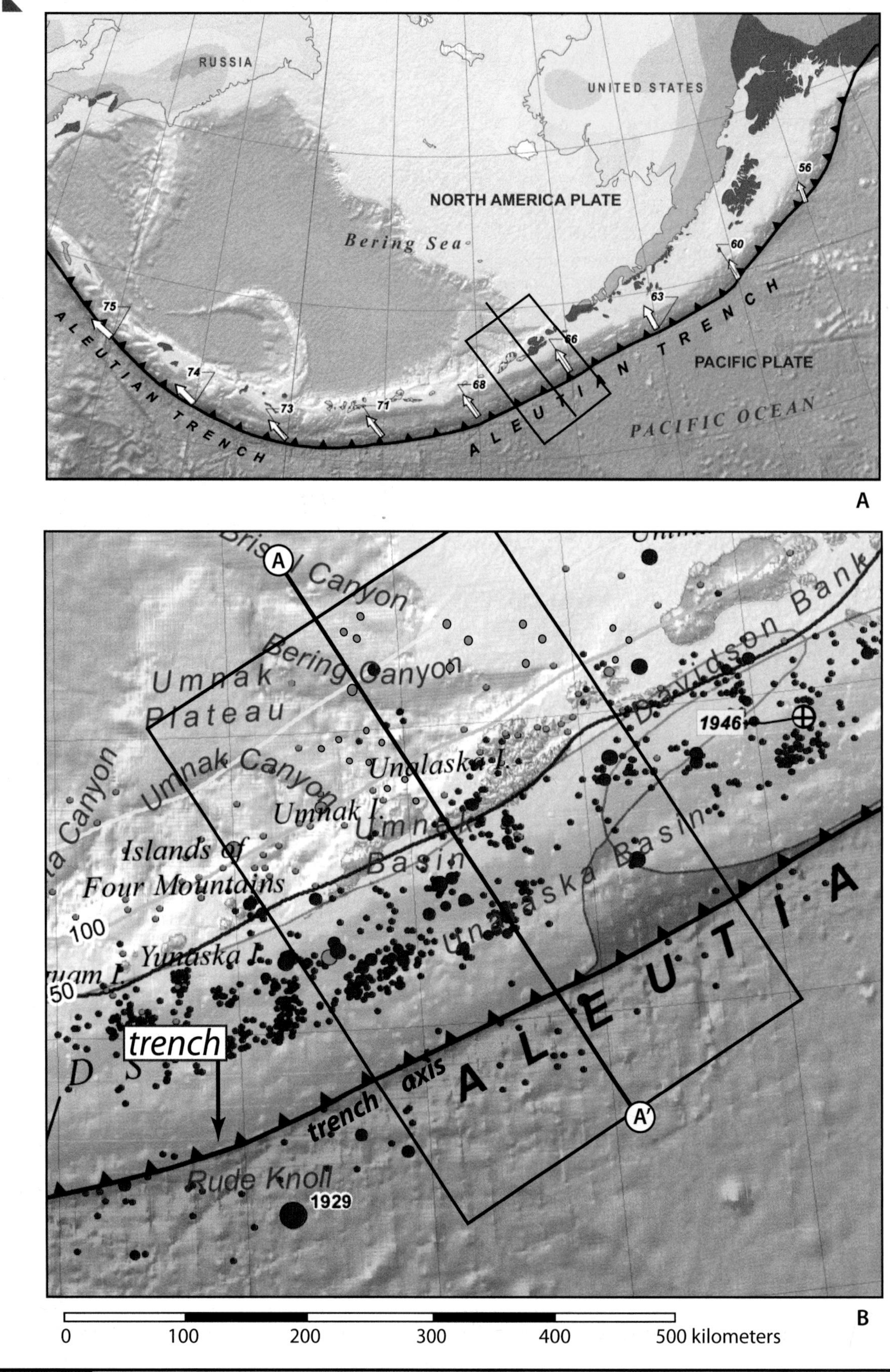

FIGURE 11.7 Maps of the Aleutian Trench and northern Pacific Ocean basin.

A. Regional map.

B. Detailed tectonic map of the central portion of the Aleutian Trench. The rectangular box bisected by line A–A′ corresponds to the rectangular box in the lower middle portion of the regional map. A–A′ corresponds to the cross section shown in figure 11.8. Scale bar applies to map B.

Table 11.4

Distance and depth data for earthquakes occurring in the central Aleutian Islands. The second column indicates the distance of the epicenter from the trench axis along the line A–A′ on figure 11.7. Positive and negative values indicate distances (in km) south and north of the trench axis, respectively. The third column lists the depth of the earthquake (focus) below the surface in kilometers. Greater values indicate greater depths of earthquake originations.

Quake	Distance	Depth
1.	-5	15
2.	+60	40
3.	-25	10
4.	-350	200
5.	-300	150
6.	-100	25
7.	-100	35
8.	-50	25
9.	+50	50
10.	-200	50
11.	-150	50
12.	-200	90
13.	+15	40
14.	-240	110
15.	-270	150
16.	-320	40
17.	-325	175

Quake	Distance	Depth
18.	+50	15
19.	-125	20
20.	-125	40
21.	-175	40
22.	-210	75
23.	-250	90
24.	-350	220
25.	-370	200
26.	-175	50
27.	-200	70
28.	-250	150
29.	-275	125
30.	-325	150
31.	-300	125
32.	-150	40
33.	-350	175
34.	-250	60

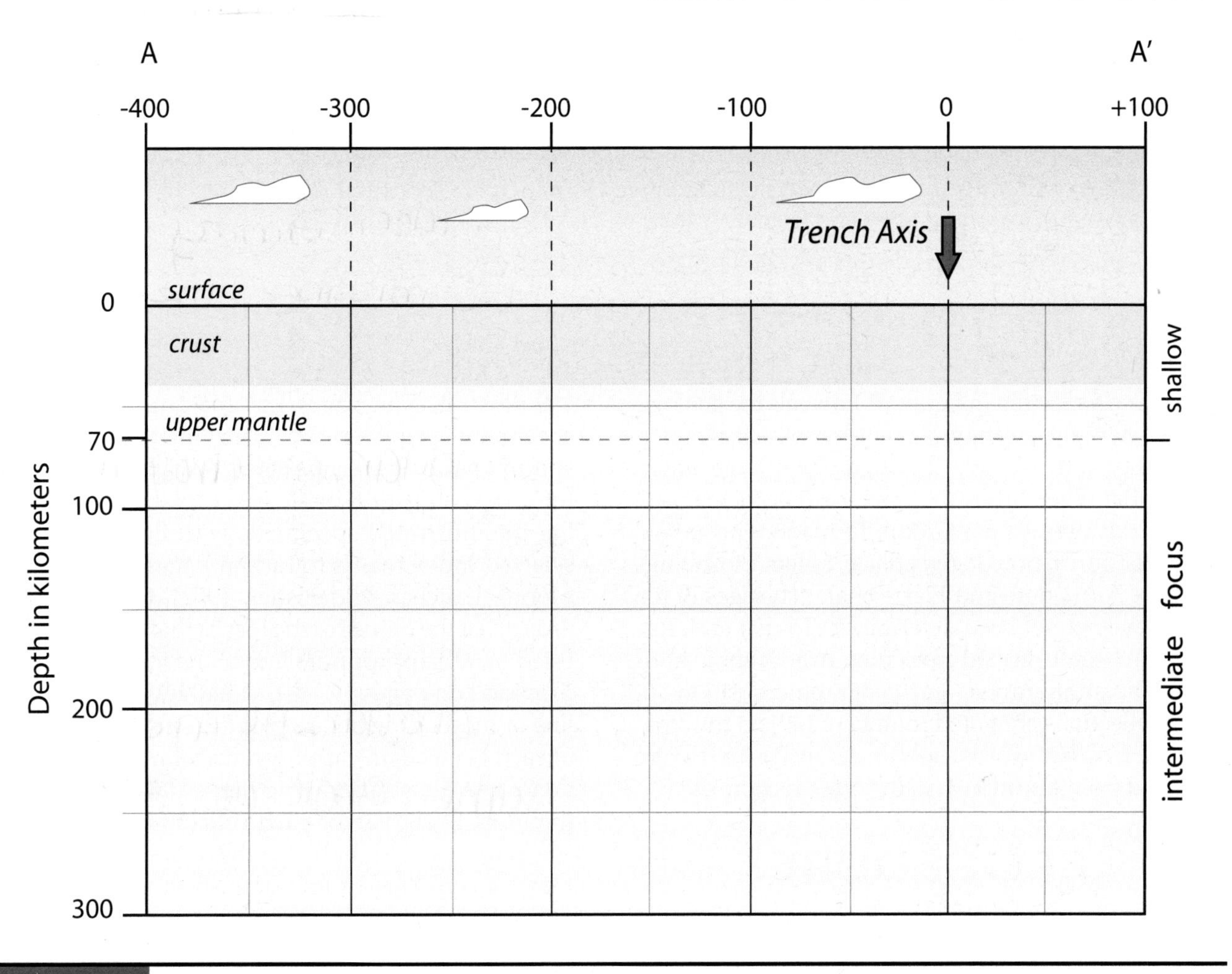

Figure 11.8 Cross section of line A–A′ on figure 11.7B. The horizontal scale is distance from the Aleutian Trench axis measured in kilometers. Vertical and horizontal scale are the same. Depths below the surface are measured in kilometers. Distances seaward of the Aleutian Trench are assigned positive values. Distances northward from the trench axis are assigned negative values. Earthquakes occurring at a depth of 0 to 70 km are called "shallow" quakes; those occurring between 70 and 299 km are considered to be "intermediate-depth" earthquakes.

Exercise 12

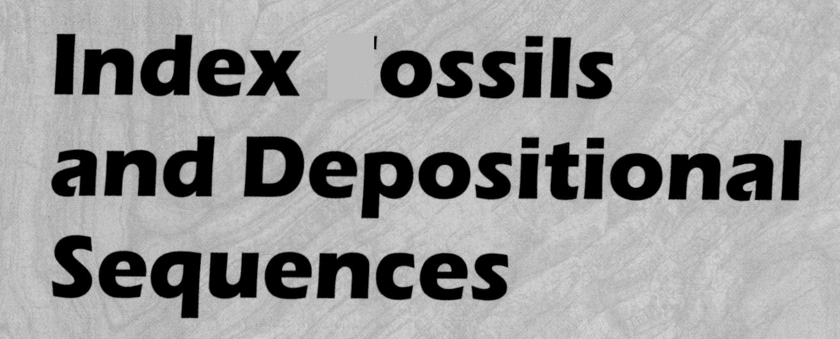

Index Fossils and Depositional Sequences

Learning Objectives

After completing this exercise, you will be:

1. acquainted with the key index fossils of selected geological systems; and
2. able to utilize fossil-based age determinations and stratigraphic/structural relationships to work out the sequence stratigraphy and tectonic history of a geologically complex region.

The cross section in figure 12.1 reveals the geological structure and stratigraphy of a hypothetical region of northeastern Canada. Stratigraphers have determined that seven lithostratigraphic units are exposed in the study area: Chuar Group, Tremadoc Formation, Labrador Formation, Percha Shale, Salem Limestone, Paradox Formation, and Pierre Shale. The ages of the units have not yet been determined. To that end, fossils have been collected from nine stratigraphic horizons. The stratigraphic and geographic positions of these collections are noted (numbers 1–9) on the cross section. In other words, specimens in Fauna 1 were collected from Location 1 on the cross section, etc.

PROCEDURE

Part A

Relative Ages

Determine the age of each formation in figure 12.1 by identifying the fossils present in Faunas 1–9 illustrated on pages 189–196 , or by identifying specimens provided by your instructor. Identify specimens to the genus level (except specimen 1c, for which a general name is acceptable) by comparing the illustrated specimens with line drawings of selected index fossils found on pages 174–188 (after Moore, Lalicker, and Fisher, 1952). For each fauna group, list the names of the fossils that you have identified on the appropriate line of the fossil data sheets provided on pages 197–201. Next, by using a vertical line or by shading boxes, plot the stratigraphic range of each genus in the appropriate column. Once this is done, use the stratigraphic range data to determine the most likely age of each collection. Indicate that age in the lower right box of the data sheet. An example of a completed fossil data sheet is illustrated in figure 12.2. Note: Some lines and columns will be left blank. For example, Fauna 4 contains only two specimens; hence the data sheet for Fauna 4 will contain information for only two genera.

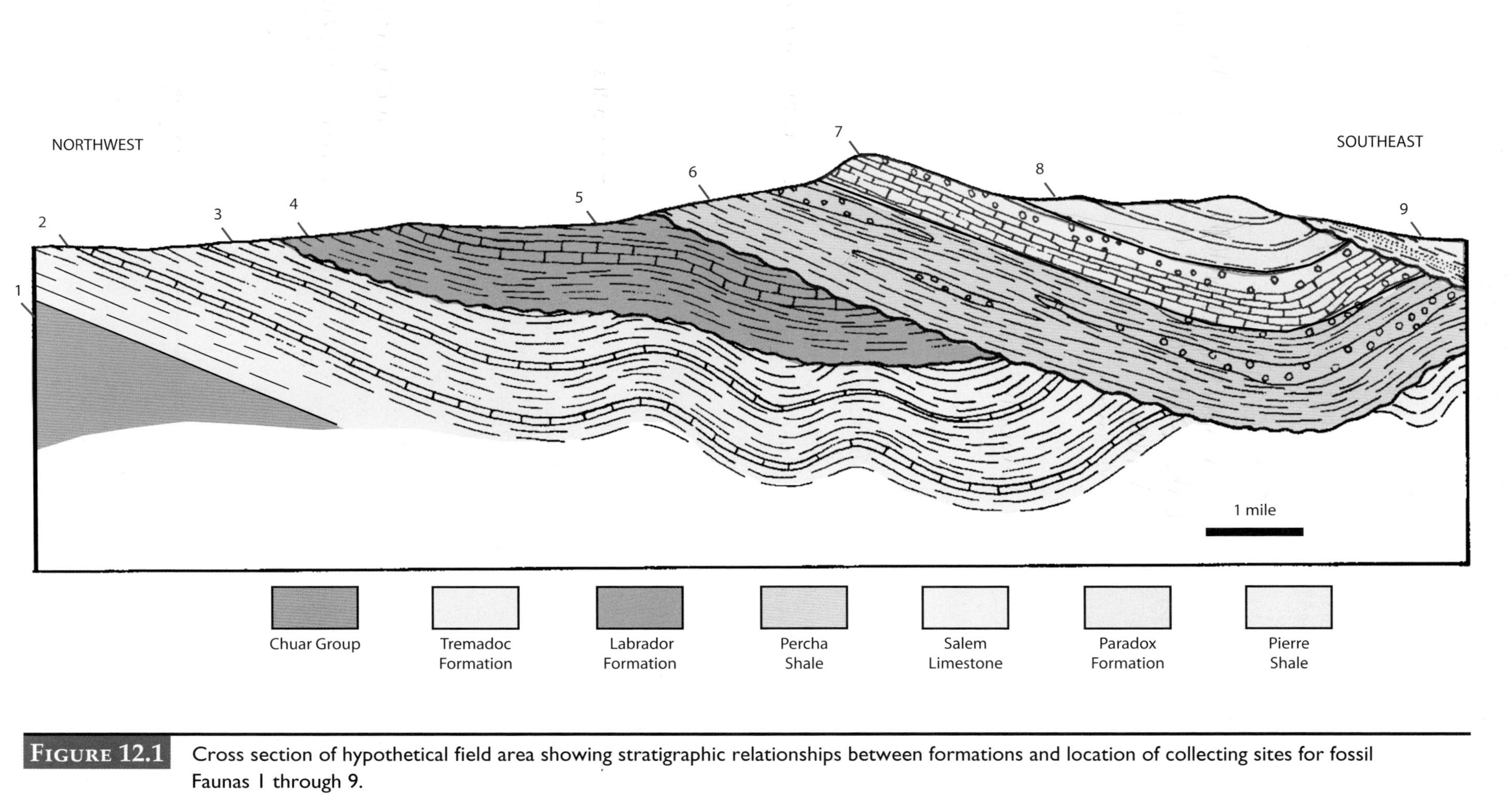

FIGURE 12.1 Cross section of hypothetical field area showing stratigraphic relationships between formations and location of collecting sites for fossil Faunas I through 9.

PART B
Chronostratigraphy

Using information on your fossil data sheets, determine and list the age of each lithostratigraphic unit.

1. Chuar Group ____________
2. Tremadoc Formation ____________
3. Labrador Formation ____________
4. Percha Shale ____________
5. Salem Limestone ____________
6. Paradox Formation ____________
7. Pierre Shale ____________

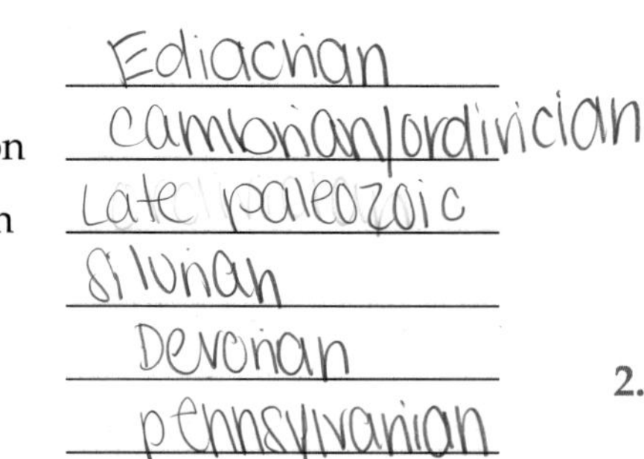

PART C
Depositional Sequences

List the formations that comprise each of the standard North American depositional sequences within this region:

1. Sauk Sequence:

2. Tippecanoe Sequence:

3. Kaskaskia Sequence:

4. Absaroka Sequence:

5. Zuni Sequence:

PART D
Significant Surfaces

Note the nature of the following surfaces:

1. Boundary between Chuar Group and Tremadoc Formation:

2. Boundary between Tremadoc and Labrador Formations:

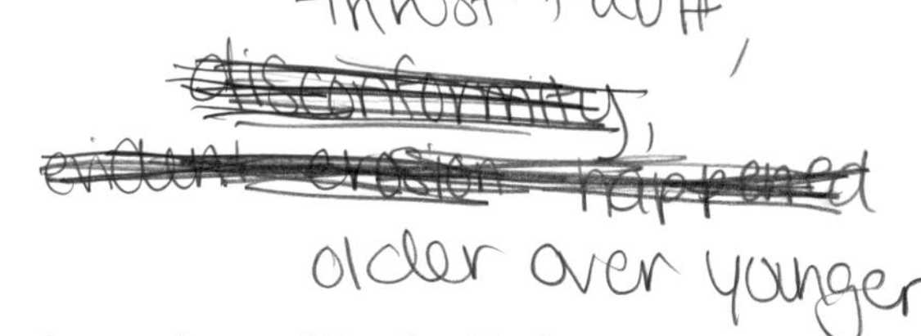

3. Surface at base of Percha Shale:

paraconformity, looks to be missing time.

4. Contact between Percha Shale and Salem Limestone:

Angular, no errosion.

5. Contact between Salem Limestone and Paradox Formation:

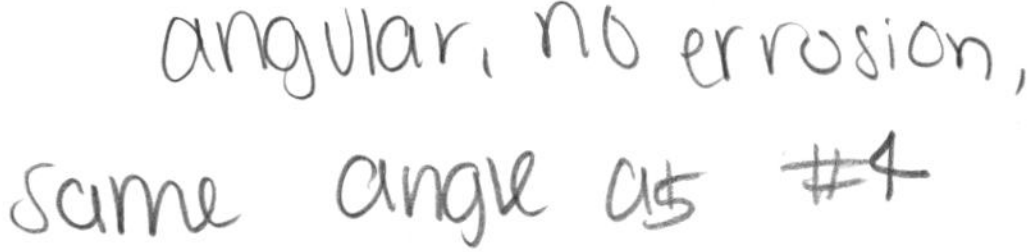

6. Surface at the base of Pierre Shale:

thrust faulted, layered in an unusual way. looks like it was cut into.

Part E
Classification

To help you become familiar with the invertebrate classification above the genus level, and to help you see the evolutionary relationships between invertebrate taxa, list the genera that you have identified in the course of this lab in the correct box on table 12.1.

Part F
Report

Write a one- or two-page summary of the geological history of the region based upon the lithostratigraphy and geometries seen on the cross section. Indicate the age and probable depositional environment of each formation. Indicate which type or types of fossils are the best index fossils for a given geological system. Discuss the nature and significance of contacts and other surfaces that separate the formations. Relate these to tectonic events and depositional sequences that you have studied. This portion of the lab permits you to synthesize the paleontological, chronological, and lithologic patterns present in this geological region. Your synthesis should be concise and complete and should reflect your ability to synthesize the data. The summary report will be evaluated by the instructor accordingly.

	FAUNA 1	A.	B.	C.	D.	E.	F.	G.	H.	I.	J.	K.	TAXON NAME
Ceno.	Neogene												A. Polygnathus
	Paleogene												B. Syringopora
Mesozoic	Cretaceous												C. Lingula
													D. Palmatolepis
	Jurassic												E. Atrypa
	Triassic												F. Phacops
Paleozoic	Permian												G.
	Pennsylvanian												H.
	Mississippian												I.
	Devonian												J.
	Silurian												K.
	Ordovician												Age of Bed: Devonian
	Cambrian												

Figure 12.2 Example of the procedure for completing the fossil data sheets for Faunas 1–9.

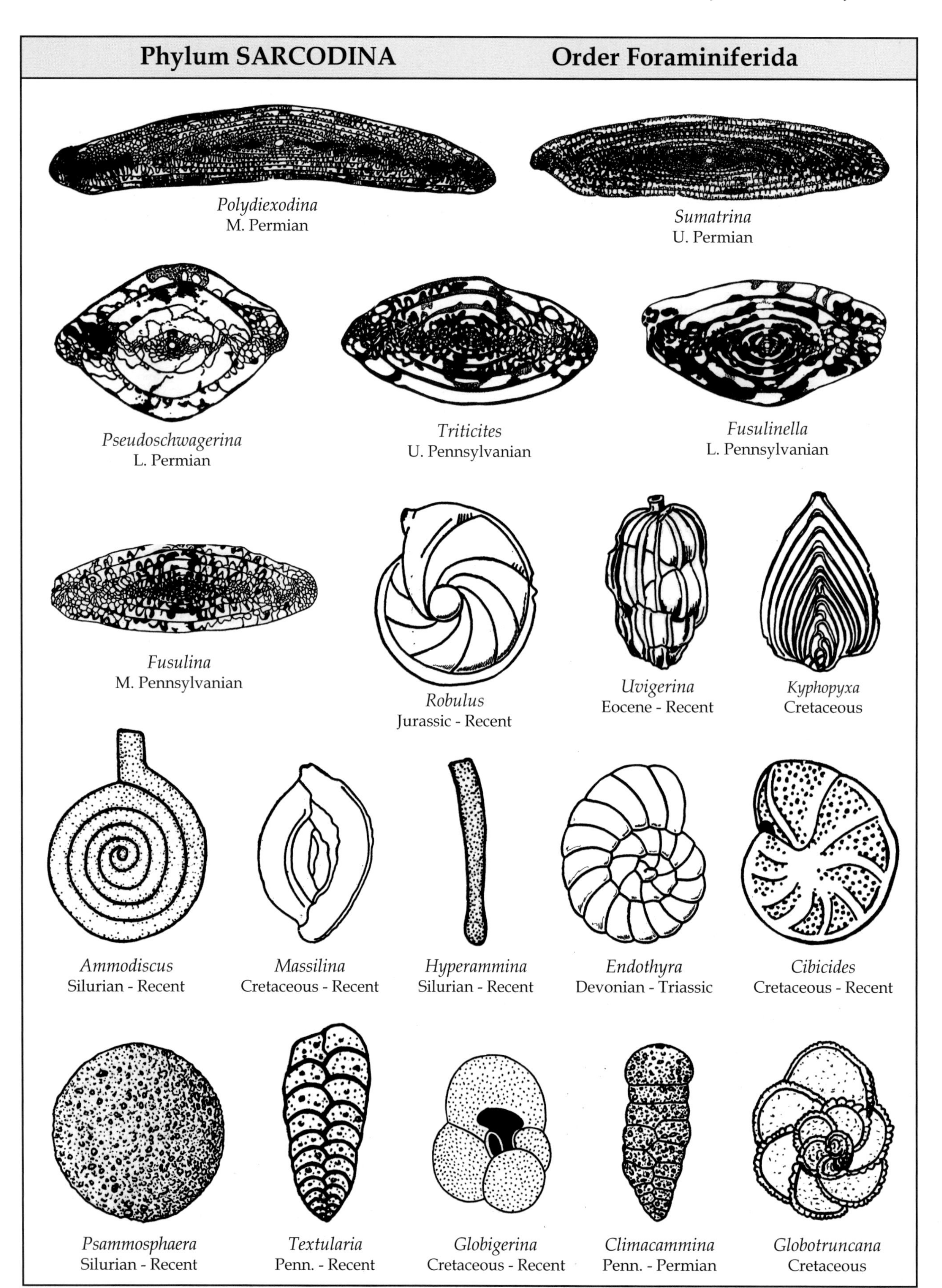
Phylum SARCODINA
Order Foraminiferida
Polydiexodina
M. Permian
Sumatrina
U. Permian
Pseudoschwagerina
L. Permian
Triticites
U. Pennsylvanian
Fusulinella
L. Pennsylvanian
Fusulina
M. Pennsylvanian
Robulus
Jurassic - Recent
Uvigerina
Eocene - Recent
Kyphopyxa
Cretaceous
Ammodiscus
Silurian - Recent
Massilina
Cretaceous - Recent
Hyperammina
Silurian - Recent
Endothyra
Devonian - Triassic
Cibicides
Cretaceous - Recent
Psammosphaera
Silurian - Recent
Textularia
Penn. - Recent
Globigerina
Cretaceous - Recent
Climacammina
Penn. - Permian
Globotruncana
Cretaceous

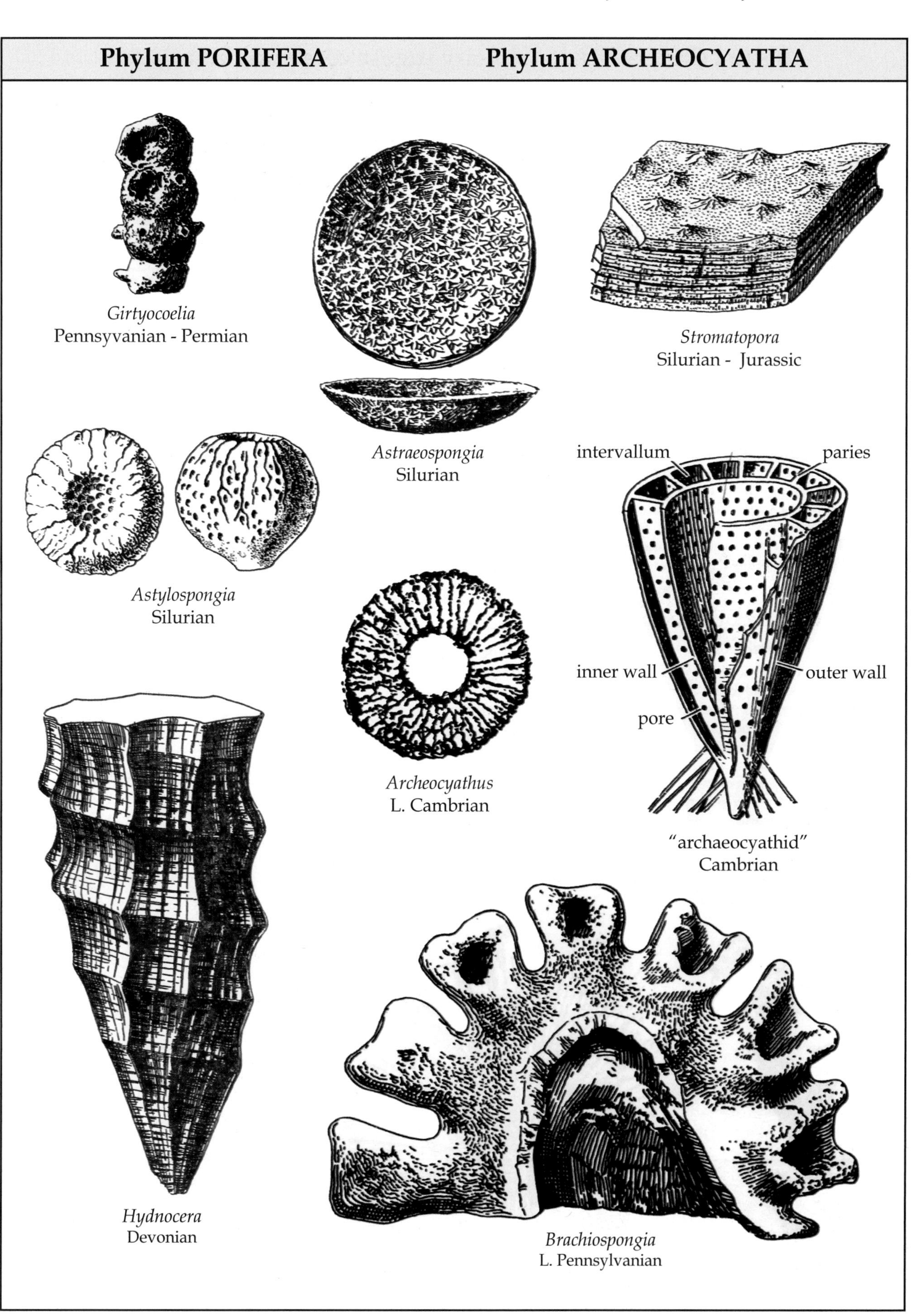

Girtyocoelia
Pennsyvanian - Permian

Astraeospongia
Silurian

Stromatopora
Silurian - Jurassic

Astylospongia
Silurian

Archeocyathus
L. Cambrian

"archaeocyathid"
Cambrian

Hydnocera
Devonian

Brachiospongia
L. Pennsylvanian

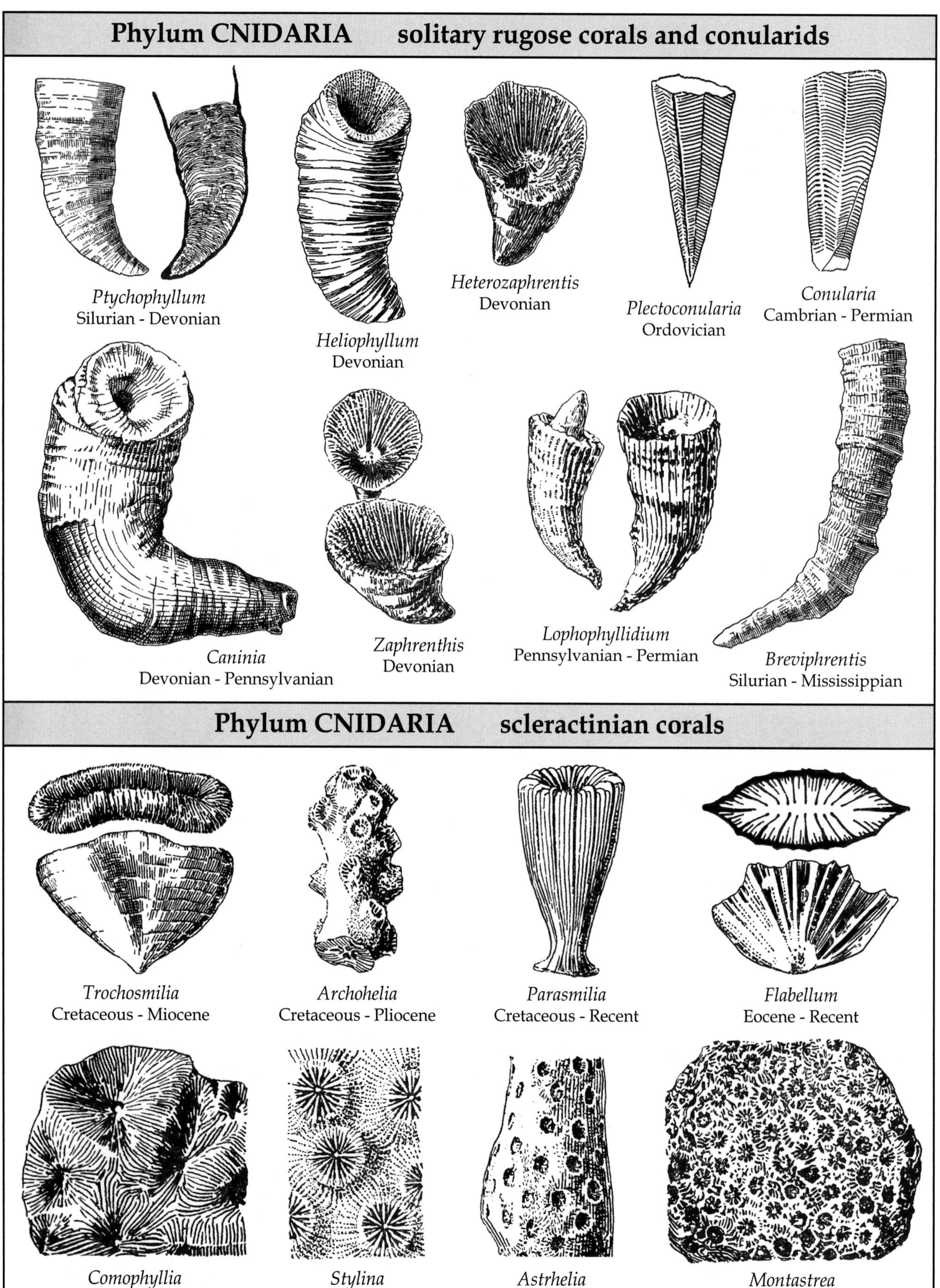
Phylum CNIDARIA solitary rugose corals and conularids
Ptychophyllum
Silurian - Devonian
Heliophyllum
Devonian
Heterozaphrentis
Devonian
Plectoconularia
Ordovician
Conularia
Cambrian - Permian
Caninia
Devonian - Pennsylvanian
Zaphrenthis
Devonian
Lophophyllidium
Pennsylvanian - Permian
Breviphrentis
Silurian - Mississippian
Phylum CNIDARIA scleractinian corals
Trochosmilia
Cretaceous - Miocene
Archohelia
Cretaceous - Pliocene
Parasmilia
Cretaceous - Recent
Flabellum
Eocene - Recent
Comophyllia
Jurassic
Stylina
Triassic - Cretaceous
Astrhelia
Miocene
Montastrea
Jurassic - Recent

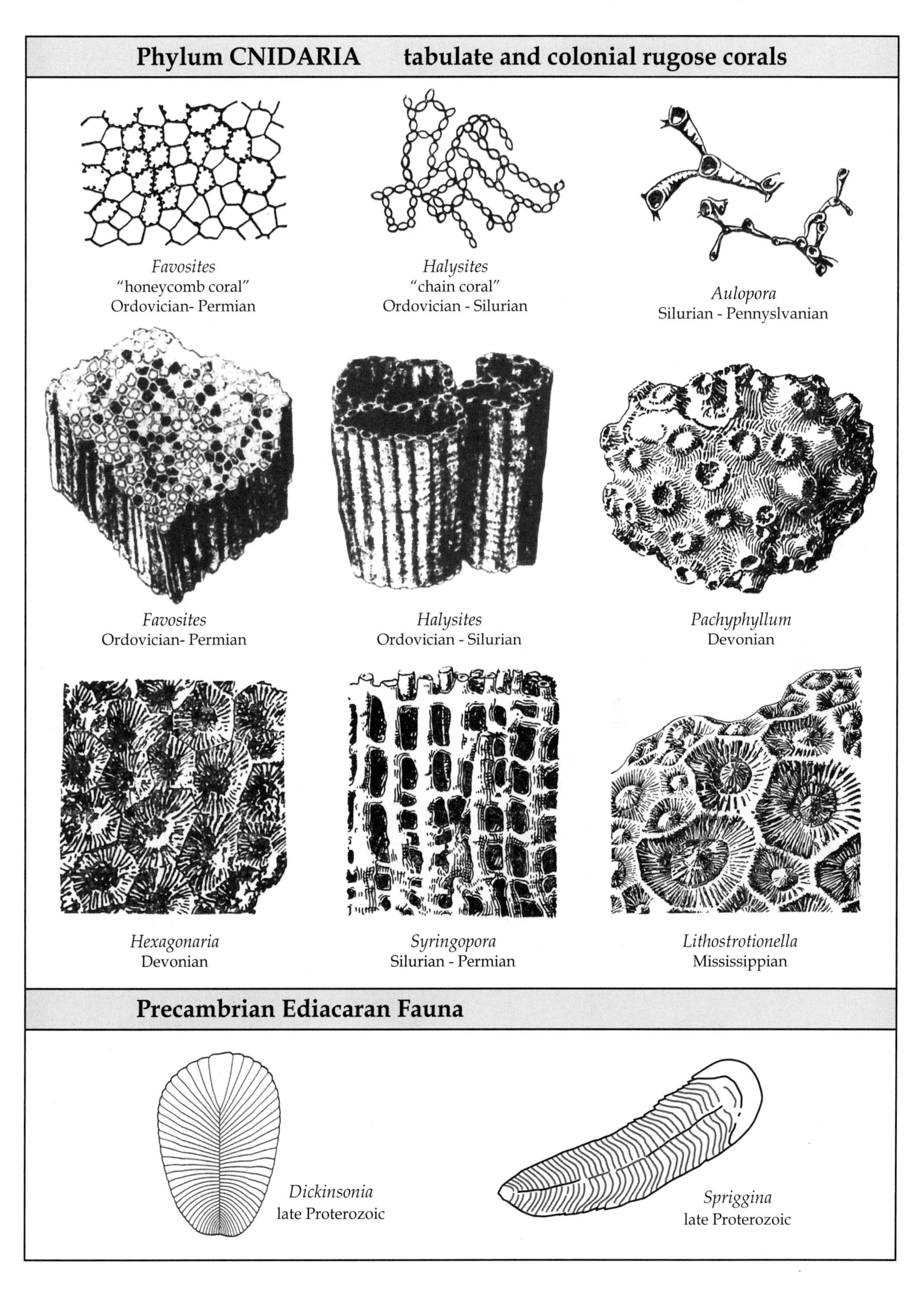

Phylum CNIDARIA tabulate and colonial rugose corals

Favosites
"honeycomb coral"
Ordovician- Permian

Halysites
"chain coral"
Ordovician - Silurian

Aulopora
Silurian - Pennyslvanian

Favosites
Ordovician- Permian

Halysites
Ordovician - Silurian

Pachyphyllum
Devonian

Hexagonaria
Devonian

Syringopora
Silurian - Permian

Lithostrotionella
Mississippian

Precambrian Ediacaran Fauna

Dickinsonia
late Proterozoic

Spriggina
late Proterozoic

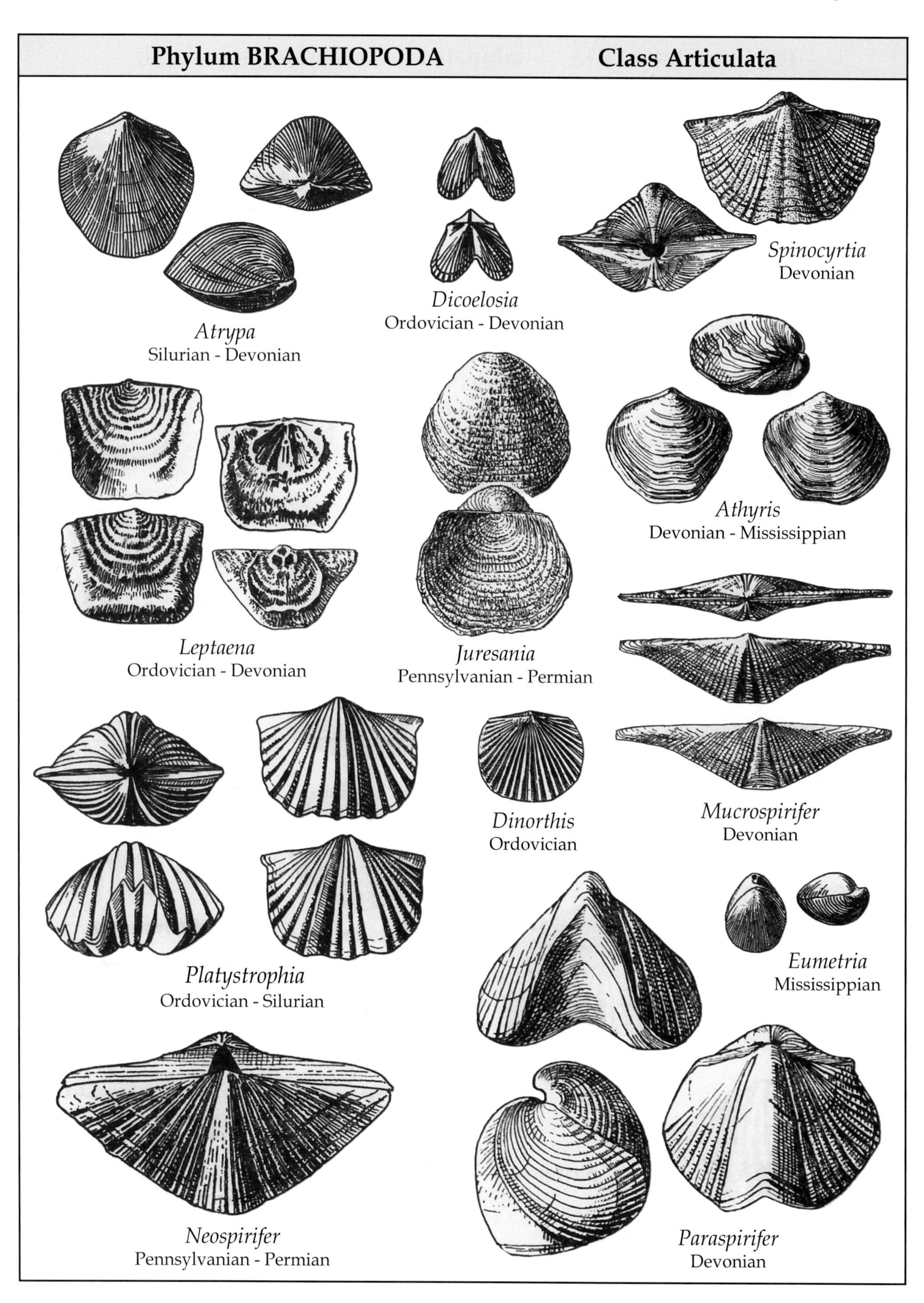
Phylum BRACHIOPODA
Class Articulata
Atrypa
Silurian - Devonian
Dicoelosia
Ordovician - Devonian
Spinocyrtia
Devonian
Leptaena
Ordovician - Devonian
Juresania
Pennsylvanian - Permian
Athyris
Devonian - Mississippian
Platystrophia
Ordovician - Silurian
Dinorthis
Ordovician
Mucrospirifer
Devonian
Eumetria
Mississippian
Neospirifer
Pennsylvanian - Permian
Paraspirifer
Devonian

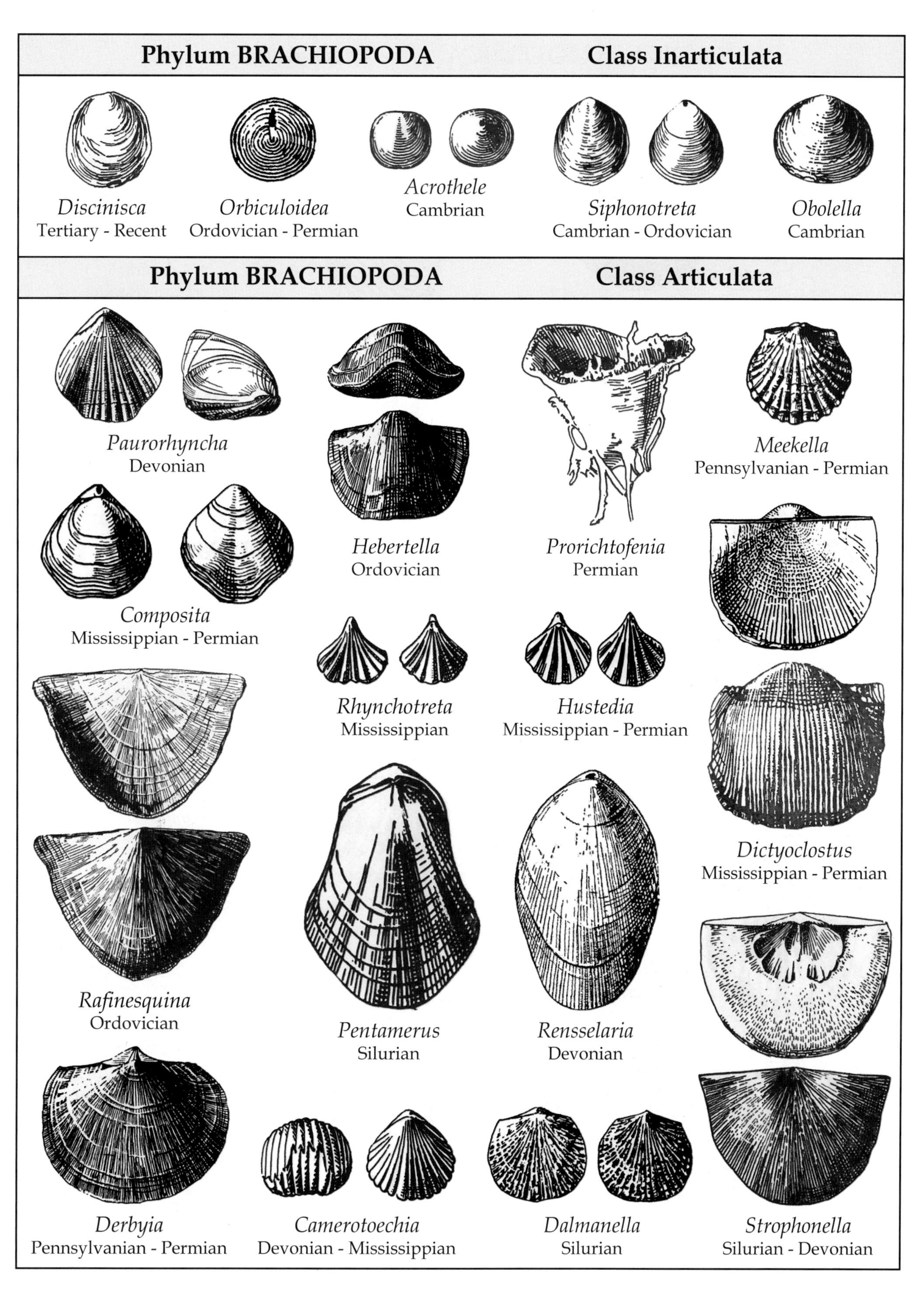
Phylum BRACHIOPODA Class Inarticulata
Discinisca
Tertiary - Recent
Orbiculoidea
Ordovician - Permian
Acrothele
Cambrian
Siphonotreta
Cambrian - Ordovician
Obolella
Cambrian
Phylum BRACHIOPODA Class Articulata
Paurorhyncha
Devonian
Hebertella
Ordovician
Prorichtofenia
Permian
Meekella
Pennsylvanian - Permian
Composita
Mississippian - Permian
Rhynchotreta
Mississippian
Hustedia
Mississippian - Permian
Dictyoclostus
Mississippian - Permian
Rafinesquina
Ordovician
Pentamerus
Silurian
Rensselaria
Devonian
Derbyia
Pennsylvanian - Permian
Camerotoechia
Devonian - Mississippian
Dalmanella
Silurian
Strophonella
Silurian - Devonian

Phylum MOLLUSCA Class Bivalvia
Pychnodonte
Jurassic - Eocene
Arctica
Jurassic - Recent
Crassatellites
Cretaceous - Recent
Corbula
Triassic - Recent
Monopleura
Cretaceous
"rudist"
Ostrea
Jurassic - Recent
Cardium
Triassic - Recent
Idonearca
Cretaceous
Barbatia
Cretaceous - Recent
Paranomia
Cretaceous
Trigonia
Jurassic - Recent
Exogyra
Jurassic - Cretaceous
Exogyra
Jurassic - Cretaceous
Inoceramus
Jurassic - Cretaceous

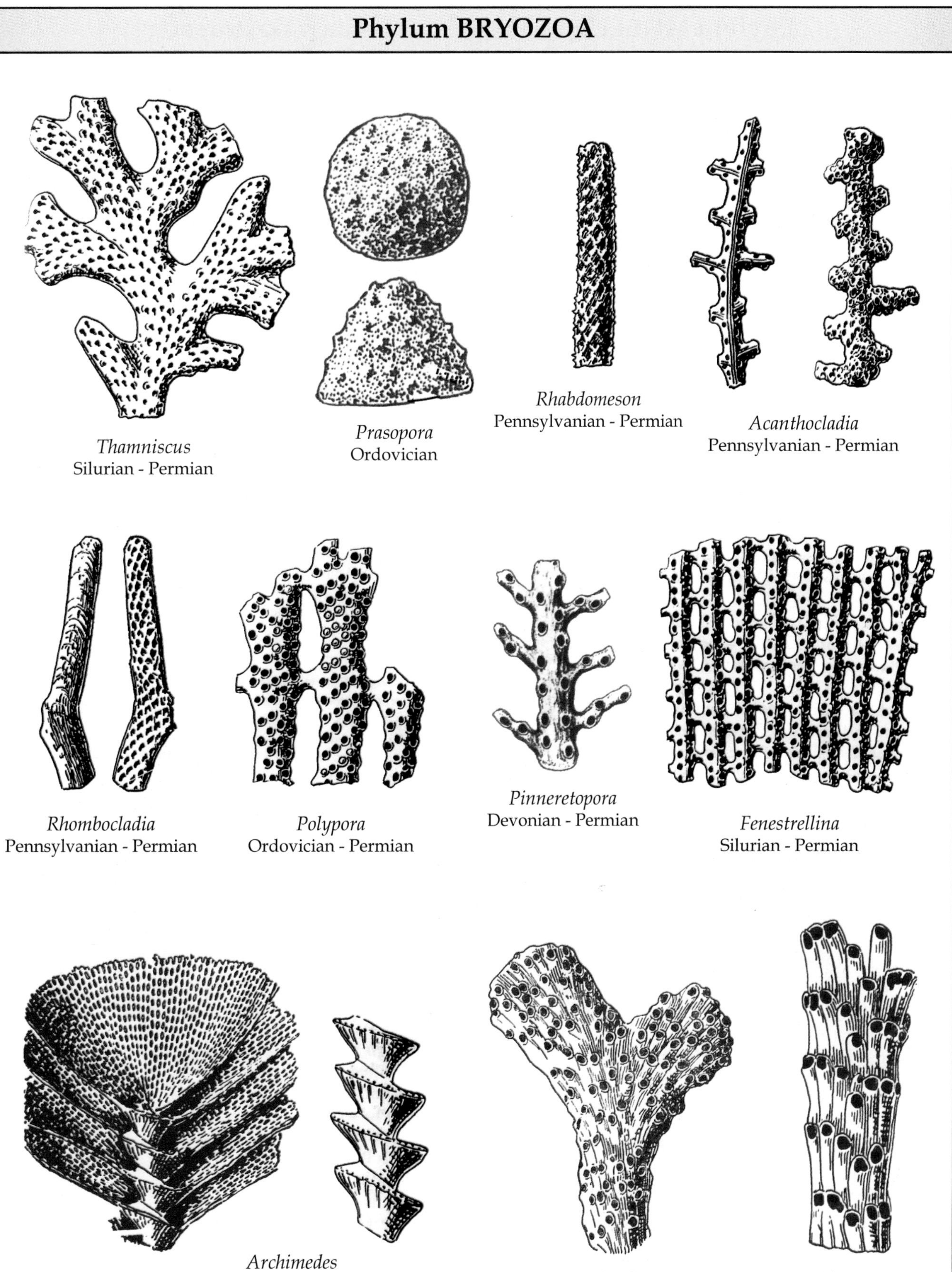
Phylum BRYOZOA
Thamniscus
Silurian - Permian
Prasopora
Ordovician
Rhabdomeson
Pennsylvanian - Permian
Acanthocladia
Pennsylvanian - Permian
Rhombocladia
Pennsylvanian - Permian
Polypora
Ordovician - Permian
Pinneretopora
Devonian - Permian
Fenestrellina
Silurian - Permian
Archimedes
Mississippian - Permian
Psilosolen
Pleistocene - Recent
Idmonea
Jurassic - Recent

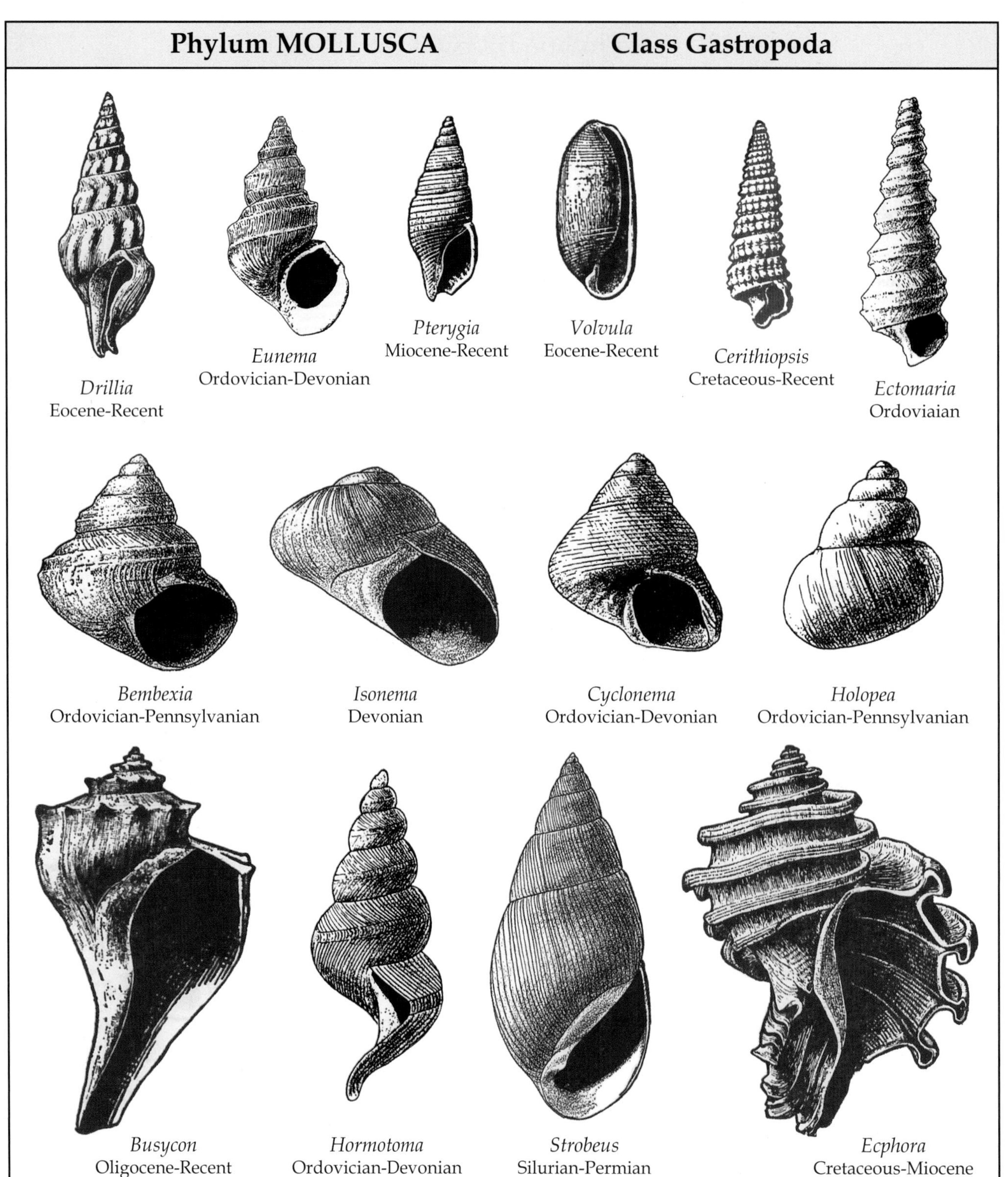
Phylum MOLLUSCA
Class Gastropoda
Drillia
Eocene-Recent
Eunema
Ordovician-Devonian
Pterygia
Miocene-Recent
Volvula
Eocene-Recent
Cerithiopsis
Cretaceous-Recent
Ectomaria
Ordoviaian
Bembexia
Ordovician-Pennsylvanian
Isonema
Devonian
Cyclonema
Ordovician-Devonian
Holopea
Ordovician-Pennsylvanian
Busycon
Oligocene-Recent
Hormotoma
Ordovician-Devonian
Strobeus
Silurian-Permian
Ecphora
Cretaceous-Miocene

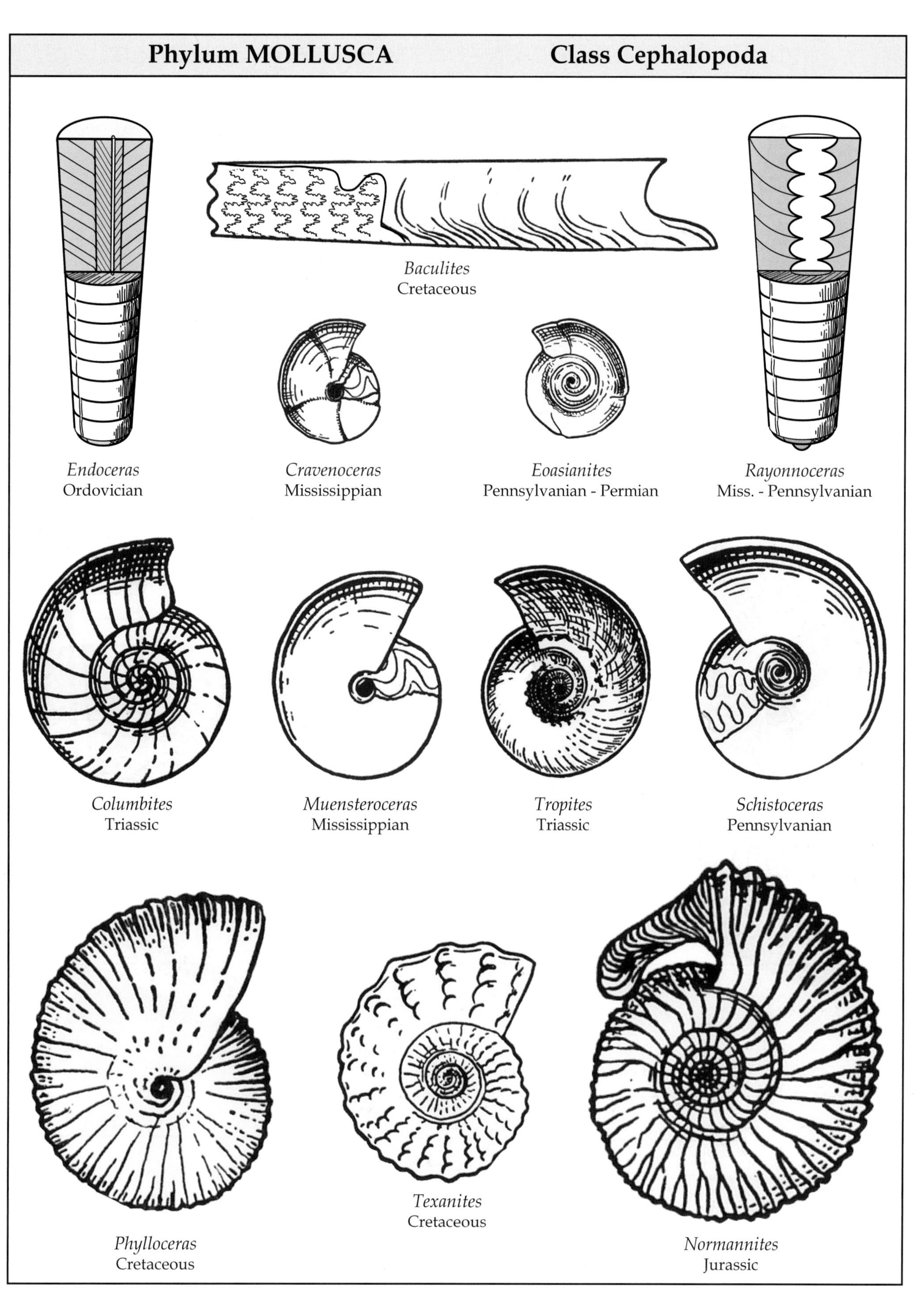
Phylum MOLLUSCA Class Cephalopoda
Baculites
Cretaceous
Endoceras
Ordovician
Cravenoceras
Mississippian
Eoasianites
Pennsylvanian - Permian
Rayonnoceras
Miss. - Pennsylvanian
Columbites
Triassic
Muensteroceras
Mississippian
Tropites
Triassic
Schistoceras
Pennsylvanian
Texanites
Cretaceous
Phylloceras
Cretaceous
Normannites
Jurassic

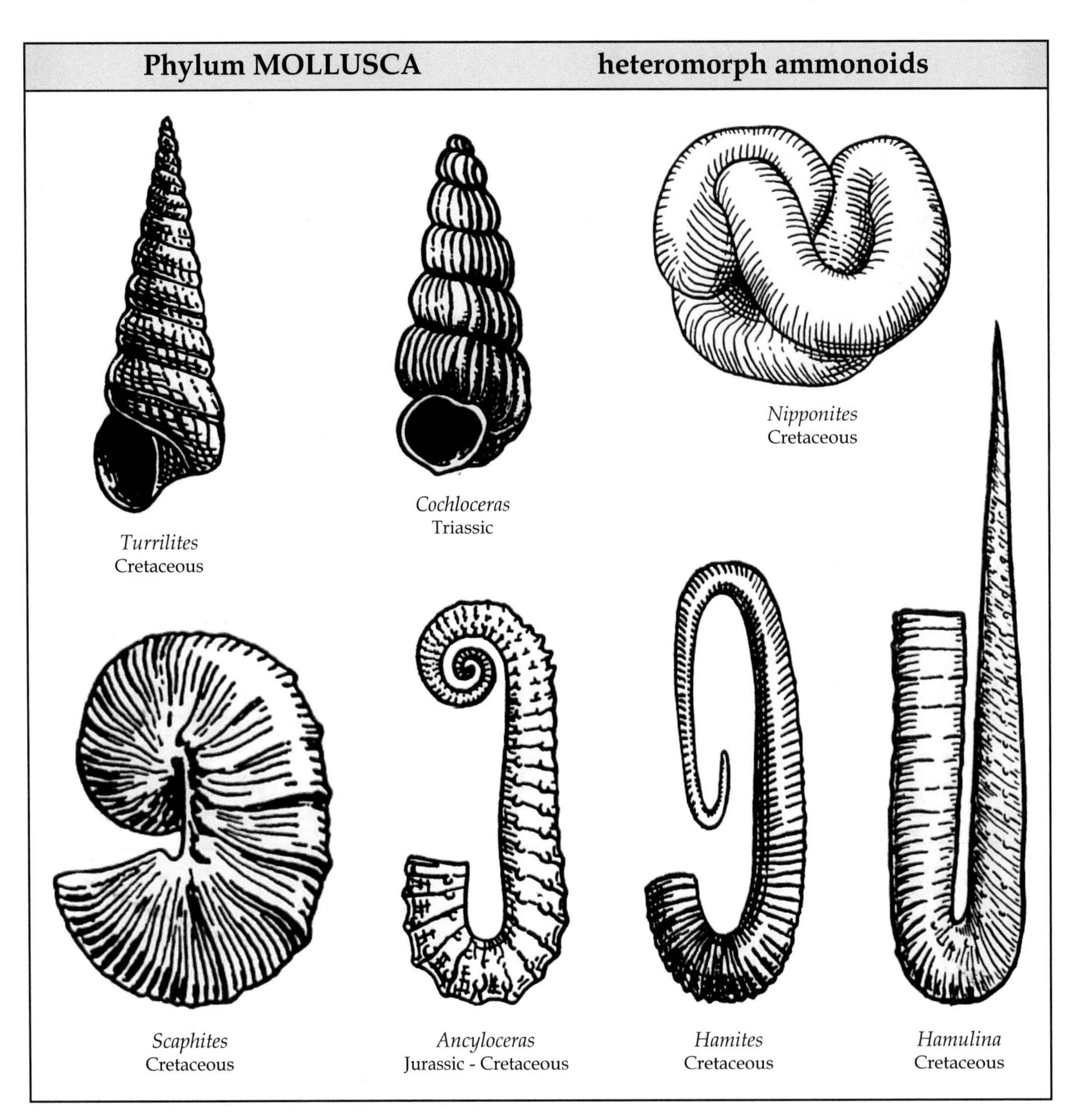
Phylum MOLLUSCA
heteromorph ammonoids
Turrilites
Cretaceous
Cochloceras
Triassic
Nipponites
Cretaceous
Scaphites
Cretaceous
Ancyloceras
Jurassic - Cretaceous
Hamites
Cretaceous
Hamulina
Cretaceous

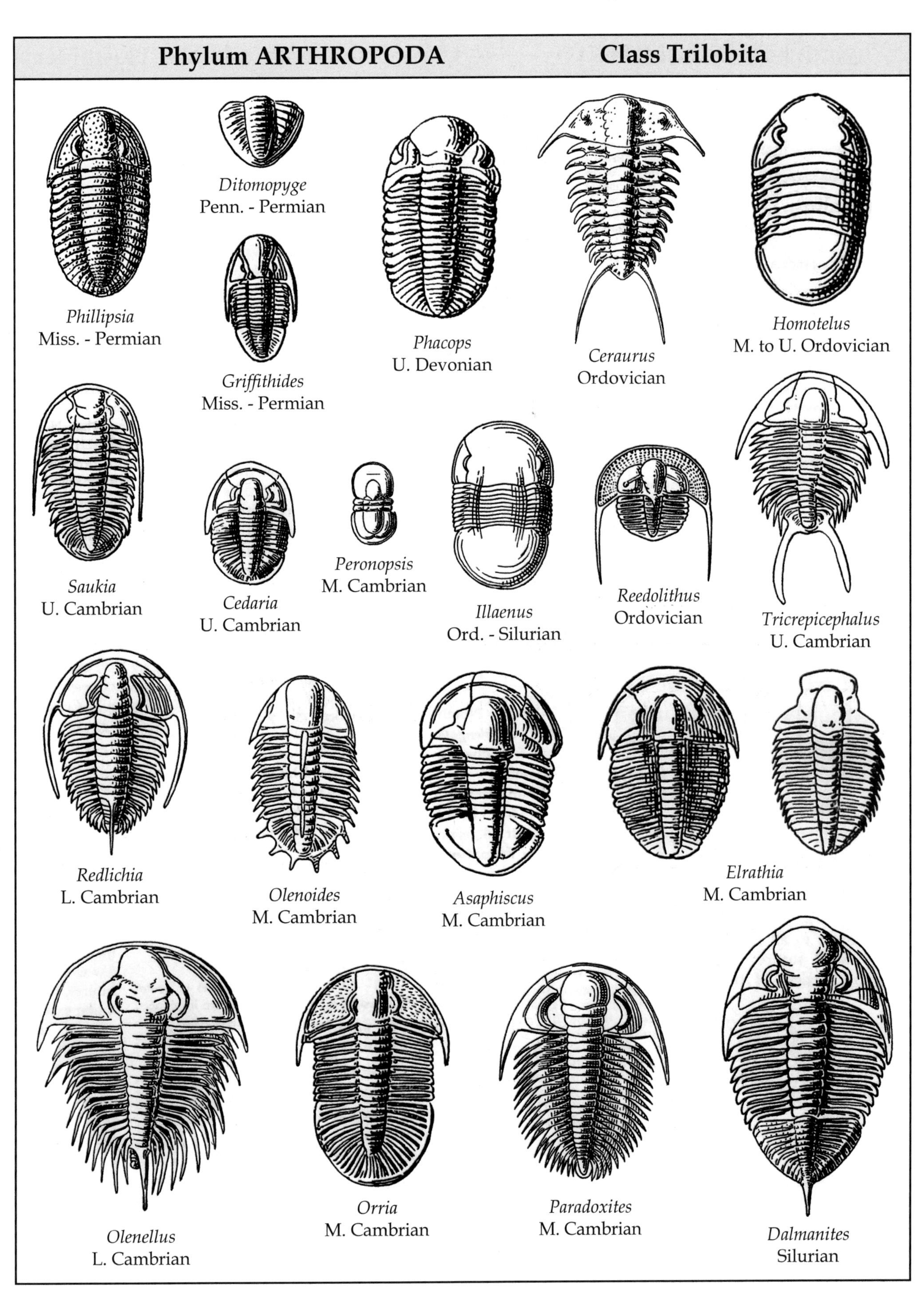
Phylum ARTHROPODA Class Trilobita
Phillipsia
Miss. - Permian
Ditomopyge
Penn. - Permian
Griffithides
Miss. - Permian
Phacops
U. Devonian
Ceraurus
Ordovician
Homotelus
M. to U. Ordovician
Saukia
U. Cambrian
Cedaria
U. Cambrian
Peronopsis
M. Cambrian
Illaenus
Ord. - Silurian
Reedolithus
Ordovician
Tricrepicephalus
U. Cambrian
Redlichia
L. Cambrian
Olenoides
M. Cambrian
Asaphiscus
M. Cambrian
Elrathia
M. Cambrian
Olenellus
L. Cambrian
Orria
M. Cambrian
Paradoxites
M. Cambrian
Dalmanites
Silurian

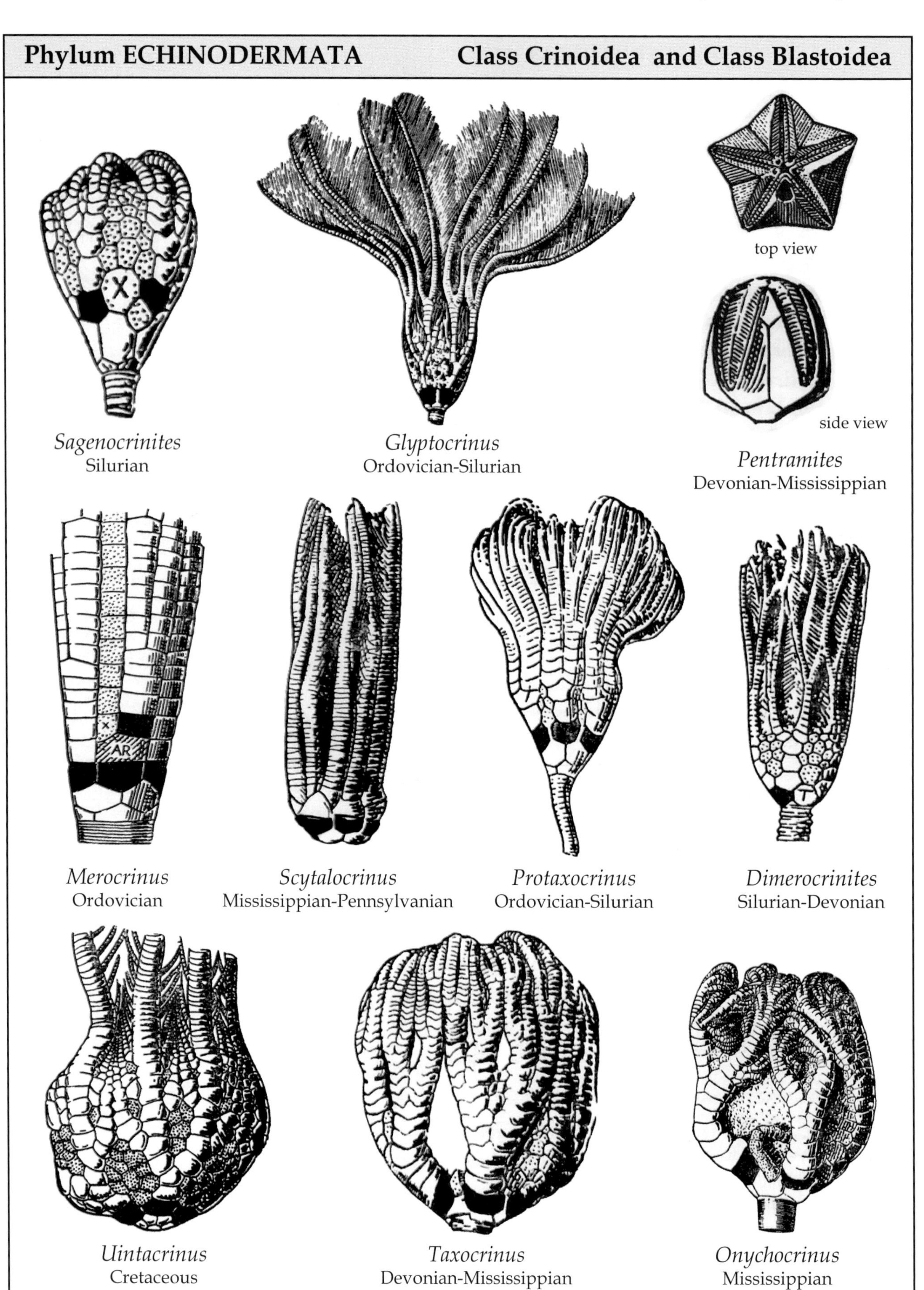
Phylum ECHINODERMATA
Class Crinoidea and Class Blastoidea
X
top view
side view
AR
T
Sagenocrinites
Silurian
Glyptocrinus
Ordovician-Silurian
Pentramites
Devonian-Mississippian
Merocrinus
Ordovician
Scytalocrinus
Mississippian-Pennsylvanian
Protaxocrinus
Ordovician-Silurian
Dimerocrinites
Silurian-Devonian
Uintacrinus
Cretaceous
Taxocrinus
Devonian-Mississippian
Onychocrinus
Mississippian

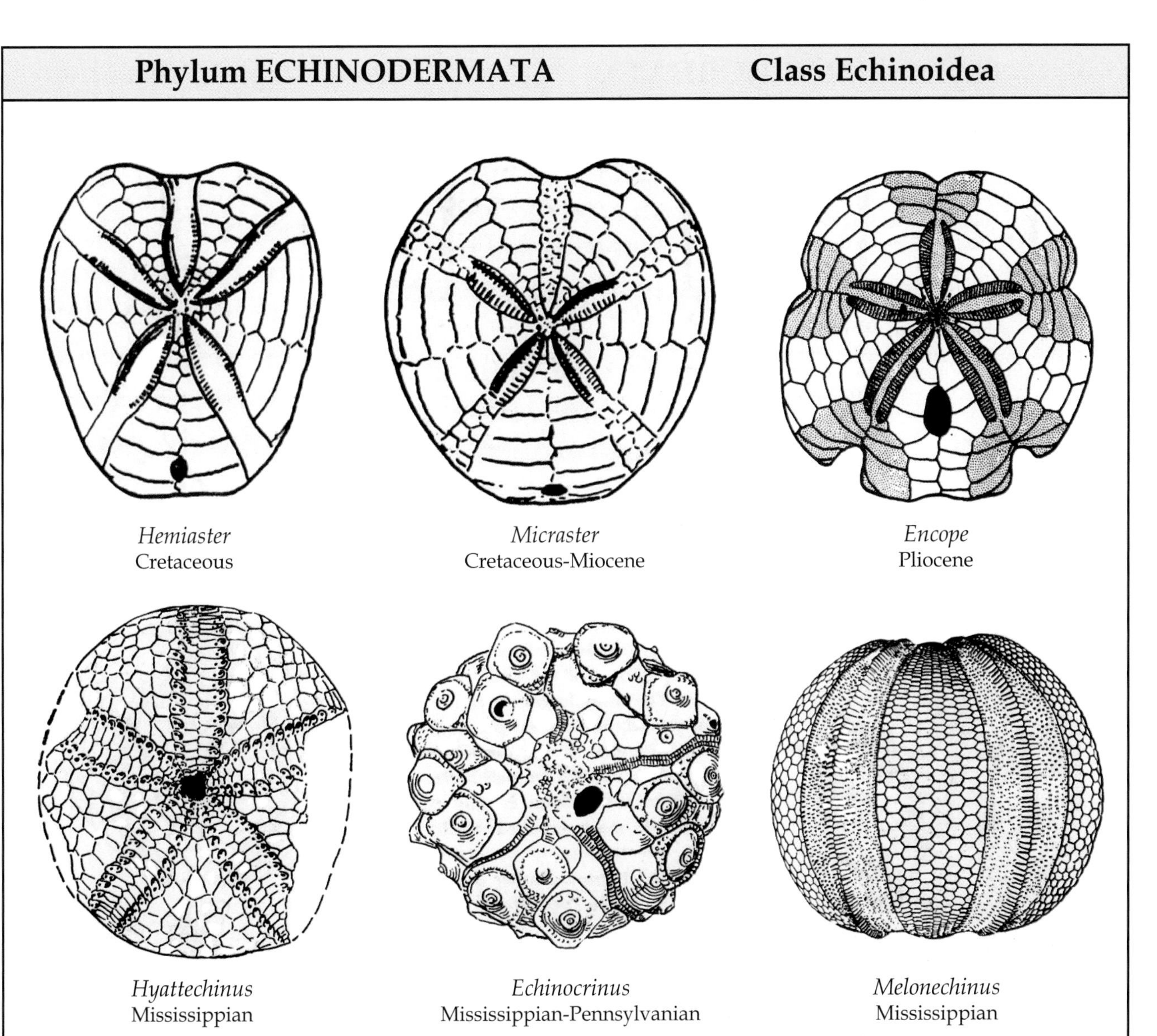
Phylum ECHINODERMATA
Class Echinoidea
Hemiaster
Cretaceous
Micraster
Cretaceous-Miocene
Encope
Pliocene
Hyattechinus
Mississippian
Echinocrinus
Mississippian-Pennsylvanian
Melonechinus
Mississippian

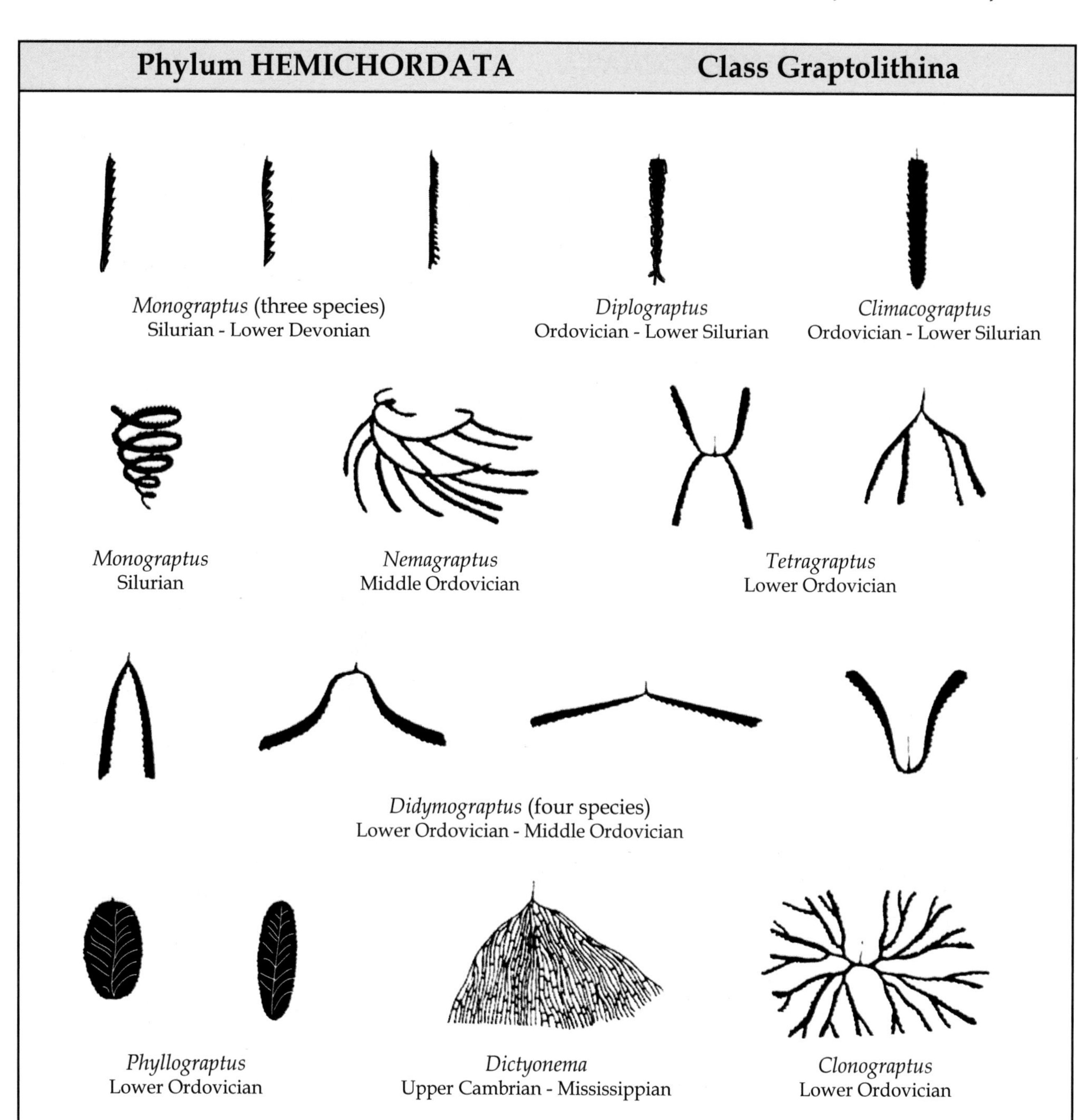

Phylum HEMICHORDATA
Class Graptolithina
Monograptus (three species)
Silurian - Lower Devonian
Diplograptus
Ordovician - Lower Silurian
Climacograptus
Ordovician - Lower Silurian
Monograptus
Silurian
Nemagraptus
Middle Ordovician
Tetragraptus
Lower Ordovician
Didymograptus (four species)
Lower Ordovician - Middle Ordovician
Phyllograptus
Lower Ordovician
Dictyonema
Upper Cambrian - Mississippian
Clonograptus
Lower Ordovician

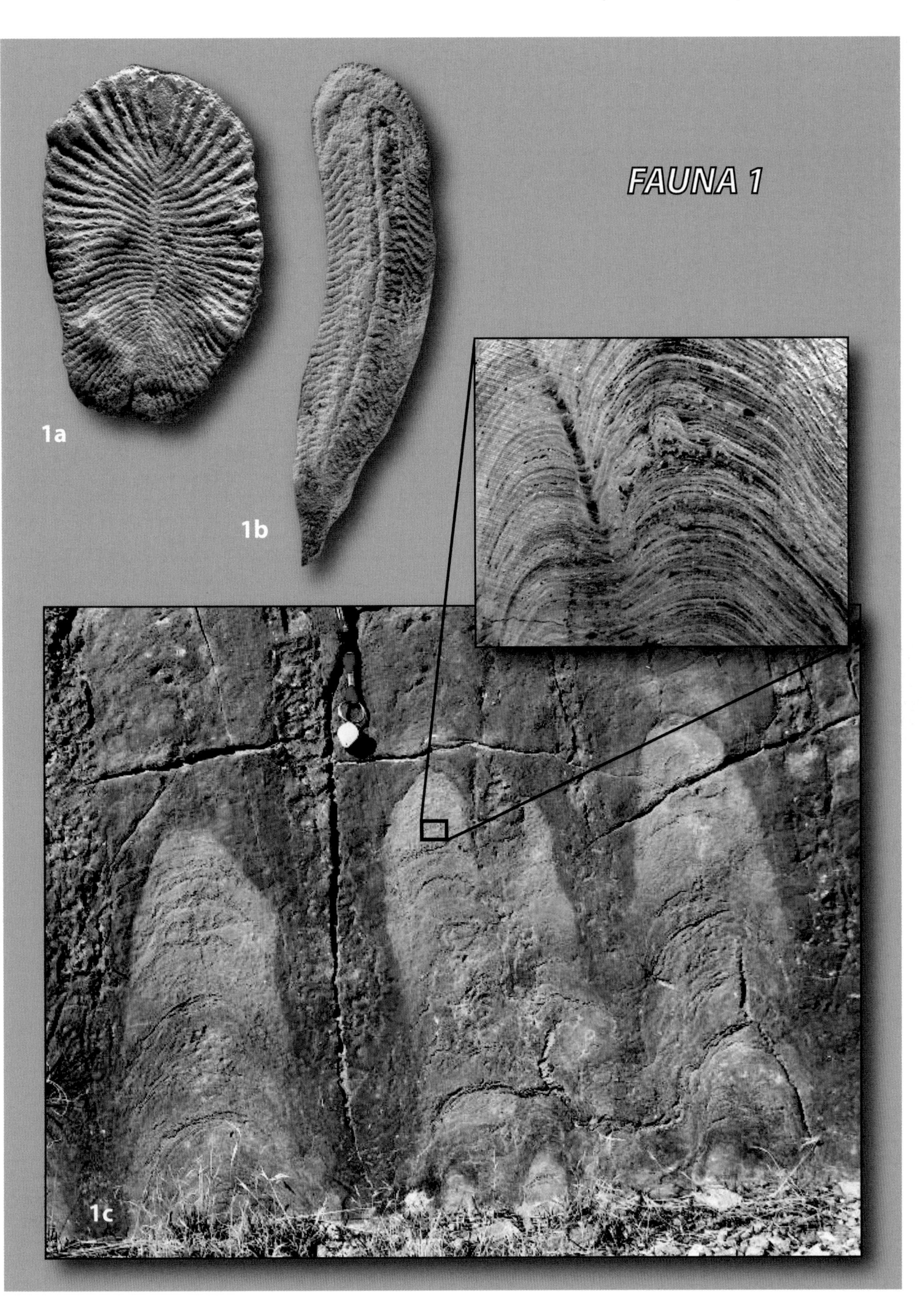
FAUNA 1
1a
1b
1c

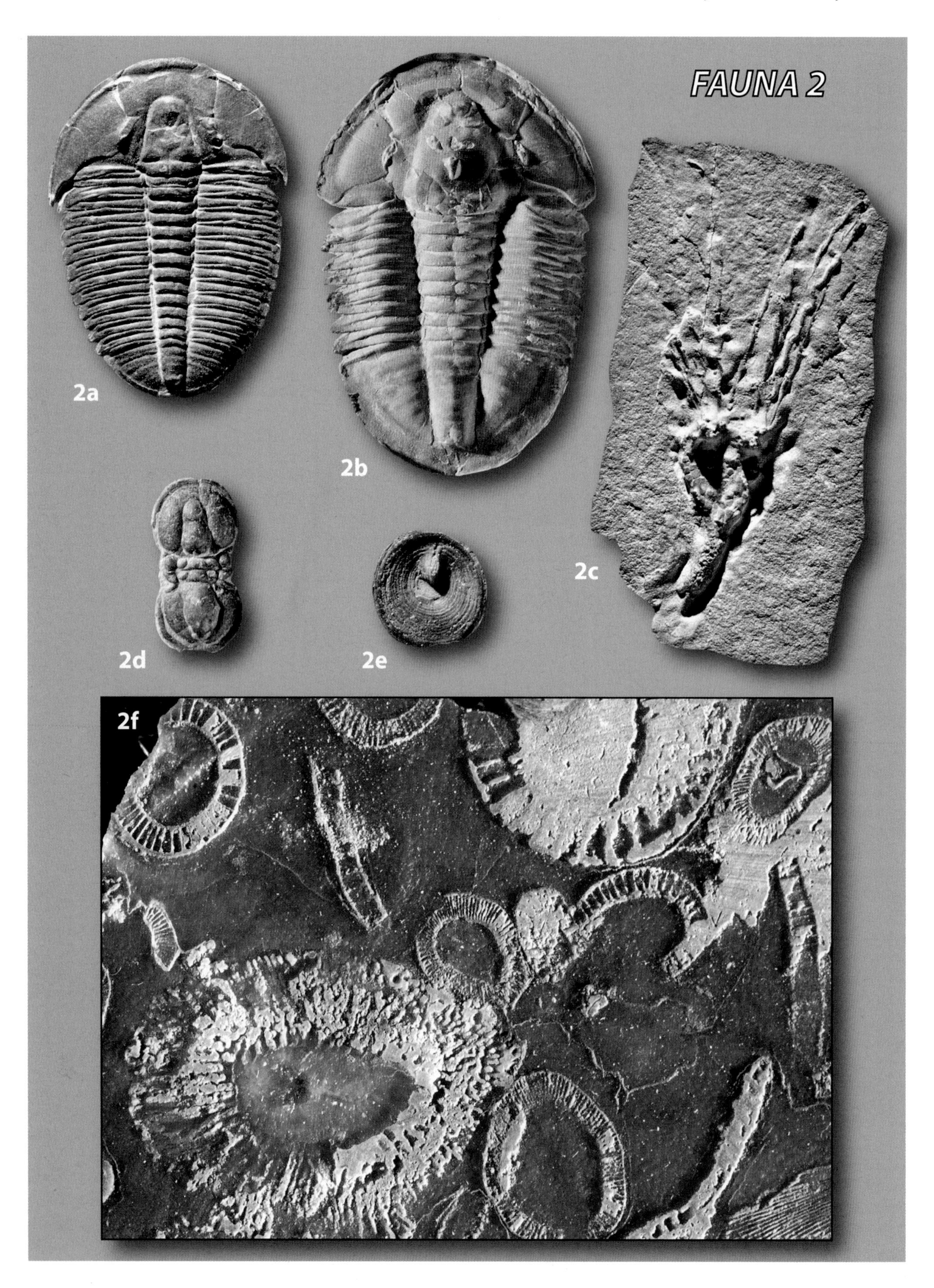
FAUNA 2
2a
2b
2c
2d
2e
2f

FAUNA 3
3a
3b
3c
3d
top
3d
front
3e
3f
LIBERTY
1982

FAUNA 4
4a
4b

FAUNA 5
5a
5b

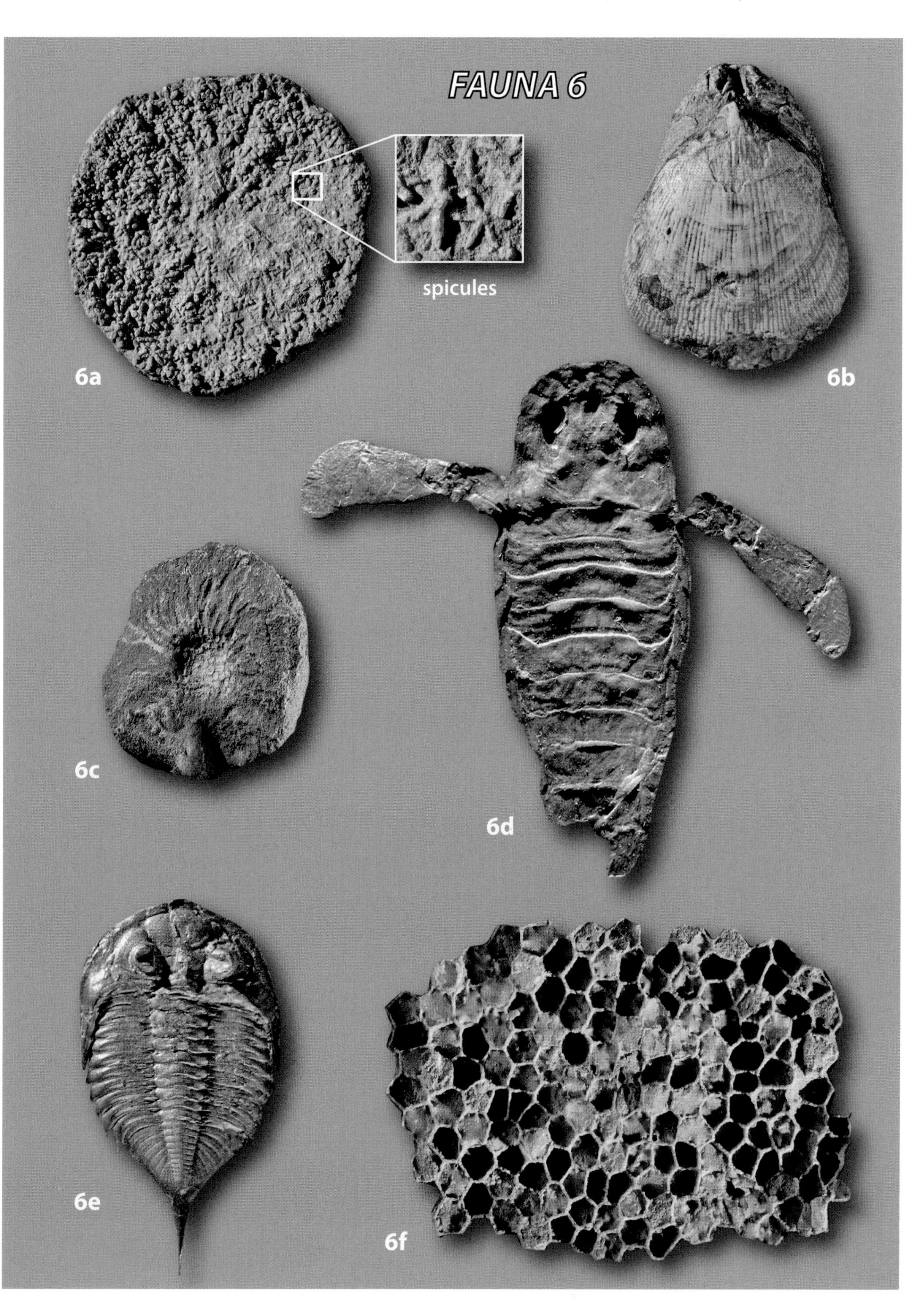
FAUNA 6
spicules
6a
6b
6c
6d
6e
6f

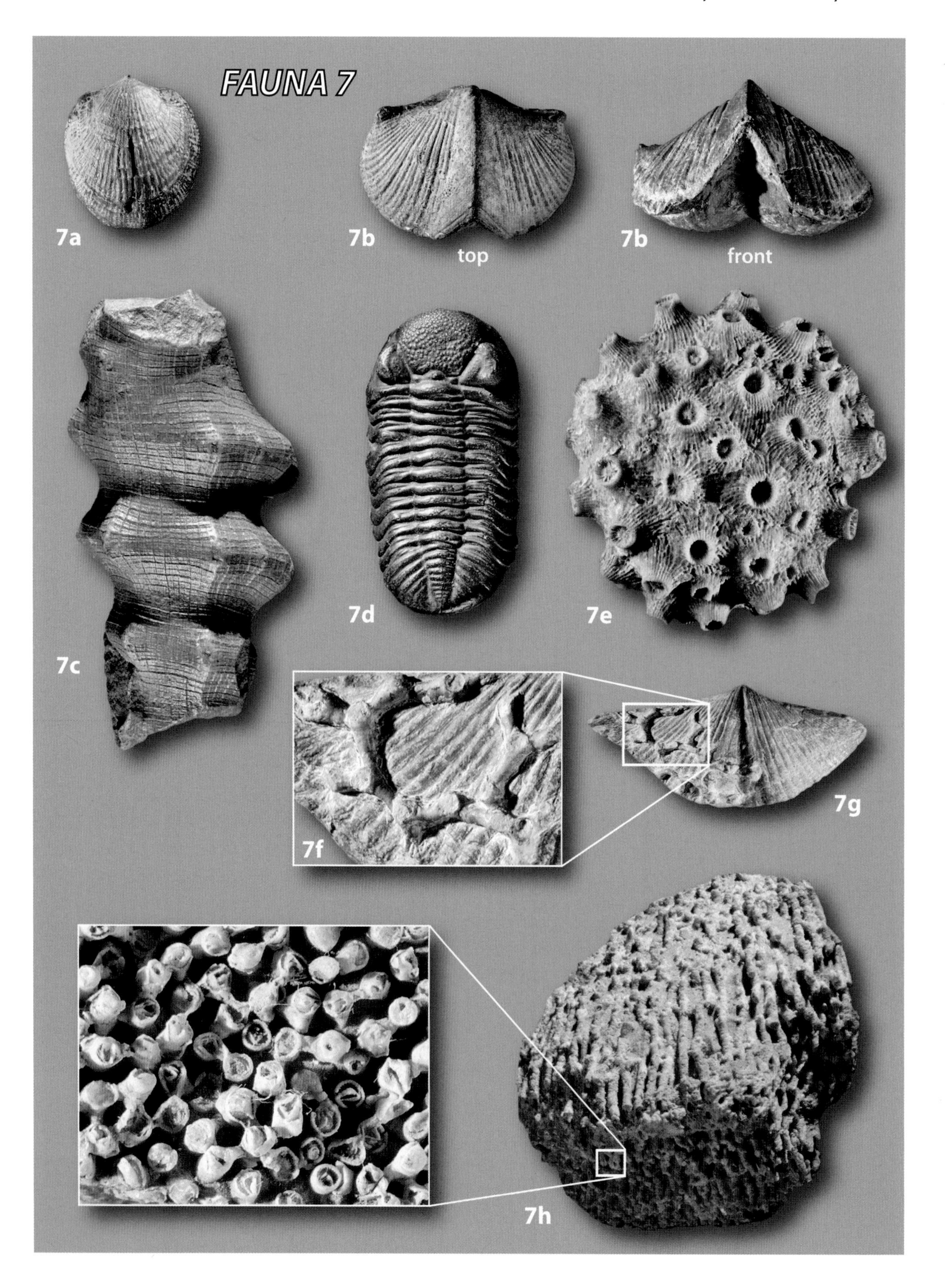
FAUNA 7
7a
7b
top
7b
front
7c
7d
7e
7f
7g
7h

FAUNA 8
8a
top
side
8b
8c
8d
8e
8f
8g
thin section image

FAUNA 9
9a
9b
9c
9d
9e
9f
9g

	FAUNA 1	A.	B.	C.	D.	E.	F.	G.	H.	I.	J.	K.
Ceno.	Neogene			X								
	Paleogene			X								
Mesozoic	Cretaceous			X								
	Jurassic			X								
	Triassic			X								
Paleozoic	Permian			X								
	Pennsylvanian			X								
	Mississippian			X								
	Devonian			X								
	Silurian			X								
	Ordovician			X								
	Cambrian			X								
	Ediacaran	X	X	X								

TAXON NAME

A. Dickinsonia
B. Spriggina
C. stromatolite

Age of Bed: Ediacaran

	FAUNA 2	A.	B.	C.	D.	E.	F.	G.	H.	I.	J.	K.
Ceno.	Neogene											
	Paleogene											
Mesozoic	Cretaceous											
	Jurassic											
	Triassic											
Paleozoic	Permian											
	Pennsylvanian											
	Mississippian											
	Devonian											
	Silurian											
	Ordovician											
	Cambrian	X	X	X		X						
	Ediacaran											

TAXON NAME

A. Elrathia
B. Asaphiscus
C. Gogia
D.
E. Acrothele
F. archeocyathus

Age of Bed: cambrian

FAUNA 3		A.	B.	C.	D.	E.	F.	G.	H.	I.	J.	K.	TAXON NAME
Ceno.	Neogene												A. Reedolithus
	Paleogene												B. rafinesquina
Mesozoic	Cretaceous												C. Ravonnoceras D. platystrophia
	Jurassic												E. Leptaena
	Triassic												F. Leperditia
Paleozoic	Permian												
	Pennsylvanian			X									
	Mississippian			X			X						
	Devonian						X						
	Silurian				X	X	X						Age of Bed:
	Ordovician	X	X		X	X	X						ordivician
	Cambrian						X						

FAUNA 4		A.	B.	C.	D.	E.	F.	G.	H.	I.	J.	K.	TAXON NAME
Ceno.	Neogene												A. Olenellus
	Paleogene												B. Peronopsis
Mesozoic	Cretaceous												
	Jurassic												
	Triassic												
Paleozoic	Permian												
	Pennsylvanian												
	Mississippian												
	Devonian												
	Silurian												Age of Bed:
	Ordovician												cambrian
	Cambrian	X	X										

FAUNA 5		A.	B.	C.	D.	E.	F.	G.	H.	I.	J.	K.
Ceno.	Neogene											
	Paleogene											
Mesozoic	Cretaceous											
	Jurassic											
	Triassic											
Paleozoic	Permian											
	Pennsylvanian											
	Mississippian											
	Devonian											
	Silurian											
	Ordovician	X	X									
	Cambrian		X									
	Edicarian		X									

TAXON NAME

A. Tetragraptes

B. phyllograptus

Age of Bed: ~~late~~ ~~paleozoic~~ ordivician

FAUNA 6		A.	B.	C.	D.	E.	F.	G.	H.	I.	J.	K.
Ceno.	Neogene											
	Paleogene											
Mesozoic	Cretaceous											
	Jurassic											
	Triassic											
Paleozoic	Permian				X		X					
	Pennsylvanian				X		X					
	Mississippian				X		X					
	Devonian				X		X					
	Silurian	X	X	X	X	X	X					
	Ordovician				X		X					
	Cambrian											

TAXON NAME

A. Astraaospongia

B. pentamerus

C. Astylospongia

D. Eurypteridia

E. Dalanites

F. Favasites

Age of Bed: silurian

	FAUNA 7	A.	B.	C.	D.	E.	F.	G.	H.	I.	J.	K.
Ceno.	Neogene											
	Paleogene											
Mesozoic	Cretaceous											
	Jurassic											
	Triassic											
Paleozoic	Permian			X	X				X			
	Pennsylvanian			X	X		X		X			
	Mississippian			X	X		X		X			
	Devonian	X		X		X	X	X	X			
	Silurian	X	X				X		X			
	Ordovician		X									
	Cambrian											

TAXON NAME

A. Atrypa
B. platystrophia
C. Hydnoceras
D. Phillipsia
E. pachyphyllum
F. Aulopor
G. microspirifer
H. Synngopora

Age of Bed:
Devonian

	FAUNA 8	A.	B.	C.	D.	E.	F.	G.	H.	I.	J.	K.
Ceno.	Neogene											
	Paleogene											
Mesozoic	Cretaceous											
	Jurassic											
	Triassic											
Paleozoic	Permian			X	X		X					
	Pennsylvanian	X	X	X	X	X	X	X				
	Mississippian	X	X				X					
	Devonian											
	Silurian											
	Ordovician		X									
	Cambrian											

TAXON NAME

A. pentrimites
B. pletoconvlara
C. birtyocoelia
D. neospirifer
E. brachiospongia
F. archimedes
G. Fusolinella

Age of Bed:
pennsylvannian

FAUNA 9		A.	B.	C.	D.	E.	F.	G.	H.	I.	J.	K.
Ceno.	Neogene		X					X				
	Paleogene		X		X			X				
Mesozoic	Cretaceous	X	X	X	X	X	X	X				
	Jurassic	X			X		X	X				
	Triassic	X										
Paleozoic	Permian											
	Pennsylvanian											
	Mississippian											
	Devonian											
	Silurian											
	Ordovician											
	Cambrian											

TAXON NAME
A. Belemnite
B. Micraster
C. Monopleura
D. Pychnodonte
E. Scaphites
F. Exogyra
G. Trigonia
Age of Bed: Cretaceous

TAXONOMIC CATEGORY	LIST OF GENERA REPRESENTED IN FAUNAS
Phylum Archeocyatha	
Phylum Porifera	
Phylum Cnidaria Order Rugosa	
Order Scleractinia	
Order Tabulata	
Phylum Brachiopoda Class Articulata	
Class Inarticulata	
Phylum Bryozoa	
Phylum Mollusca Class Bivalvia	
Class Gastropoda	
Class Cephalopoda	
Phylum Arthropoda Class Trilobita	
Order Ostracoda	
Eurypterida	
Phylum Echinodermata Class Crinoidea	
Class Blastoidea	
Class Echinoidea	
Phylum Hemichordata Class Graptolithina	

TABLE 12.1 Classification of genera identified in exercise 12.

Interpretation of Geological Maps

Exercise 13

Learning Objectives

After completing this exercise, you will:

1. be able to recognize and understand the fundamental elements of a geological map;
2. be familiar with basic geological map symbols;
3. know how to trace contacts, faults, or other planar features on a geological map;
4. be able to construct a topographic profile;
5. be able to construct a geological cross section; and
6. be able to recognize the basic types of folds, faults, and unconformities from patterns portrayed on geological maps.

Introduction

A map is a graphic representation of a geographic area. It is a scale model that uses symbols to depict relationships between objects that occur in the map area. For example, highway maps use one set of symbols to depict cities and towns, another set of symbols to represent roads or waterways connecting these municipalities, and yet a third set to represent the scale of the map. Highway maps are constructed with a specific purpose in mind—to help you efficiently and safely move between two points of interest. **Topographic maps** have a different purpose—to depict relationships between topographic elements such as valleys, hills, mountains, etc. The fundamental symbol on a topographic map is the **contour line**. These lines depict points of equal elevation. Contour lines that are closely spaced indicate steep topography. Flatter areas are represented by more widely spaced contour lines. Historically, humans have spent much time producing maps because these scale models permit us to visualize and interpret patterns that are otherwise too large to discern and decipher.

Geological maps show the distribution of rock bodies and the traces of geological structures exposed at the Earth's surface (figure 13.1). These features, once they're carefully documented in the field, are depicted by tracing their distribution on a topographic base map using the symbols shown in figure 13.2. The fundamental symbols on most geological maps are formational **contacts** and **faults**.

Contacts are the planar boundaries between adjacent rock formations. These are depicted using solid black lines. Faults are depicted using somewhat heavier solid black lines along with symbols that show the direction of offset and nature of the fault. The attitude of bedding and other planar features is indicated using the **strike and dip** symbol. If strata are not horizontal, then there exists a line created by the intersection of the dipping stratum or fault plane and an imaginary horizontal plane (figure 13.3). The azimuth of the line of intersection is the strike direction. Dip is the angle of tilt below the imaginary horizontal plane oriented 90° to strike in the direction in which the beds are inclined.

Geological maps have great practical significance. Indeed, the first geological maps were constructed by William "Strata" Smith to expedite the construction of coal canals in early nineteenth cen-

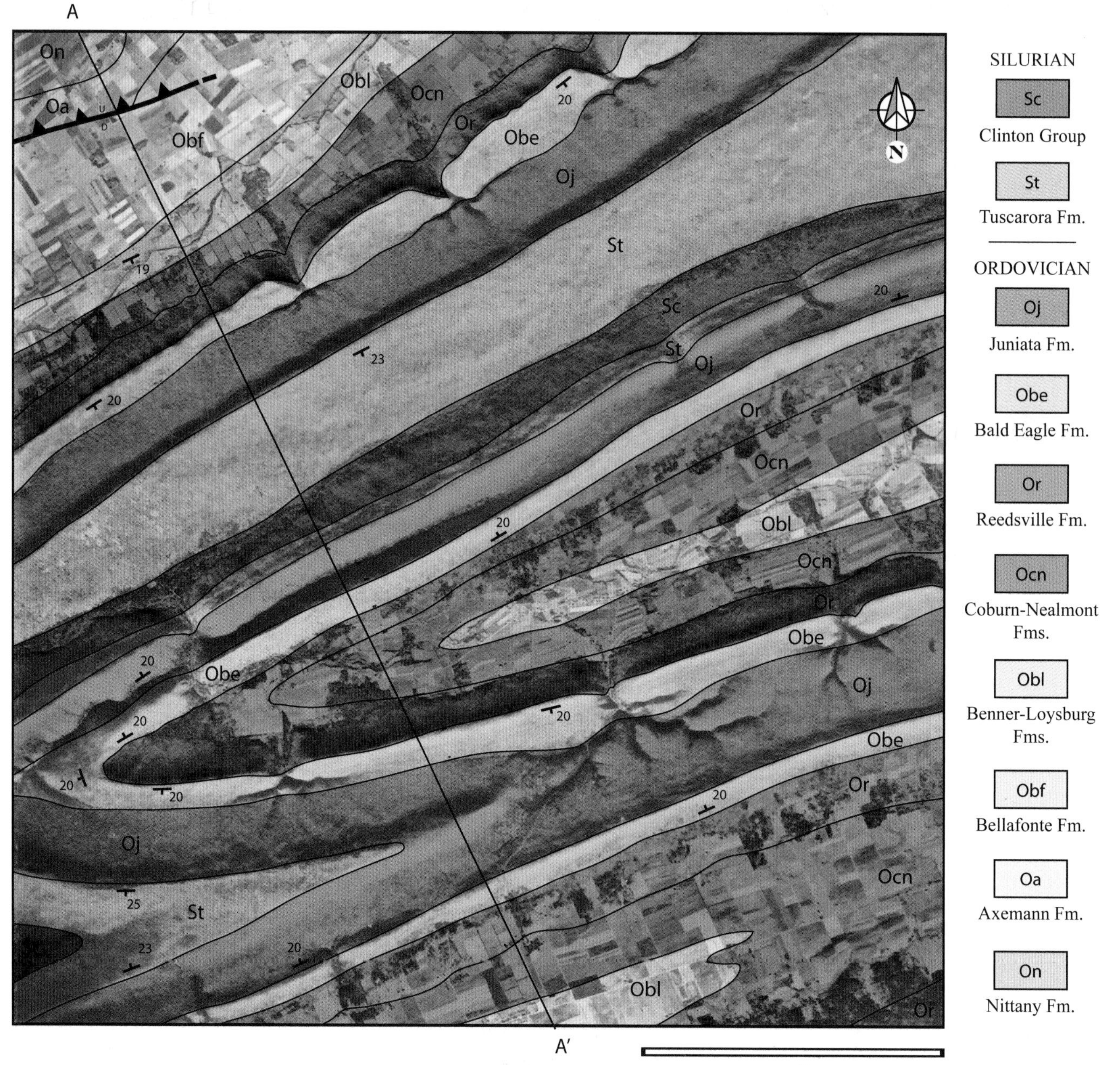

FIGURE 13.1 Geological map of the Mill Hall region of Pennsylvania showing the relationship of geology and topography.

tury England. Nowadays they are used to delineate the existence of resource-bearing strata and trace the occurrence of geological hazards. They are also indispensable tools in reconstructing Earth history on local, regional, and global scales. The patterns or relationships depicted on these scale models permit us to resolve the timing and nature of geological processes that affected the map area. The purpose of this exercise is to help you develop your geological map interpretation skills. Several subsequent labs are based upon map interpretations, which will provide critical hands-on experience with these important geological tools.

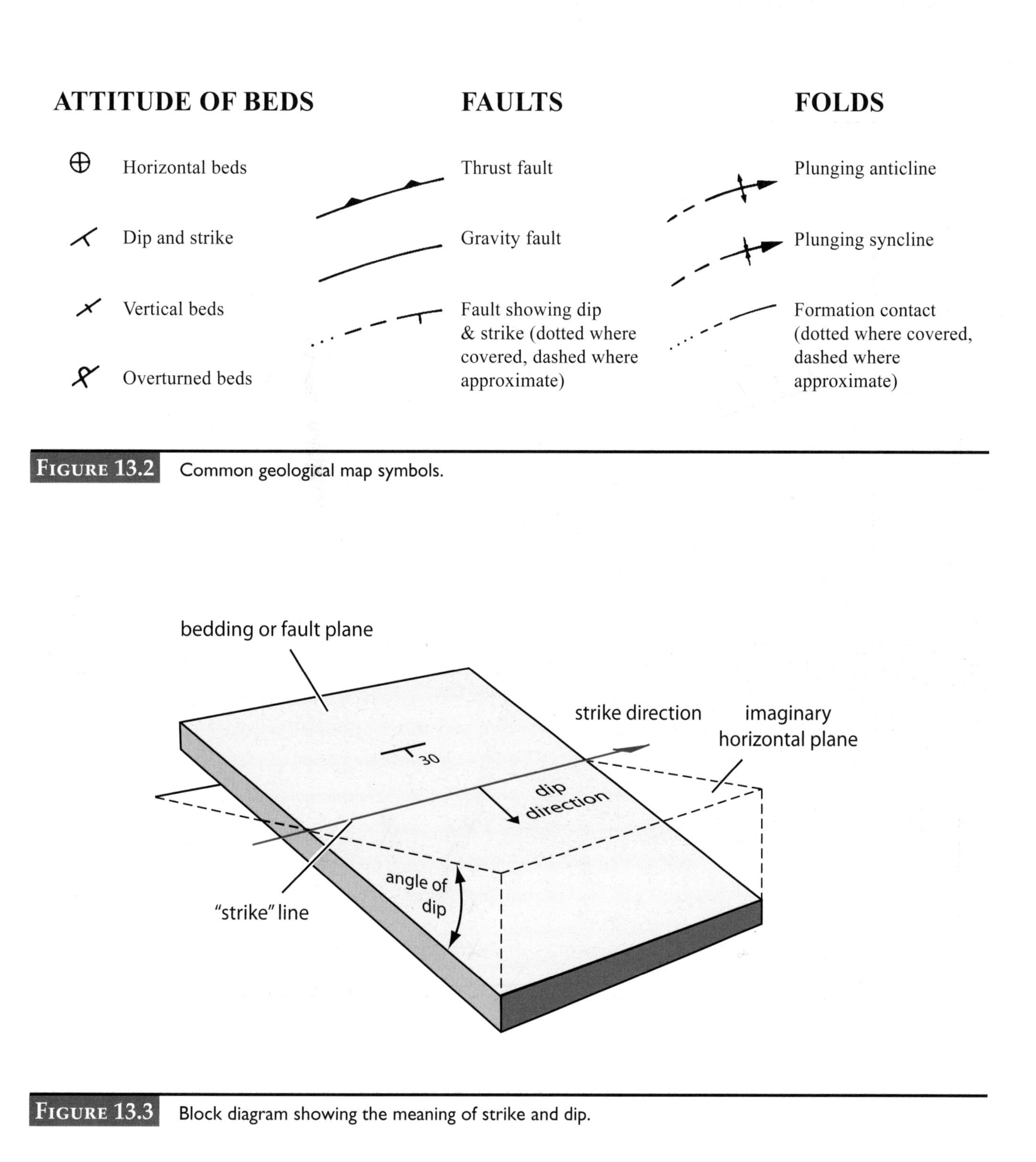

FIGURE 13.2 Common geological map symbols.

FIGURE 13.3 Block diagram showing the meaning of strike and dip.

Map Interpretation

Geological maps are comprised of colored (or patterned) areas separated by black lines representing formation-bounding contacts and faults (figure 13.1). These colored areas represent the distribution of rock formations (defined in exercise 5) across the map area. The distribution of the formation within the map area is determined by a number of factors, including the thickness of the formation, the attitude (strike and dip) of the formation, structural complexities such as faults and folds, and the topography of the surface over which the formation is exposed. Figure 13.4 illustrates the effect of strike and dip on the distribution of a 40-ft thick formation in a hilly region. Notice that contacts between formations, represented by black lines, are parallel to topographic contours when the stratum is horizontal (figure 13.4A). Conversely, in figure 13.4B, where the same layer strikes to the northwest and is oriented vertically, the contacts are not deflected around topographic features, but cut directly across the topography (contour lines).

Figure 13.4A illustrates the effect of topography on the outcrop width of a given formation. The central feature depicted on this topographic map is

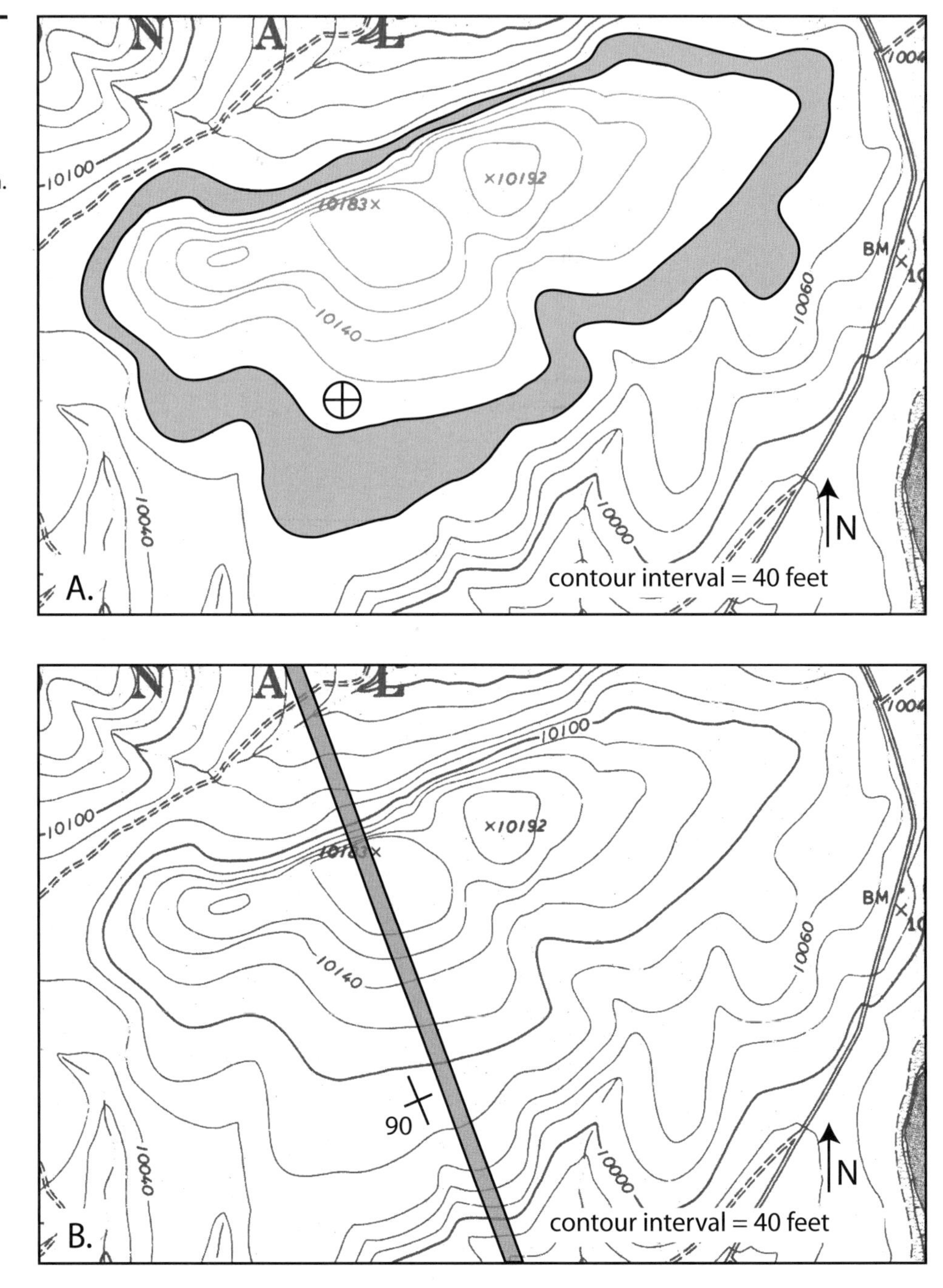

Figure 13.4

A. Geological map showing the relationship between topography and the outcrop pattern of a horizontal stratum.

B. Relationship between topography and a vertically oriented bed.

a small hill with a steep northwest face and gently inclined southeast side. The outcrop width of the bed shaded in blue is narrower on the steep northwest face than it is on the gentler southeast side. On figure 13.5 we have removed the topography and can see that the outcrop width of a particular stratum is also determined by the degree of structural tilt (dip). As a rule, outcrop width of a particular stratum decreases with increasing degree of tilt.

Tracing Planar Features across Topography

Figure 13.6 illustrates a simple procedure for determining the trace of a formation contact or other planar feature (fault plane, coal seam, ash bed, etc.) across topography portrayed on a contour map. This is done by using strike and dip data to draw a projection of the planar feature, beginning at a point where the target layer intersects the land surface (outcrop at point A on figure 13.6). To create this projection, orient a piece of scaled graph paper (use the map scale as reference) so that the strike of the planar feature is perpendicular to the cross section. Project the position of point A (known outcrop) on the map to the 9000-ft line on the cross section by drawing a line parallel to the measure's strike (see strike and dip symbol adjacent to point A). This line is the 9000-ft structural contour line. Draw a circle on the cross section where the 9000-foot structural contour intersects the line of equal elevation (in this case 9000) on the cross section. Draw a circle at the point of intersection of these two lines. Using the measured dip (35°), draw a line showing the projection of the planar feature on the cross section. Draw circles where the dipping line intersects other lines of elevation on the cross sections. Now, draw structure contours representing various elevations across the map. In figure 13.6, we have drawn structure contours at 200-ft intervals. Place a point on the map at each place that structure contours (straight lines) intersect topographic contour lines of the same elevation. Finally, draw a line connecting the series of points. This line represents the trace of the contact (or other planar feature) across the landscape.

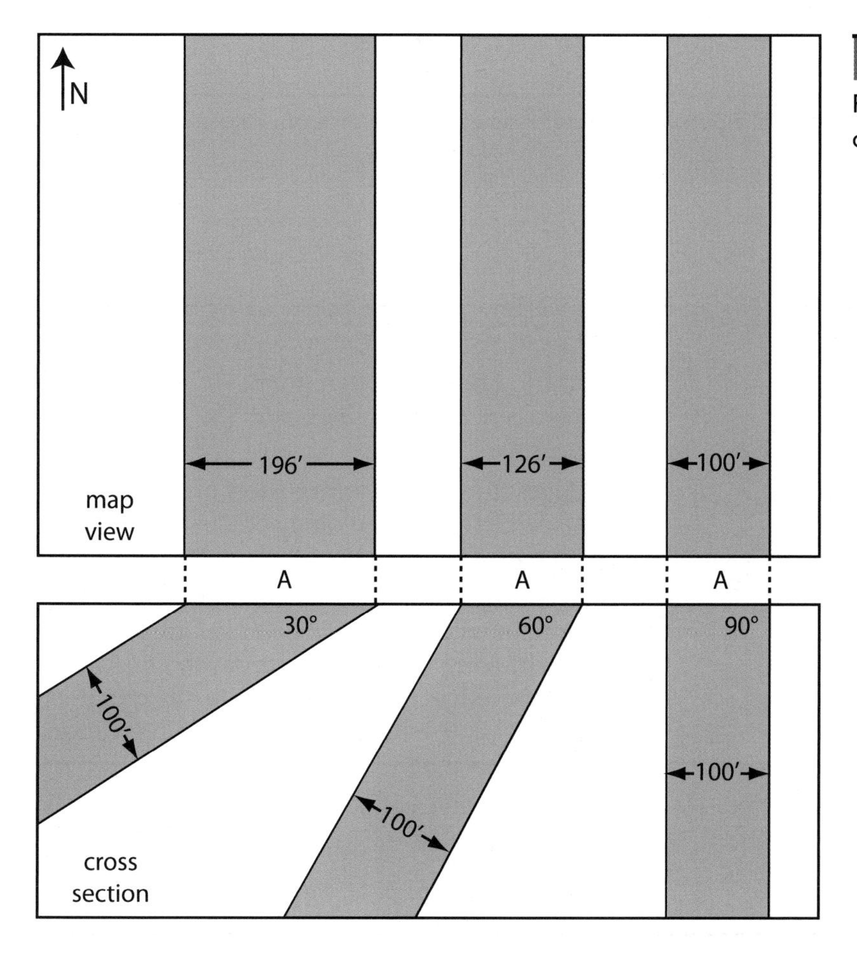

FIGURE 13.5

Relationship between dip angle and the outcrop width of a 100-ft thick rock layer.

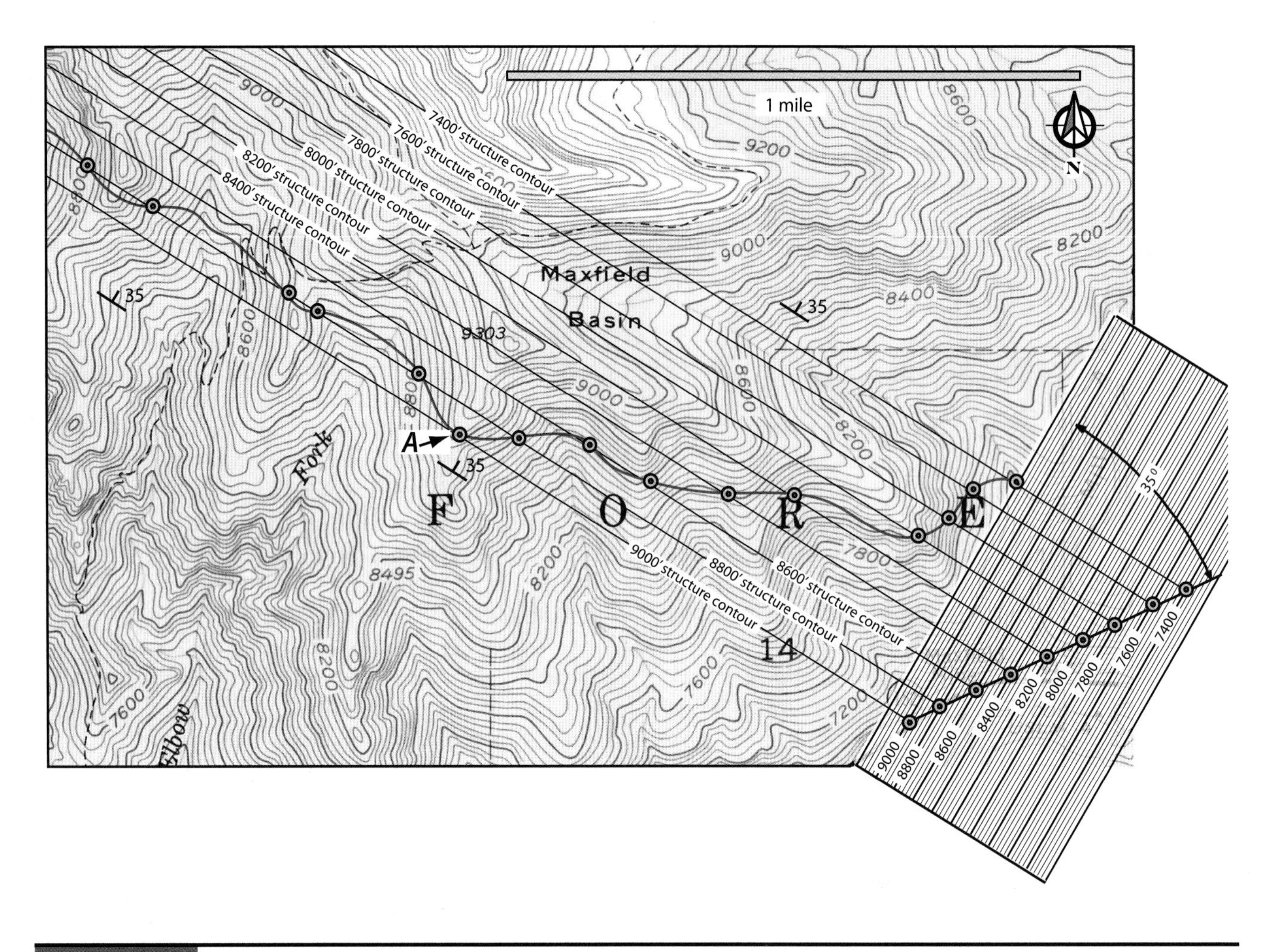

FIGURE 13.6 Technique for projecting the trace of a planar geological feature across an area of irregular topography.

Constructing a Geological Cross Section

Geological maps are two-dimensional representations of three-dimensional features. To illustrate the projection of the mapped features into the third dimension, in this case into the subsurface, geologists construct cross sections. These show the inferred distribution of formations and structures at depth based upon interpretation of relationships mapped at the surface. Geologic cross sections can be made from geologic maps by first constructing a topographic profile along a prescribed line of section (such as line A–A′ in figure 13.1), and then populating this profile with information transferred from the geological map. Figure 13.7 shows the procedure for drafting a topographic profile. The basic steps are:

1. Overlay a strip of paper along the specified line of section on the geological map.
2. Mark and label the points of intersection between contour lines and the edge of the overlay.
3. Align the overlay with a sheet of graph paper scaled to reflect the range of elevations in the map area.
4. Transfer elevation data from the overlay to horizontal elevation lines on the graph paper by drawing vertical lines from the tick mark on the edge of the overlay to the point of intersection with the line of equal elevation. Mark this with a dot.
5. Draw a line connecting the dots. This is a scale model of the topography.

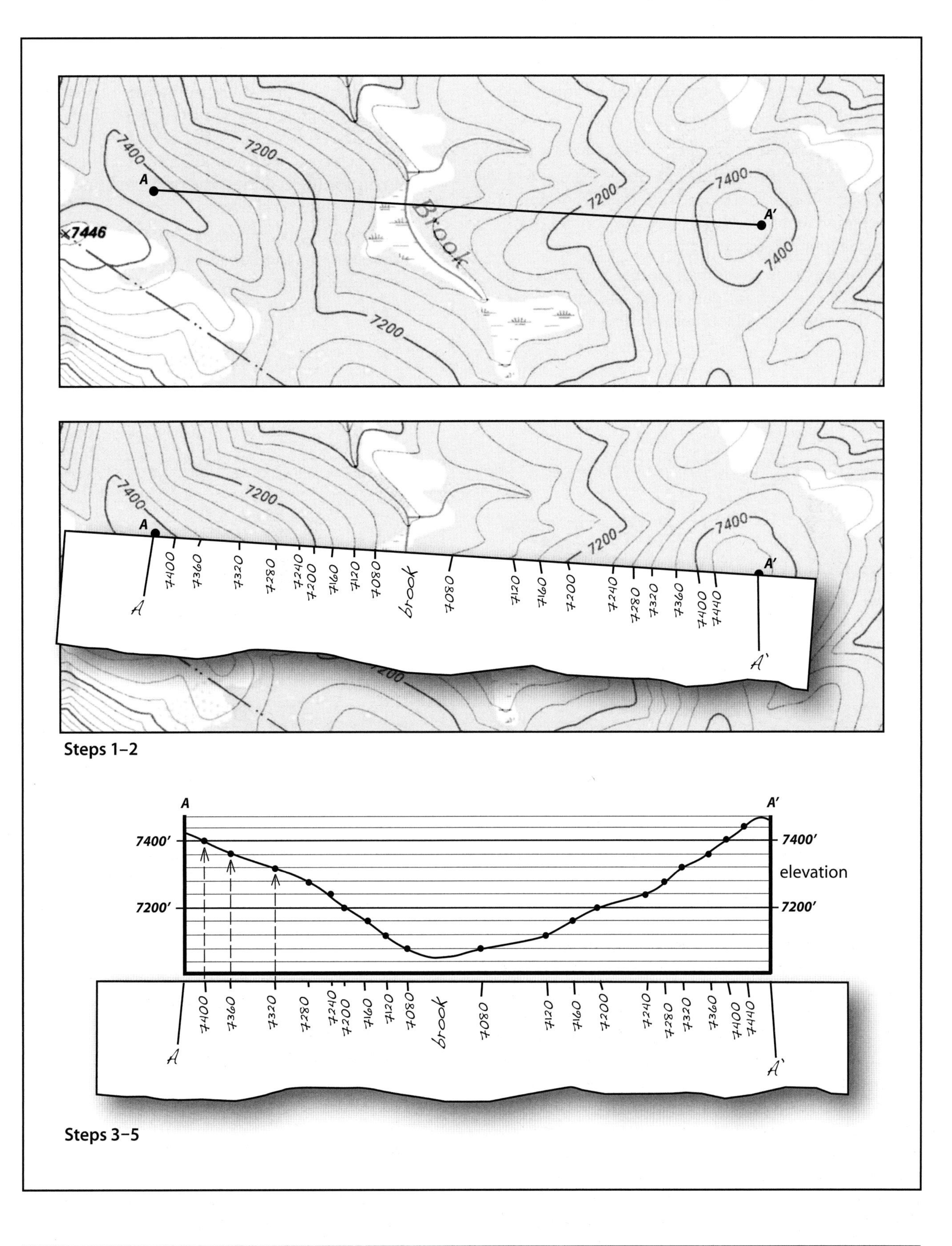

FIGURE 13.7 Technique for constructing a topographic profile from a topographic map.

The second part of the procedure is similar to the first, only this time you transfer geological information from the map to the topographic profile using the following steps:

1. Overlay a strip of paper along the specified line of section on the geological map.
2. Make a tick mark on the edge of the strip of paper wherever it intersects a contact, fault, or other geological feature.
3. Determine from the map the direction and angle of dip of each formation and fault and project these into the subsurface a short distance.
4. Connect the projected lines in the subsurface in accordance with map data. If strata are dipping gently, but are not otherwise faulted or folded, contacts will continue as straight lines into the subsurface. If the mapped area is characterized by folds, such as that illustrated in figure 13.8, connect the projected lines with curved lines as shown.
5. Be careful to maintain uniform formation thickness in the subsurface unless you have evidence to the contrary.

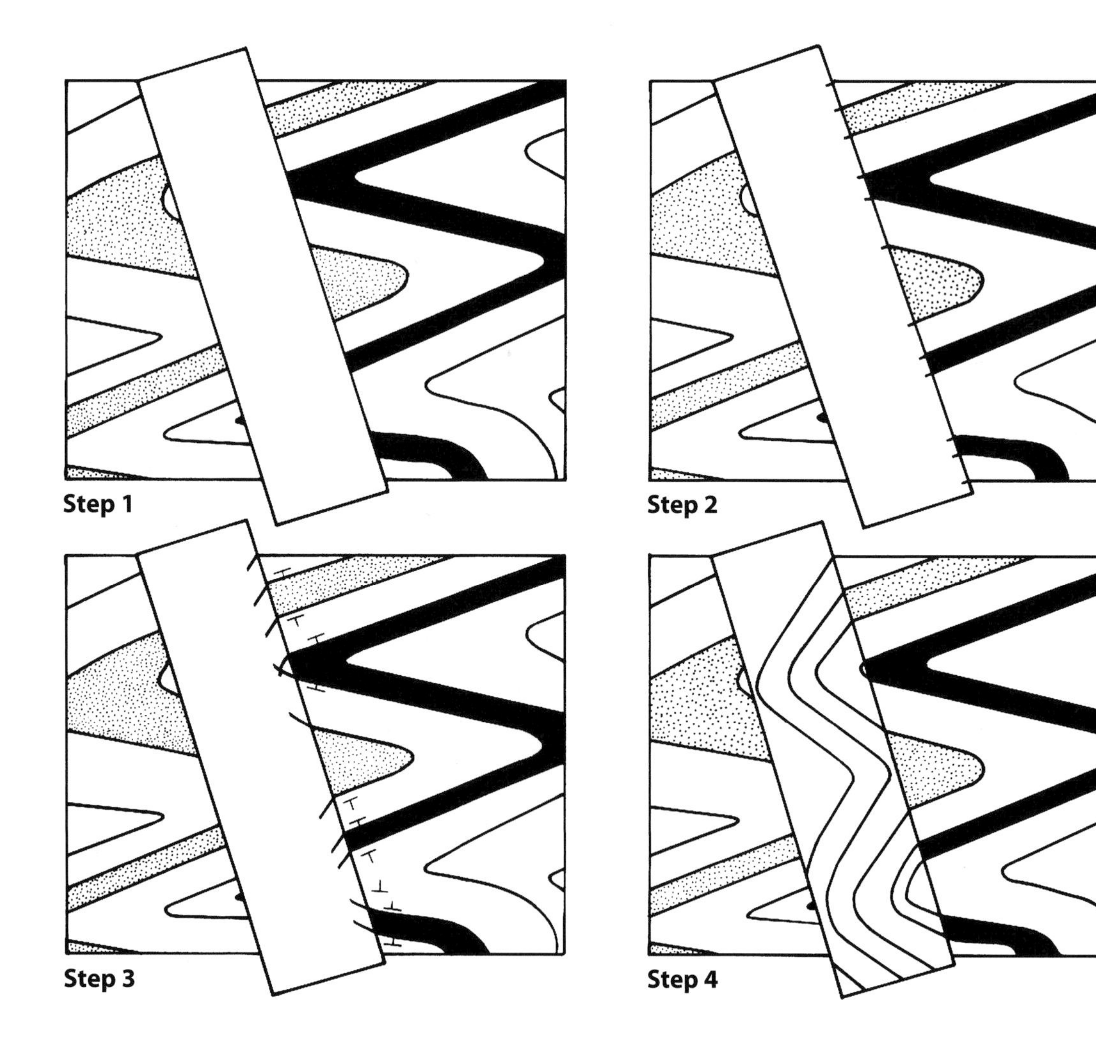

FIGURE 13.8 Drawing a structural cross section from map data.

PROCEDURE

PART A

Figure 13.9 shows the topography of an area of central Utah known for its production of coal. Geologists located the contact between the Salina Canyon Formation and the overlying Huntington Formation at point A on the east side of the ridge that trends north–south through the map area. The contact is overlain by a coal seam that comprises the lower 15 ft of the Huntington Formation. Both formations are Cretaceous (K) in age. Strata in the map area strike 0° (north) and dip 10° to the west.

1. Using the technique illustrated in figure 13.6, trace the contact between the Huntington and Salina Formations across the map of this area. Create a geological map by lightly coloring and labeling the two Cretaceous formations.

2. Is the coal seam exposed on both sides of the prominent north–south trending mountain, or just on the east side? What is the lowest elevation at which coal occurs in the map area?

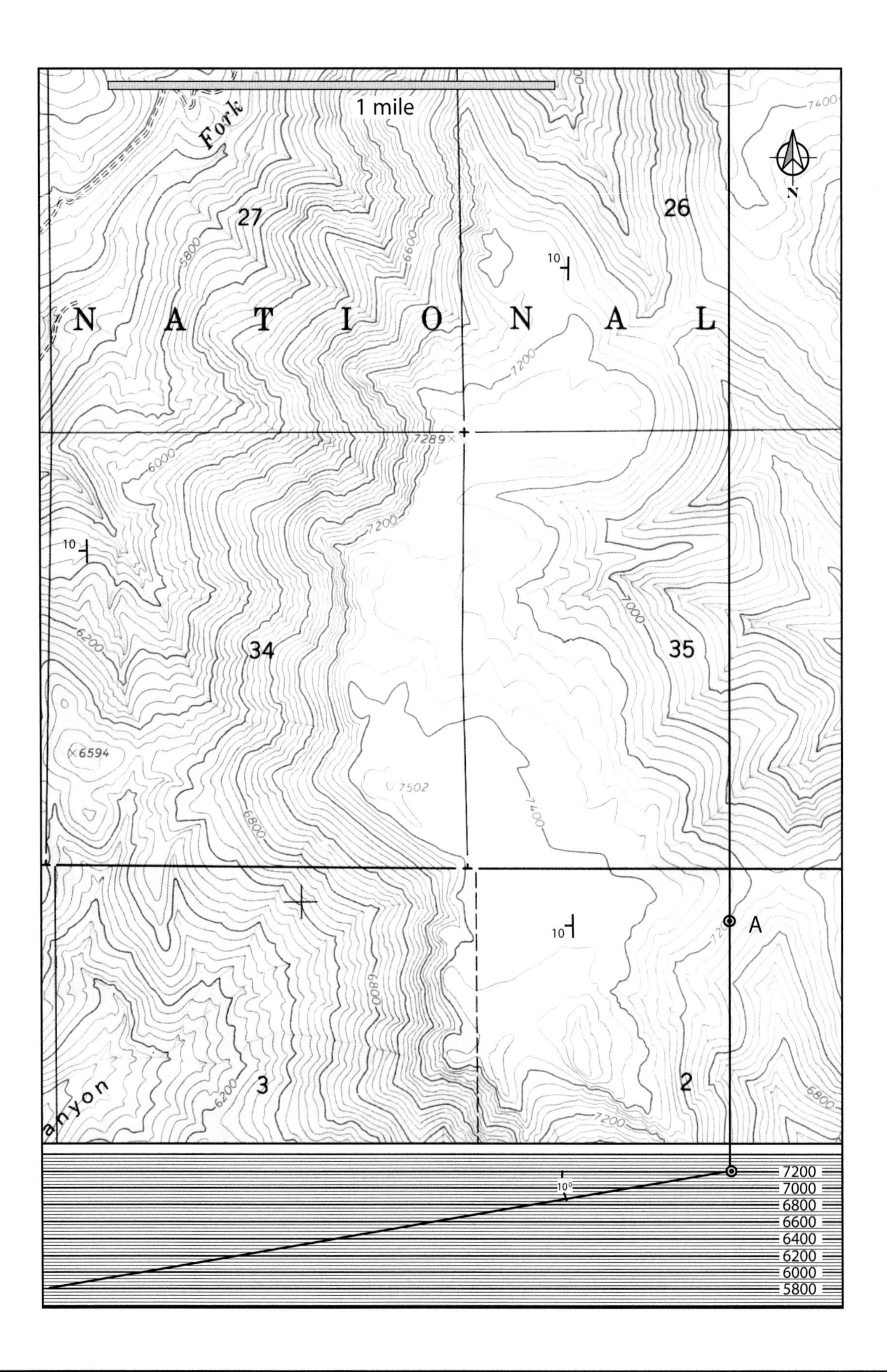

FIGURE 13.9 Topographic map showing a portion of the central Wasatch Plateau in Utah.

Part B

1. On figure 13.10, draw a geologic cross section representing line A–A′ in figure 13.1. The topographic profile has been drawn already, so in this case you need only transfer the geological data to the profile and project relationships into the subsurface.

2. What geological structures are predominant in this part of Pennsylvania?

folds and faults,
fault in lato ordincian
folds from seiunan
to middle oralician.

3. How can the pattern of a syncline on a geological map be distinguished from the pattern of an anticline?

anticline scoop up
Syclines scoop up like an arch

4. What type of tectonic stress (compressional, tensional, or shear) is indicated by the folding and faulting present in the map area?

compressional,
stress caused a
reverse fault,
more folds on left side
due to compression

5. List the two or three formations that appear most resistant to weathering and erosion.

A

A′

Part C

1. A portion of the Salem, Kentucky, geological quadrangle map is shown in figure 13.11. Carefully examine the stratigraphic and structural relationships shown on the map and construct a geological cross section along the line A–A′.

2. What geological systems are present in this map area?

3. Why is the outcrop width of the Kincaid Limestone (Mkc) greater on the eastern side of the syncline than it is on the western side?

4. What type of fault is most common in this area? Cite your evidence.

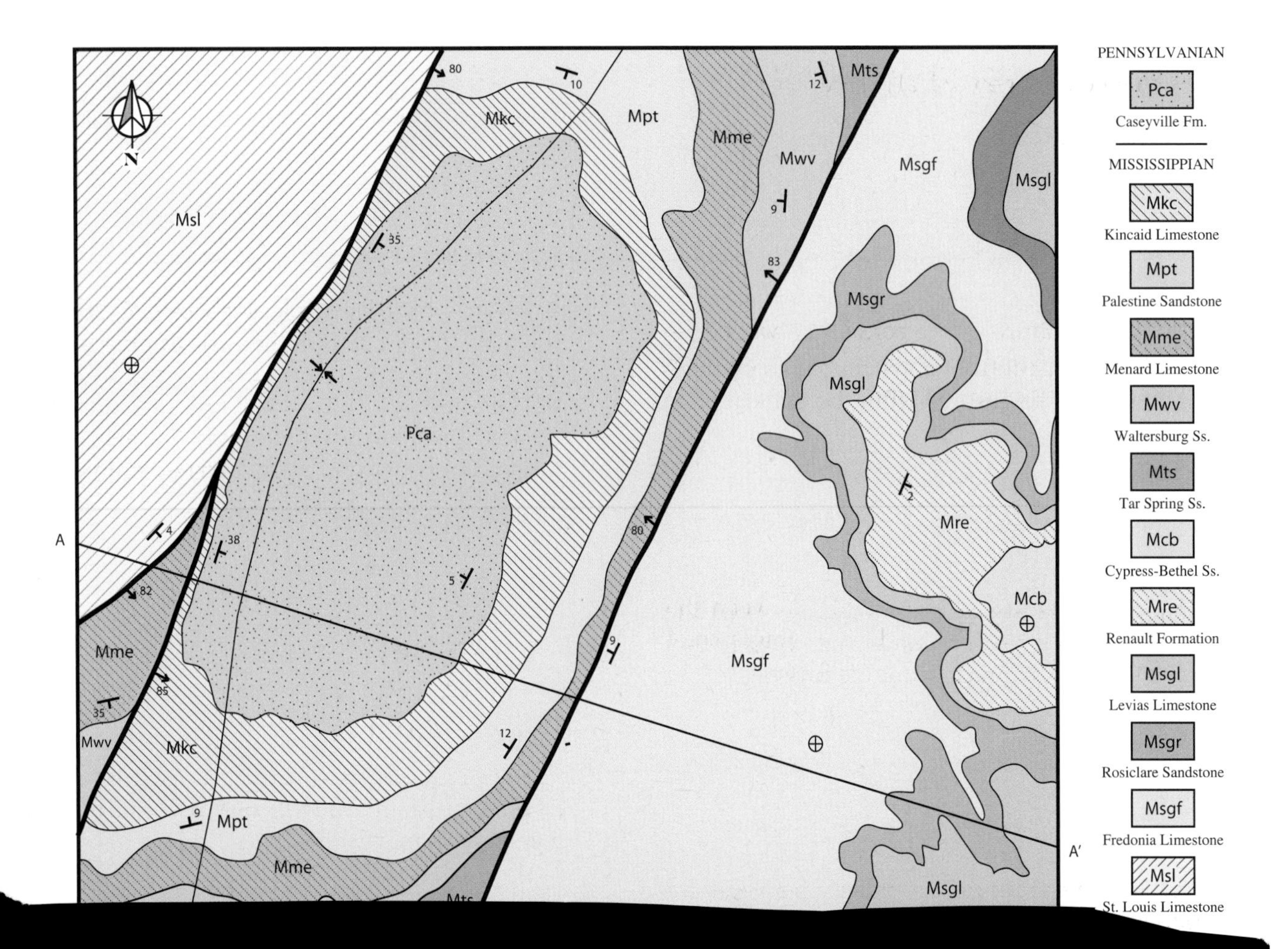

Part D

Figure 13.12 is a geological map of the southeastern Black Hills, South Dakota. This region displays a variety of unconformities. Briefly review the definitions of nonconformity, angular unconformity, and disconformity in exercise 1 and then answer the following questions.

1. Which Paleozoic and Mesozoic geological systems are represented in the map area?

paleozoic - cambrian, carbonserous, permian
mesosoic - triassic, jurrasic, certaceous

2. Which Paleozoic geological systems are not represented?

ordivician, silurian, Devonian

3. What is the nature of the contact between Precambrian rocks and the base of the Deadwood Formation? Cite your evidence.

non conformity, contact btwn sedimentary & metamorphic rock

4. What is the nature of the contact between the Deadwood Formation and the Englewood Limestone? How much "time" is missing?

disconformity, gap of time btwn 2 sedimentary ~~rock~~ layers.

5. What is the nature of the contact between the Sundance Formation and Skull Creek Shale in the southern one-third of the map area? How does this relationship change northward along the strike?

paraconformity, parallel strata. little erosion.

6. The Tertiary White River Group crops out on the eastern edge of the map. List the formations that are in contact with the base of the White River Group.

minnekanta, spearfish, sundance, Skull creek, mowry, the green horn limestone.

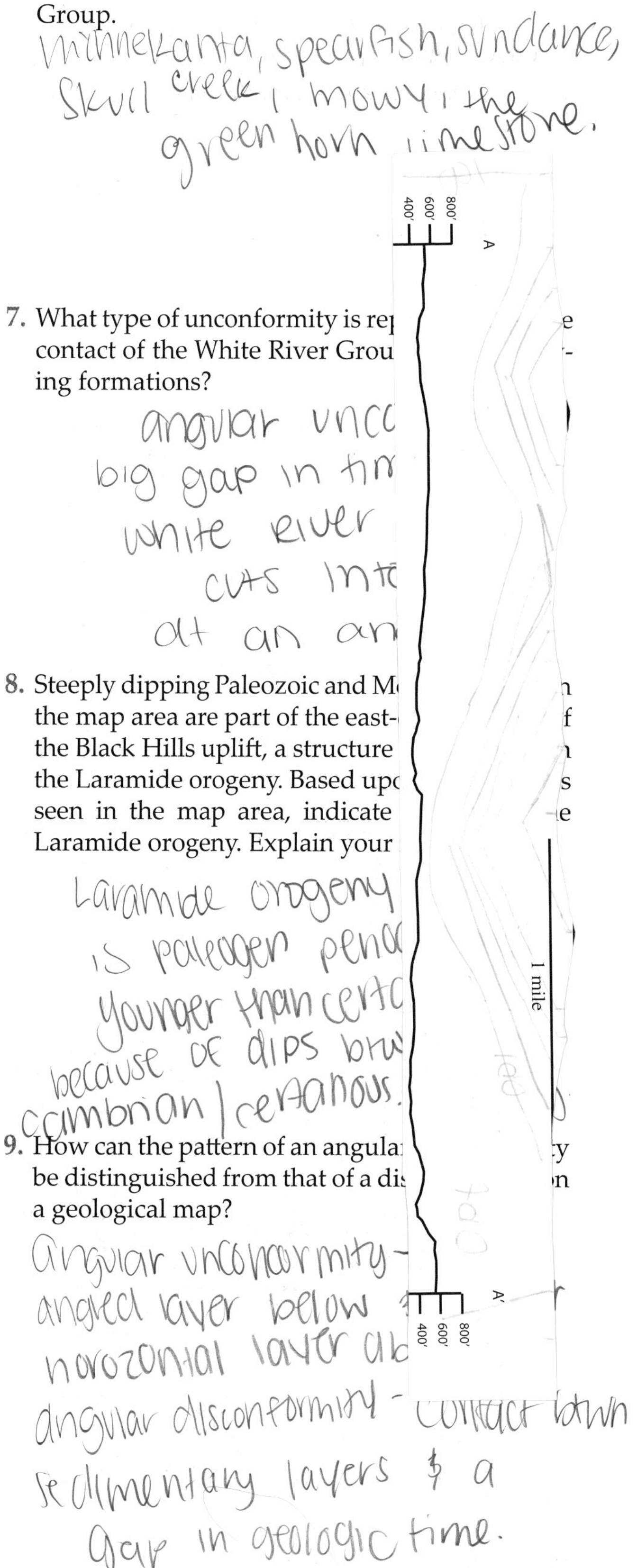

7. What type of unconformity is re[...]e contact of the White River Grou[...]- ing formations?

angular unc[...] big gap in tim[...] white river [...] cuts into [...] at an an[...]

8. Steeply dipping Paleozoic and M[...]n the map area are part of the east-[...]f the Black Hills uplift, a structure [...]n the Laramide orogeny. Based upo[...]s seen in the map area, indicate [...]e Laramide orogeny. Explain your [...]

Laramide orogeny is paleogen perio[...] younger than certa[...] because of dips btw[...] cambrian/certanous.

9. How can the pattern of an angula[...]y be distinguished from that of a dis[...]n a geological map?

angular unconformity - angled layer below [...] horozontal layer ab[...]
angular disconformity - contact btwn sedimentary layers & a gap in geologic time.

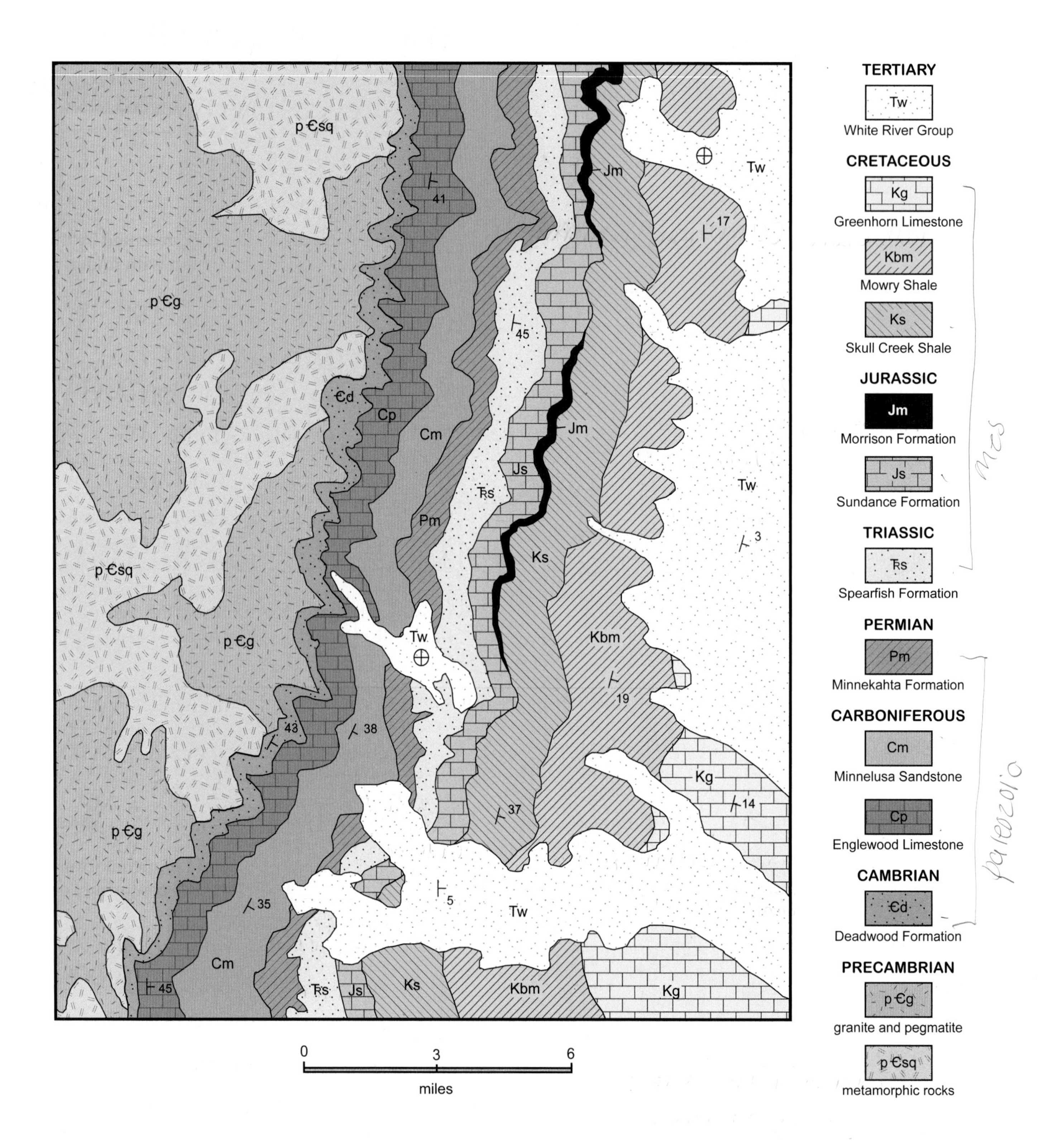

FIGURE 13.12 Colored geological map of the southeastern portion of the Black Hills, South Dakota.

Canadian Shield and Stable Platform

Exercise 14

Learning Objectives

While completing this exercise, you will:

1. become acquainted with the Precambrian evolution of the North American continent;
2. gain a greater understanding of the nature and history of the stable platform; and
3. enhance your understanding of the age and significance of strata deposited in key sedimentary basins of North America.

Each of the major continents is comprised of three structural domains known as the **shield**, **stable platform**, and **marginal mobile belt**. Marginal mobile belts are the high relief areas of active tectonism (earthquakes, volcanoes, uplift), such as the modern Andes and Himalayas. By contrast, the shield is a tectonically quiescent, topographically flat region where Precambrian-age crystalline (and to a lesser degree, sedimentary) rocks are exposed at the surface. The largest shield area on Earth today is located in northern North America and Greenland and is known as the Canadian Shield (figure 14.1). Rocks and structures exposed in the **Canadian Shield** reveal a complex story of repeated mountain building, erosion, deposition, collision, and occasional rifting. Owing to the fact that (1) events stretch back nearly four billion years, (2) selected regions of the shield have been subjected to multiple episodes of tectonic activity, and (3) fieldwork is logistically difficult, it has taken great effort to piece together the Precambrian history of the Canadian Shield. The results of these efforts are summarized in figure 14.2, which shows the major structural provinces and rock bodies that comprise the Precambrian core of North America. The proto-North American continent is called Laurentia by geologists. You will see this name used in some parts of this exercise. This term is derived from outcrops of Precambrian rocks in the Laurentian Mountains of Canada.

South and west of the Canadian Shield, Precambrian rocks are covered by sedimentary sequences of Paleozoic, Mesozoic, and Cenozoic age (figure 14.3). Such tectonically stable regions where Precambrian basement rock are masked by sedimentary strata are called **stable platforms**. Together, the shield and stable platform comprise the continent's **craton**. Notice on figure 14.3 that limited exposures of Precambrian rocks occur within the confines of the stable platform, but only in the cores of uplifted fault blocks (Rocky Mountains) and in the bottoms of deeply incised canyons (Inner Gorge of the Grand Canyon). The sedimentary cover of the North American platform is thinnest adjacent to the shield and thickens toward the marginal mobile belts. Uneven subsidence of Precambrian basement rocks has resulted in the accumulation of great thicknesses of sedimentary rocks in localized **intracratonic basins** such as the Illinois, Appalachian, and Williston basins. Intracratonic basins are frequently the sites of important reserves of coal, petroleum, natural gas, and minerals and hence, are of great economic interest. Sedimentary strata of the North American platform comprise six unconformity-bounded, transgressive-regressive depositional sequences known as the Sauk, Tippecanoe, Kaskaskia, Absaroka, Zuni, and Tejas. These are illustrated in figure 17.1 and are further explored in exercise 17 of this lab manual.

PROCEDURE

PART A

Carefully examine figure 14.2 and answer the following questions. You will need to refer to your textbook and/or the Internet to find answers to some of the questions. The purpose of part A of this exercise is to help you become acquainted with Precambrian rocks and structures of North America and with the events/processes that formed them.

1. List the Archean provinces and cratons of North America.

wyoming provence, nain, Superior provence, provence, Slave provence, Hearne provence, Rae provence

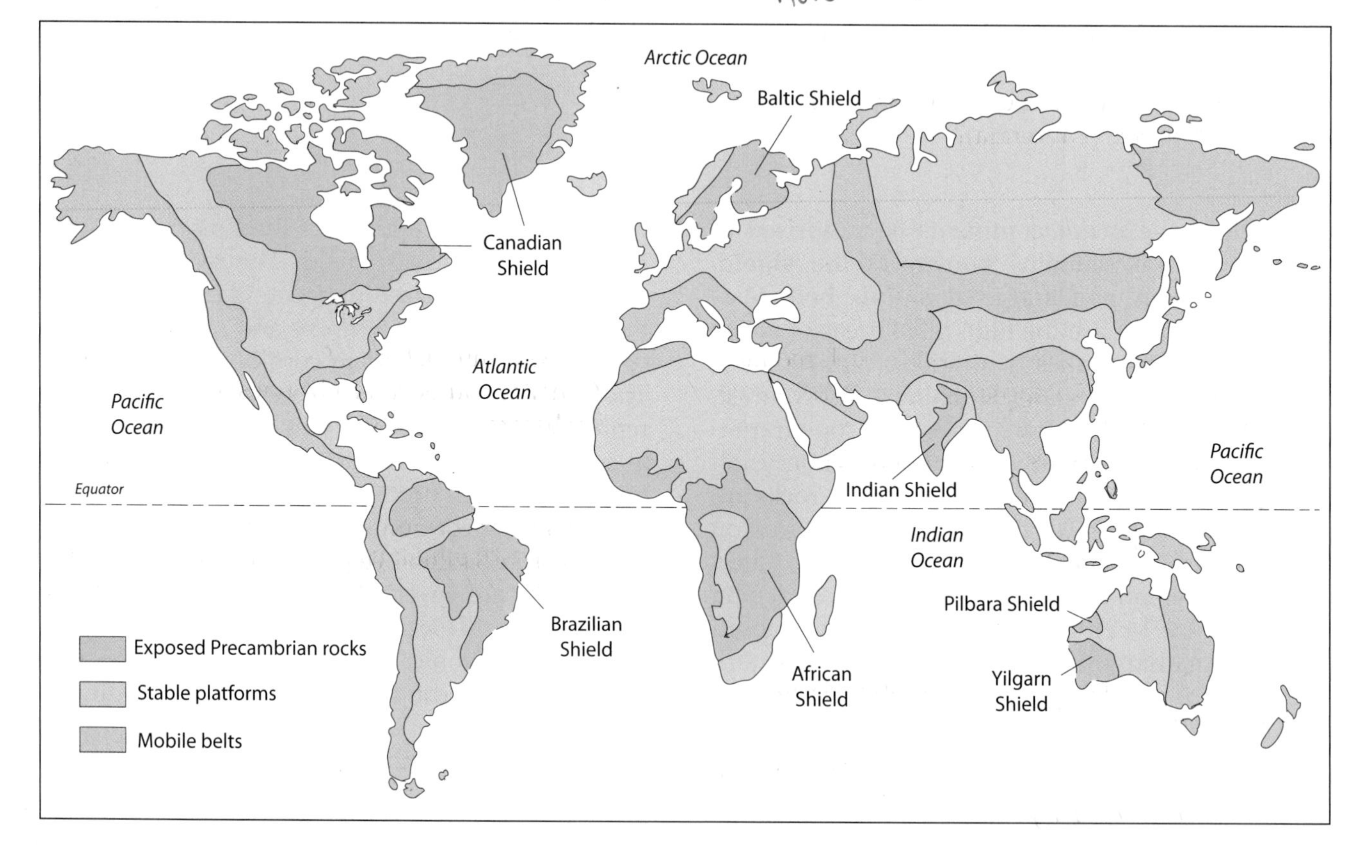

FIGURE 14.1 Distribution of shields, stable platforms, and marginal mobile belts on Earth today.

2. ~~Approximately what percentage of modern North America had formed by the end of the Archean Eon?~~

3. What is the Abitibi Belt and what age/type of rocks are exposed there?

A greenstone from ontario-quebec. volcanic rocks and its archaean (2.8-2.6 BYA) (igneous rock)

4. Discuss the nature and significance of greenstone belts. What type of rocks characterize greenstone belts and why are they restricted to Archean provinces?

mineral deposits gold/silver
carbonate rocks
granite + gneiss
Restricted because unformed prior to plate tectonism

5. What is the nature (geographic distribution, rock type, age) and significance of the Trans-Hudson orogen in the evolution of ancestral North America (Laurentia)?

largest palieoproterozoa in the world. formed 1.85 BYA. superior province collided w their pronce caused great plains flattened.

6. How do the ages of the rocks in the colored map area of figure 14.2 support the concept of continental growth by accretion? Cite specific examples of patterns that support this concept.

the older rocks are in middle w younger rocks surrounding it, law of superposition

7. What is the age, nature (rocks, fault style), and significance of the Midcontinent Rift?

1.1 billion years old, mesoproterozoic era
major reverse faults
the rifting prevented
new ocean from beling formed
when north american continent
ripped apart.

8. Where was accretion and tectonic activity most prevalent in Laurentia during the Paleo- and Mesoproterozoic Eras? What provinces were added to ancestral North America during these two eras? Approximately how much larger (percentage) was Laurentia after these provinces were accreted compared to the size of the continent at the end of Archean time?

turned into the central plains of the U.S. tectonic activity Idaho + canada. 30-35 of todays America

9. ~~The Paleoproterozoic Animikie Group crops out in the area south and west of Lake Superior. It is chiefly comprised of sedimentary and metasedimentary rocks deposited in the Animikie Basin between 2.5 and 1.8 billion years ago. These rocks are economically important because they contain vast amounts of iron ore in the form of banded iron formations (BIF). Briefly explain the nature and origin of BIF in the context of Earth's evolving atmosphere.~~

10. ~~The Gunflint Chert of northwestern Ontario (Canada) and northern Minnesota contains BIF, but became even more noteworthy when fossils were found in black, stromatolite-bearing carbonate rocks within the formation. What type(s) of organisms would have been living in shallow oceans 1.88 billion years ago? Sketch and label a specimen of fossils found in the Gunflint Chert.~~

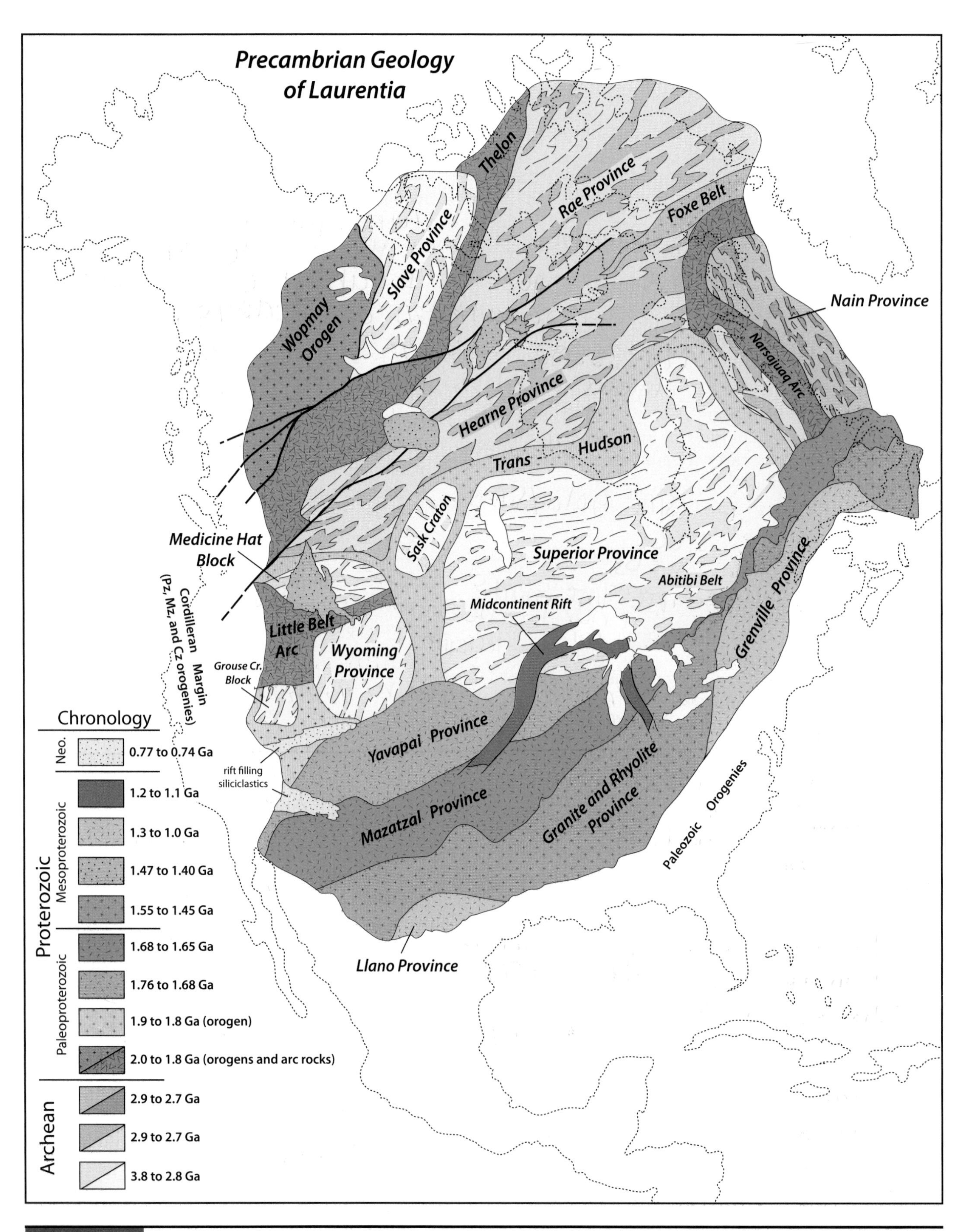

FIGURE 14.2 Major structural provinces and rock bodies that comprise the Precambrian core of North America. (After Whitmeyer and Karlstrom, 2007.)

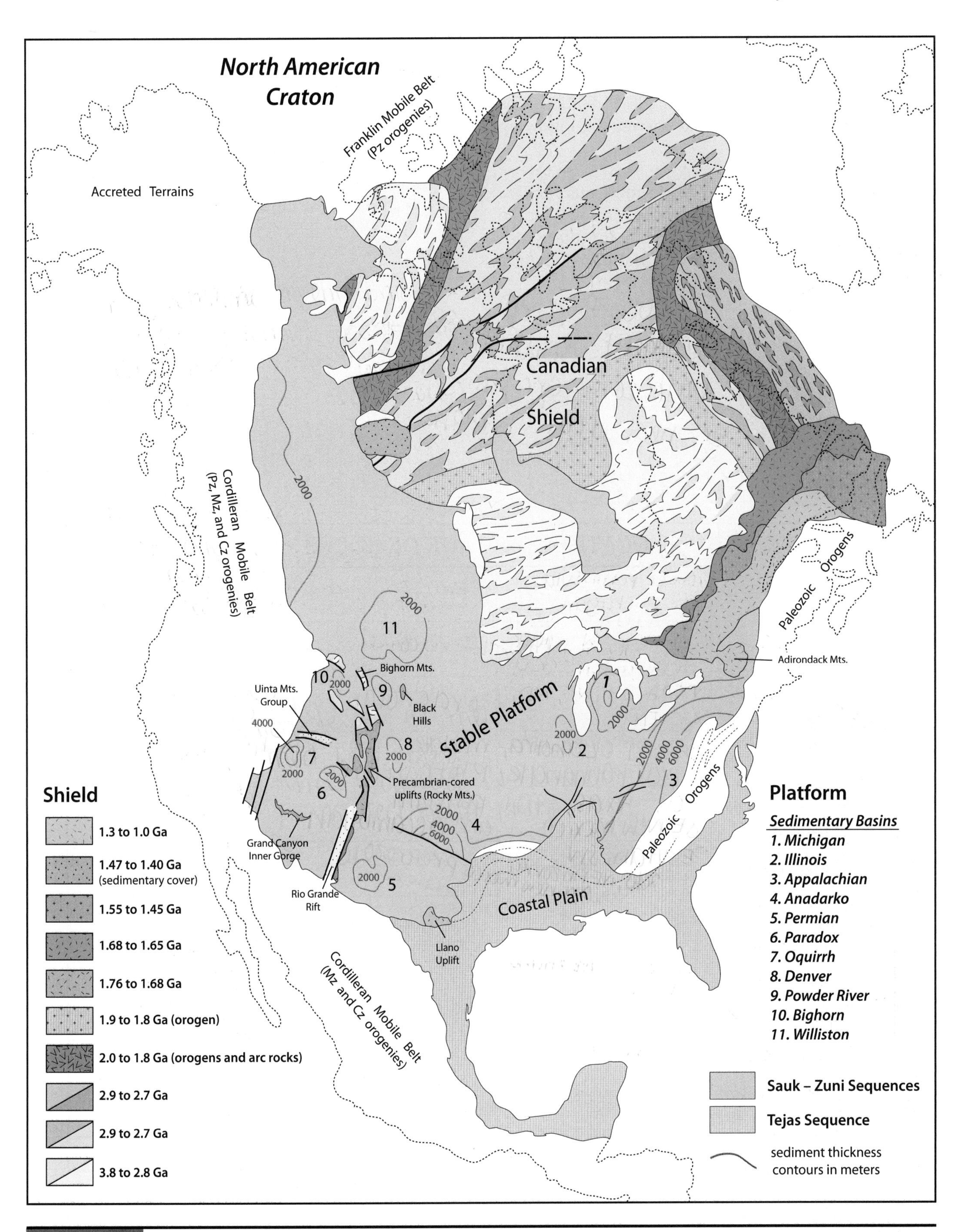
North American Craton
Franklin Mobile Belt (Pz orogenies)
Accreted Terrains
Canadian Shield
Cordilleran Mobile Belt (Pz, Mz, and Cz orogenies)
Paleozoic Orogens
Adirondack Mts.
Bighorn Mts.
Uinta Mts. Group
Black Hills
Stable Platform
Precambrian-cored uplifts (Rocky Mts.)
Grand Canyon Inner Gorge
Rio Grande Rift
Coastal Plain
Llano Uplift
Cordilleran Mobile Belt (Mz and Cz orogenies)
Shield
1.3 to 1.0 Ga
1.47 to 1.40 Ga (sedimentary cover)
1.55 to 1.45 Ga
1.68 to 1.65 Ga
1.76 to 1.68 Ga
1.9 to 1.8 Ga (orogen)
2.0 to 1.8 Ga (orogens and arc rocks)
2.9 to 2.7 Ga
2.9 to 2.7 Ga
3.8 to 2.8 Ga
Platform
Sedimentary Basins
1. Michigan
2. Illinois
3. Appalachian
4. Anadarko
5. Permian
6. Paradox
7. Oquirrh
8. Denver
9. Powder River
10. Bighorn
11. Williston
Sauk – Zuni Sequences
Tejas Sequence
sediment thickness contours in meters

FIGURE 14.3 The North American craton.

PART B

Carefully study the geological relationships portrayed in figure 14.3 and answer the following questions.

1. Note the thickness of the sedimentary strata covering the stable platform as indicated by the contour lines. How thick is the sedimentary rock cover throughout most of the platform? Where is it the thickest?

   2000-6000 ft deep of sediment. thickest by paleozoic orogins or near coastal plane

2. The stable platform of North America contains a number of basins that have resulted from increased subsidence and sedimentary infilling. Some of the major basins are numbered 1 through 11 on the map. Using your textbook and other reference materials (such as Wikipedia or other websites), fill in the requested information for each of these intracratonic basins in table 14.1.

BASIN	LOCATION	AGE OF ROCKS	RESOURCES
Michigan	lower peninsula of Michigan	Paleozoic time	water
Illinois	Illinois, SW indiana, western kentucky	paleozoic	coal
Appalachian	NY, PA, OH, WV, MD, KY, TN	ordovician	coal, iron, natural gas
Anadarko	western oklahoma, texas panhandle	middle proterozoic	crude oil
Permian	west texas, SE new mexico	permain and precambrian	oil and gas
Paradox	SE Utah, SW colorado, NE Arizona, NW new mexico	proterozoic	uranium
Oquirrh	North-central Utah and southern Idaho	25,000 feet of Pennsylvanian and Permian strata	no major resource production
Powder River	southeast montana, NE wyoming	cambrian	coal
Bighorn	north central wyoming	2.8 BYA	wildlife
Denver	North colorado	Cretateous	gold, sand, gravel
Williston	NW north dakota	proterozoic	crude oil, uranium

TABLE 14.1 Phanerozoic basins of North America.

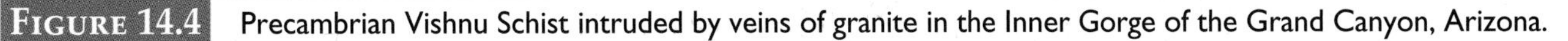

FIGURE 14.4 Precambrian Vishnu Schist intruded by veins of granite in the Inner Gorge of the Grand Canyon, Arizona.

3. The Adirondack Mountains of New England are comprised of Precambrian rocks. What is the age and nature of these Precambrian rocks? To which Precambrian province do they belong?

   1300-1400 mil years, Brenville province

4. The Bighorn Mountains of Montana and the Black Hills of South Dakota are both cored by Precambrian crystalline rocks that have been exposed by uplift and erosion. If you studied rocks in the two areas during an extended field trip, would you find them to be similar in lithology and in age or would you have examined rocks from two distinct Precambrian provinces? Be specific in your answer. You may wish to refer to figure 14.2. though dif. ages, Rocks similar. from cambrian to late cretaceous. blackhills uplifted during Hudson orogeny

5. Figure 14.4 shows a portion of the Inner Gorge of the Grand Canyon. Precambrian rocks are overlain nonconformably by the Cambrian Tapeats Sandstone. The crystalline rocks beneath the nonconformity form part of which Precambrian province illustrated on figure 14.2?

   yavapai province 1.70-1.68 BYA

Exercise 15

Paleozoic Orogenies of Ancestral North America

Learning Objectives

After completing this exercise, you will:

1. have gained additional confidence in your ability to read and interpret geological maps;
2. be better acquainted with rocks and structures that formed during the Taconic, Acadian, and Allegheny orogenies;
3. recognize the tectonic significance of formations that were deposited in the Taconic, Acadian, and Allegheny foreland basins;
4. be able to recognize the effects of the Ancestral Rocky Mountain orogeny in the Gateway, Colorado, area; and
5. enhance your ability to draw a structural cross section.

During the Paleozoic Era, the eastern and southern margins of ancestral North America (Laurentia) bordered the Iapetus Ocean. The approach, then collision of Baltica resulted in two mountain-building events called the Taconic (Ordovician) and Acadian (Devonian) orogenies. During the Pennsylvanian and Early Permian Periods, Laurentia collided with the massive continent of Gondwana, a step in the assembly of the supercontinent of Pangea. The resulting mountain-building event, known as the Allegheny–Ouachita orogeny, deformed the coast of ancestral North America from Newfoundland to Texas. Each of these collisional events produced a crystalline core complex, fold and thrust belt, and foreland basin. The latter received vast amounts of debris shed from the adjacent highlands (figure 15.1). Tectonic stresses also deformed the interior of ancestral North America resulting in the formation of fault-bounded mountains in modern-day Colorado, New Mexico, Wyoming, and Utah during the Late Paleozoic Era. Geologists refer to these mountains as the Ancestral Rockies. In this lab you will examine geological maps or areas affected by different phases of these Paleozoic orogenies.

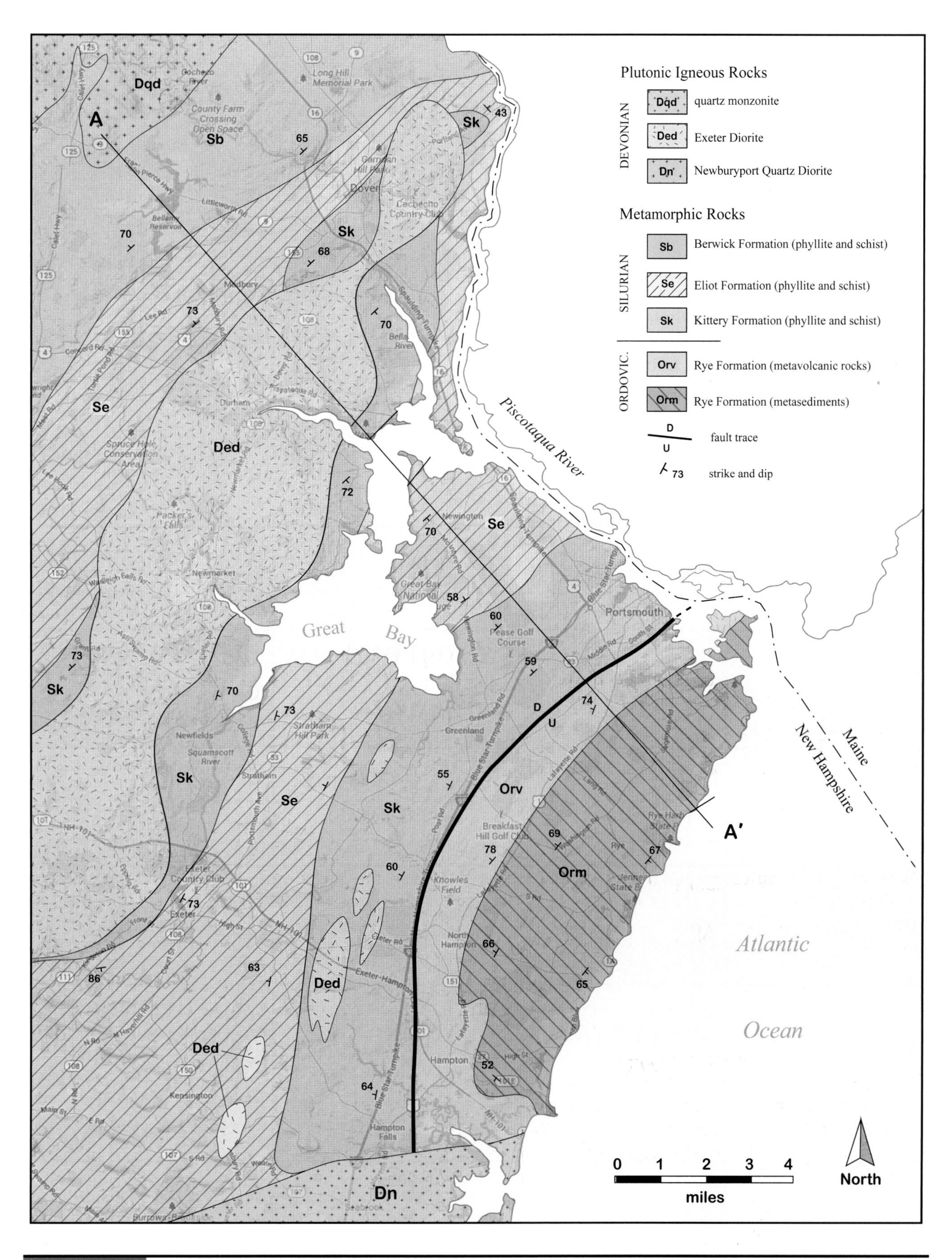

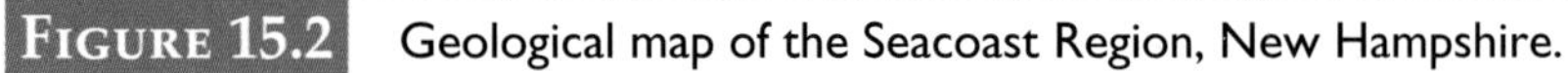
FIGURE 15.2 Geological map of the Seacoast Region, New Hampshire.

PART B

Geology of the Southern Appalachians and Coastal Plain

Scale: 1 inch = 40 miles

1. The Appalachians comprise parts of four eastern US physiographic provinces known as the Piedmont, Blue Ridge, Valley and Ridge, and Appalachian (Allegheny) Plateaus Provinces. The rocks and structures of these provinces disappear beneath nearly horizontal Cretaceous and Tertiary sediments of the Coastal Plain Province in central Alabama and Georgia. Using your textbook or web resources, fill in table 15.1 and familiarize yourself with the rocks and structures that characterize these provinces.

2. Draw boundaries between these provinces on the geological map provided on page 230 (figure 15.3).

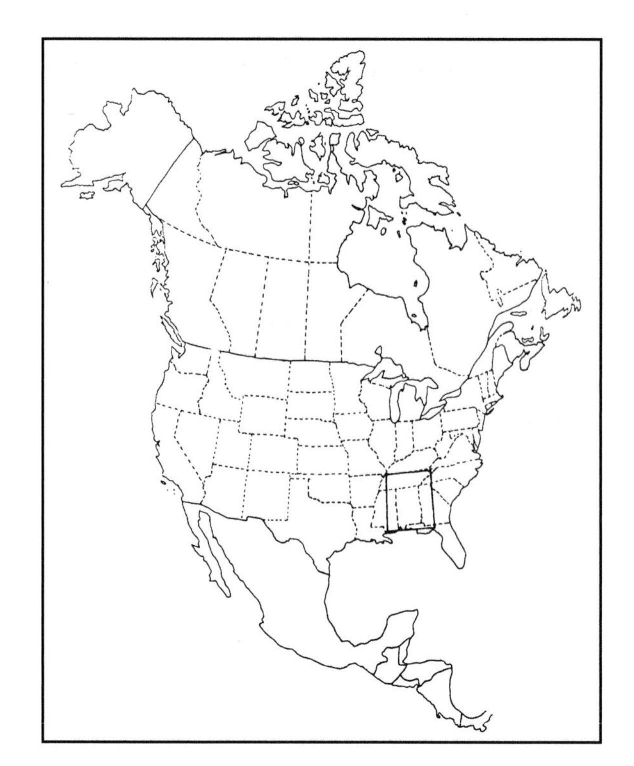

Province	Location	Age of Rocks	Structural Style
Allegheny Plateaus	western most appalachian mtns	Early mid paleozoic	underformed, deformed, highly incised
Valley and Ridge	NY, PA, VA, TN, AL	Early, mid paleozoic	fold & thrust belt deformed by foreland basin, deposits of tectonic Acadian orogeny
Blue Ridge	TN, VA, WV, NC, SC, PA, GA	precambrian	Precambrian rocks faulted during Allegheny orogeny.
Piedmont	NC, SC, GA, AL, VA, MD, NJ	paleozoic	plateau region formed by erosion of complex structures.
Coastal Plain	coastal portions of Alabama, Georgia, Carolinas, up to New Jersey	Cretaceous and Tertiary sedimentary strata	strata dip gently to east or southeast; reflects onlapping strata from transgression of coastal regions; Tejas Sequence

TABLE 15.1 Physiographic provinces of the Appalachian Mountains, eastern United States.

Geology of the Southern Appalachians and Coastal Plain

Quaternary		**Q**
Qs	Light blue	Coastal and estuarine sand and gravel
Pliocene		**P**
Pc	Tan with orange circles	Fluvial sediments
Miocene		**M**
Mt	Light orange	Tampa and Catahoula Formations Marl and sand sediments
Oligocene		**ϕ**
ϕv	Dark and light yellow diagonal	Vicksburg Group
Eocene		**E**
Ej	Light yellow	Jackson Group Marl
Ec	Dark and light yellow, horizontal	Claiborne Group
Ew	Fine light and dark yellow, horizontal	Wilcox Group
Em	Yellow-orange	Midway Group
Cretaceous		**K**
Kr	Fine green vertical	Ripley Formation
Ks	Coarse green vertical	Selma Chalk
Ke	Coarse green horizontal	Eutaw Formation
Kt	Coarse green diagonal	Tuscaloosa Formation
Pennsylvanian		**IP**
Cpv	Light gray	Pottsville Group
Mississippian		**C**
Cm	Dark blue gray	Mississippian undivided
Cmu	Light blue gray diagonal	Upper Mississippian undivided
Cmm	Medium blue stipple	Middle Mississippian undifferentiated
Cml	Medium blue horizontal, with stipple	Lower Mississippian undifferentiated
Cg	Purple triangles	Middle Paleozoic granite
Devonian		**D**
D	Medium purple	Devonian

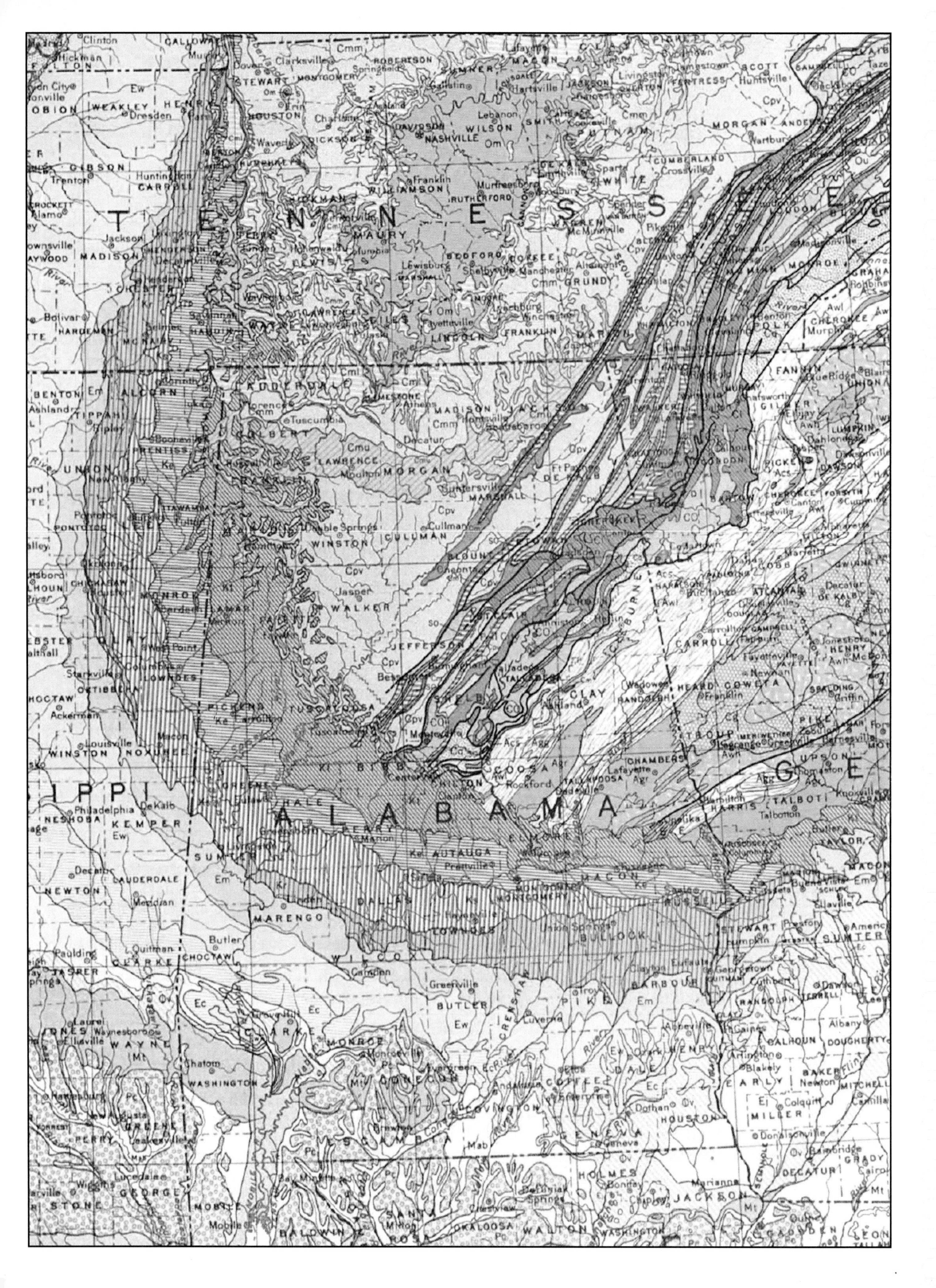

T E N N E S S E
A L A B A M A
G E
Clarksville
STEWART
MONTGOMERY
ROBERTSON
SUMNER
HOUSTON
DAVIDSON
NASHVILLE
WILSON
Lebanon
SMITH
PUTNAM
JACKSON
OVERTON
FENTRESS
SCOTT
MORGAN
CUMBERLAND
WHITE
GIBSON
CARROLL
WEAKLEY
HENRY
Dresden
Trenton
Franklin
WILLIAMSON
RUTHERFORD
Murfreesboro
HICKMAN
MAURY
MADISON
Jackson
BEDFORD
COFFEE
GRUNDY
Manchester
Lewisburg
MARSHALL
LINCOLN
FRANKLIN
Winchester
Fayetteville
HARDEMAN
Bolivar
MCNAIRY
HARDIN
WAYNE
LAWRENCE
GILES
Pulaski
ALCORN
BENTON
TIPPAH
Ripley
PRENTISS
Booneville
Ashland
UNION
New Albany
LAUDERDALE
COLBERT
Florence
Tuscumbia
Athens
LIMESTONE
MADISON
JACKSON
Huntsville
Scottsboro
Decatur
LAWRENCE
MORGAN
Moulton
MARSHALL
Guntersville
DE KALB
Ft Payne
FRANKLIN
WINSTON
CULLMAN
Cullman
BLOUNT
Jasper
WALKER
JEFFERSON
Bessemer
PONTOTOC
LEE
ITAWAMBA
MONROE
Aberdeen
LAMAR
FAYETTE
CHICKASAW
CLAY
West Point
Columbus
LOWNDES
OKTIBBEHA
Starkville
CHOCTAW
Ackerman
WEBSTER
PICKENS
Carrollton
TUSCALOOSA
Tuscaloosa
WINSTON
Louisville
NOXUBEE
Macon
NESHOBA
Philadelphia
KEMPER
DeKalb
GREENE
HALE
BIBB
Centerville
CHILTON
Clanton
SHELBY
Columbiana
TALLADEGA
Talladega
COOSA
Rockford
CLAY
Ashland
TALLAPOOSA
Dadeville
CHAMBERS
Lafayette
RANDOLPH
CLEBURNE
CALHOUN
Anniston
ST CLAIR
CHEROKEE
ETOWAH
Gadsden
FANNIN
UNION
GILMER
LUMPKIN
DAWSON
PICKENS
FORSYTH
CHEROKEE
Canton
BARTOW
COBB
Marietta
GWINNETT
DE KALB
Decatur
ATLANTA
DOUGLAS
CARROLL
Carrollton
HEARD
COWETA
Newnan
Franklin
FAYETTE
Fayetteville
HENRY
SPALDING
Griffin
TROUP
Lagrange
MERIWETHER
Greenville
PIKE
Zebulon
UPSON
Thomaston
HARRIS
Hamilton
TALBOT
Talbotton
TAYLOR
Butler
MACON
Opelika
LEE
Columbus
RUSSELL
Seale
MARION
Buena Vista
SCHLEY
Ellaville
STEWART
Lumpkin
SUMTER
Americus
PERRY
Marion
AUTAUGA
Prattville
ELMORE
Wetumpka
MONTGOMERY
MACON
Tuskegee
SUMTER
Livingston
Decatur
NEWTON
LAUDERDALE
Meridian
MARENGO
Linden
DALLAS
Selma
LOWNDES
Hayneville
BULLOCK
Union Springs
WILCOX
Camden
BUTLER
Greenville
CRENSHAW
Luverne
PIKE
Troy
BARBOUR
Clayton
Eufaula
QUITMAN
Georgetown
RANDOLPH
Cuthbert
TERRELL
Dawson
CLAY
Fort Gaines
CALHOUN
DOUGHERTY
Albany
EARLY
Blakely
Arlington
BAKER
Newton
MITCHELL
Camilla
MILLER
Colquitt
Donalsonville
Bainbridge
DECATUR
GRADY
Cairo
SEMINOLE
Butler
Quitman
CHOCTAW
CLARKE
Paulding
JASPER
Laurel
JONES
Waynesboro
WAYNE
Ellisville
CLARKE
Grove Hill
MONROE
Monroeville
Chatom
WASHINGTON
CONECUH
Evergreen
COVINGTON
Andalusia
COFFEE
Elba
Enterprise
DALE
Ozark
HENRY
Abbeville
HOUSTON
Dothan
GENEVA
Geneva
ESCAMBIA
Brewton
Hattiesburg
GREENE
Leakesville
PERRY
Augusta
GEORGE
Lucedale
STONE
Wiggins
MOBILE
Mobile
BALDWIN
Bay Minette
SANTA ROSA
Milton
OKALOOSA
Crestview
WALTON
DeFuniak Springs
HOLMES
Bonifay
WASHINGTON
Chipley
JACKSON
Marianna
GADSDEN
Quincy
POLK
Benton
BRADLEY
Cleveland
MURRAY
WHITFIELD
Dalton
GORDON
Calhoun
MONROE
Madisonville
CHEROKEE
Murphy
Hartsville
Gallatin
Springfield
Livingston
Jamestown
Huntsville
Wartburg
Crossville
McMinnville
Sparta
Tullahoma
Shelbyville
Columbia
Centerville
Lexington
Huntingdon
Paris
Linden
Waynesboro
Savannah
Dover
Erin
Charlotte
Waverly
Camden
Hohenwald
Lawrenceburg

PART C

A Portion of the Geologic Map of Pennsylvania

Scale: 1 inch = 4 miles

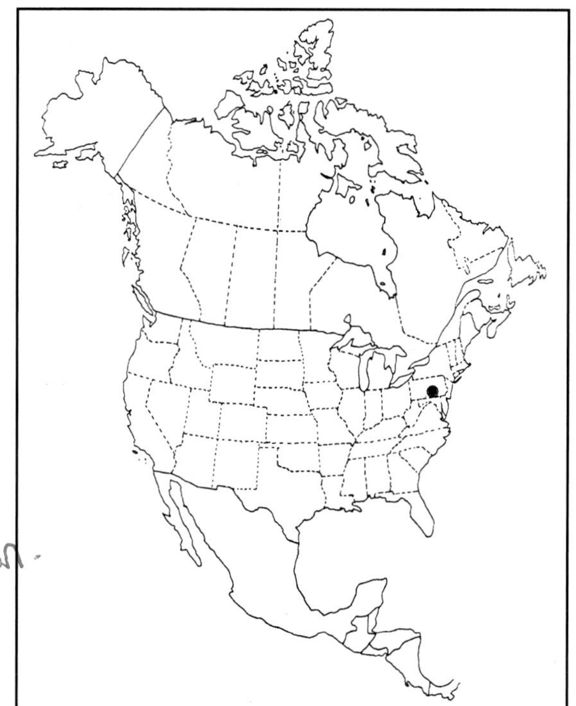

1. What is the likely source area for the Ordovician Martinsburg Formation and the Devonian Catskill Formation (figure 15.1)? What mountain-building events are associated with these formations? Cite your evidence.

   Oraivician plate convergnce closed Iapetus at basin, developed thick muds, lithifying into the formation.

2. Using the geologic map found on page 234 (figure 15.4), account for the difference in the structural patterns north and south of the Pennsylvania Turnpike, which extends east–west through Carlisle, Pennsylvania.

   Southern rocks- cumbrian, olgo, subduction of continental in orgn
   north- rocks formed during & after Silurian fuel and basin cuias

3. Date the Appalachian folding in this area. Summarize the evidence.

   devonian through mississippian. evidence in DM

4. Explain the origin and distribution of the igneous rocks in this area. With what episode of mountain building are they associated?

5. What is the reason for the present course of the Susquehanna River cutting across the folds indiscriminately and across all different rock types?

   rising sea levels fill chanel forming a shallow bay & depositing sediments.

6. In terms of plate tectonics, what type of plate interaction (extensional, strike-slip, compressional) best explains the geology of the map area? Cite evidence.

   compression, orogeny usually indicates mountain building.

Geologic Map of the Appalachians

Triassic		**TR**
TRd	Red diamonds and red lines	Dark gray igneous sills and dikes
TRlc		
TRg	Bright green with red patterns	Brunswick or Gettysburg Formation
TRb		Red sandstone and shale with minor conglomerate
TRh		
TRac		
Pennsylvanian		**IP**
IPp	Blue green	Pottsville Group Sandstone, conglomerate, shale and coals
Mississippian		**M**
Mmc	Pink	Mauch Chunk Formation Gray shale and red sandstone
Mp	Blue purple	Pocono Group Conglomerate, sandstone, and shale
Devonian		**D**
Dck	Orange	Catskill Formation Red sandstone and shale
Dm	Light orange	Marine beds, shale, sandstone, and limestone
Dho	Gray, green diagonal	Hamilton Group and Onendaga Formation Shale, sandstone, and limestone
Doh	Orange and red diagonal	Oriskany and Helderberg Formations Fossiliferous sandstone and shale, and fossiliferous limestone
Silurian		**S**
Skt	Purple and blue diagonal	Keyser Formation and Tonoloway Formations Limestone
Sw	Dark and light blue, horizontal	Wills Creek Formation Shale and limestone with local sandstone
Sc	Light green, red stipples	Clinton Group Red iron-rich shale and sandstone
Sbm	Blue and purple, diagonal	Bloomsburg and McKenzie Formation Red to green shale and sandstone
St	Red and brown, diagonal	Tuscarora Formation Light coarse sandstone

Geologic Map of the Appalachians *(continued)*

Ordovician		O
Ojb	Orange and brown, diagonal	Juniata Formation and Bald Eagle Formation Red sandstone, shale, and conglomerate
Om	Gray and pink, horizontal	Martinsburg Formation Marine shale
Oc		
Ohm	Blue grayish pink, diagonal	Chambersburg Formation, Hershey and Myerstown Formations
Osp	Light blue	St. Paul Group and Annville Formation Limestone with chert
Ob	Coarse pink and purple, diagonal	Beekmantown Group Limestone
Oor	Fine pink and purple, diagonal	Ontelaunee Formation, Epler Formation Rickenback Formation Limestone and dolomite
Os	Pink with blue, diagonal	Stonehenge Formation Limestone and limestone conglomerate

Cambrian		Є
Єc	Red and bright green, diagonal	Conococheague Group Limestone and dolomite
Єe	Grayish green, orange, crossed lines	Elbrook Formation Limestone and dolomite
Єwb	Green and orange vertical	Waynesboro Formation Red and purple shale with sandstone beds
Єt	Red and light green, diagonal	Tomstown Formation Dolomite, thin shaley beds
Єa	Pink and red squares	Antietam Formation Quartzite and schist
Єma	White and orange diagonal	
Єh	Orange and red vertical	Harpers Formation Phyllite and schist
Єwl	Pink and blue diagonal	Chickies Formation or Weverton Formation Quartzite and schist

Igneous Rocks		
mr	Yellow, red dots	Metarhyolite
vs	Green, red stippled	Greenstone schist
mb	Brown, red stippled	Metabasalt

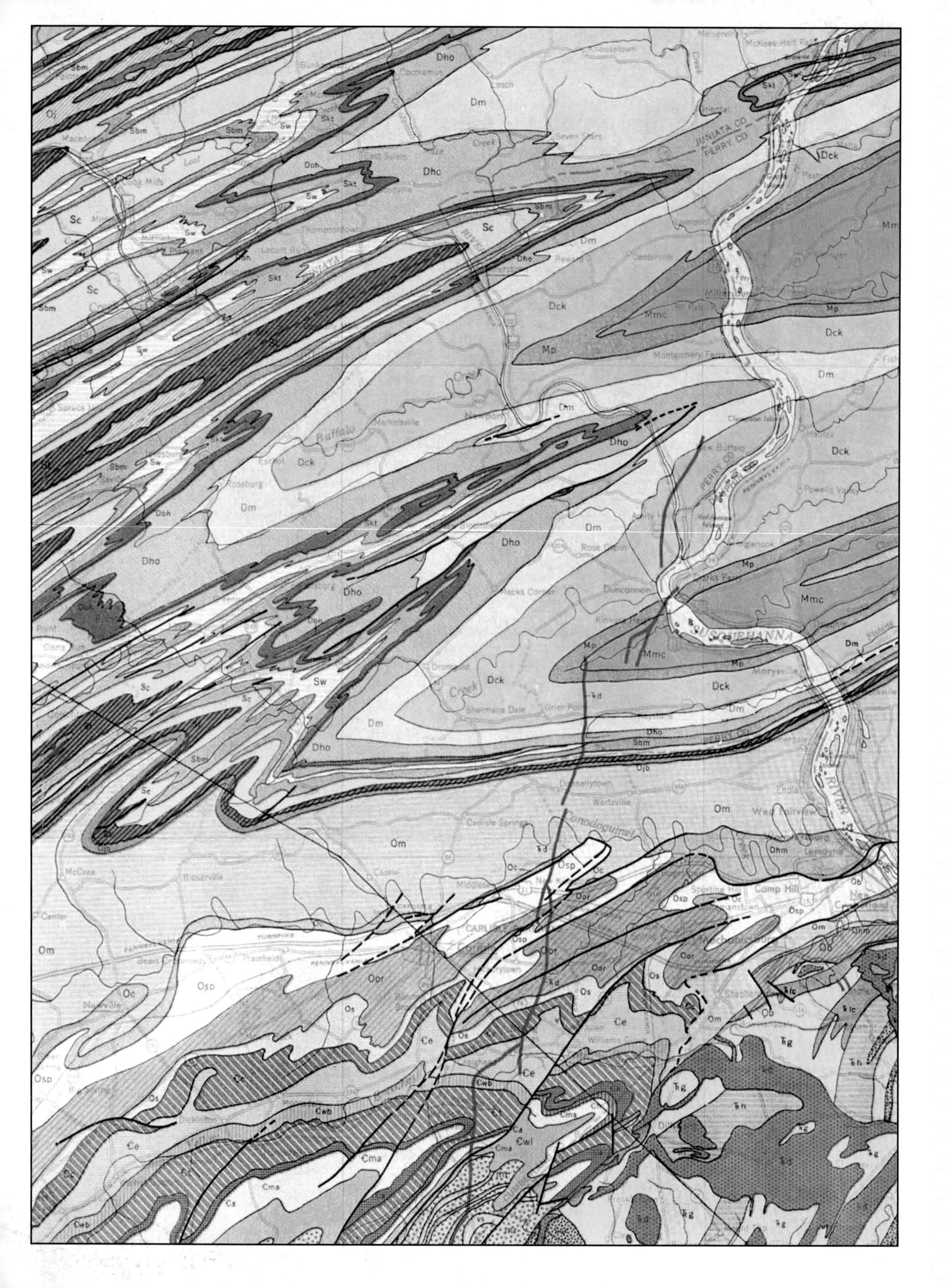

PART D

Geology of the Gateway, Colorado, Quadrangle

Scale: 1 inch = 2,000 feet (approximately 0.4 miles)

Examine figures 15.5 and 15.6 and answer the following questions about the geology of the Gateway, Colorado, area. Figure 15.5 (page 238) is a USGS geological map of the area. Figure 15.6 (page 239) shows the regional topography along with the three structural domains (areas underlain by Precambrian, Permian, and Mesozoic rocks) that occur in this area. Quaternary deposits have been omitted from figure 15.6 to emphasize the relationships between the structural domains.

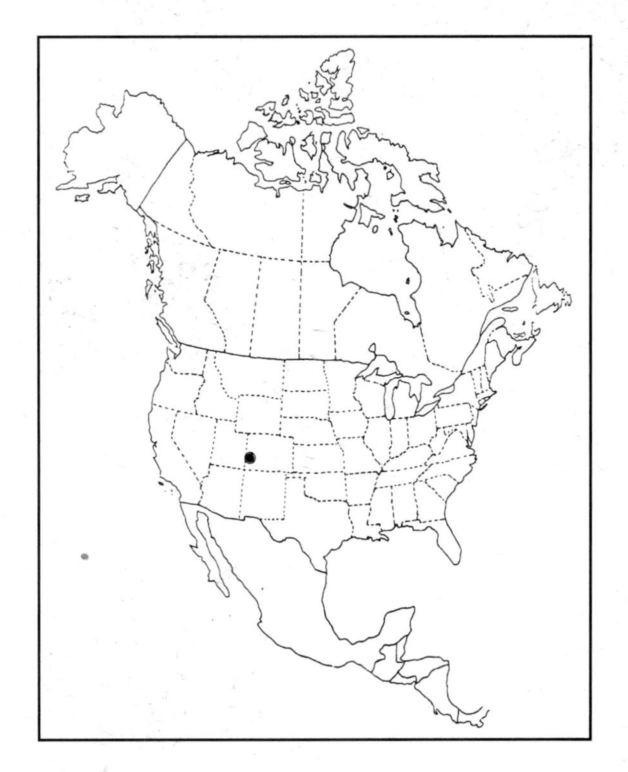

1. Using the topographic profile provided on page 236, draw a geologic cross section along the diagonal line (A to A′) from the northeastern corner to the lower edge of the geological map. Pay careful attention to strikes and dips.

2. Note the unconformable contact between the Permian Cutler Formation and the Precambrian gneisses and schists. What type of unconformity is represented by this contact?

   paraconformity, time missing (precambrian – pennsylvanian)

3. Explain the absence of the Cutler and Moenkopi Formations beneath the Chinle Formation in the northeastern corner of the map area.

   erosion, and nonconformity

4. There are no rocks of Paleozoic age in the map area except for the Cutler Formation. In this area it ranges from zero feet thick in the northern portion of the map to a few thousand feet thick just south of the map area, where several hundred feet of Pennsylvanian Age sandstone and conglomerate is also present. The Cutler consists of maroon, purple, and red crossbedded, arkosic, and lithic sandstone and conglomerate that represent deposition on an alluvial fan (fanglomerate). In the vicinity of Wright Draw and Ute Creek, the formation rests on highly irregular hills of Precambrian rocks. Grains of quartz, fresh feldspar, and pebbles and boulders of granite, gneiss, and quartzite, derived almost entirely from Precambrian rocks, are mixed together in poorly sorted, rudely crossbedded layers and lenses. How do the absence of pre-Permian Paleozoic rocks, development of unconformities, and the sedimentology of the Cutler Formation relate to the concept of the Ancestral Rocky Mountains?

   the ancestrial Rocky mountains were formed due to the collision of the North American plate, leading to a convergent boundary, water flow, erosion, weathe-ring and non disruption could explain formations.

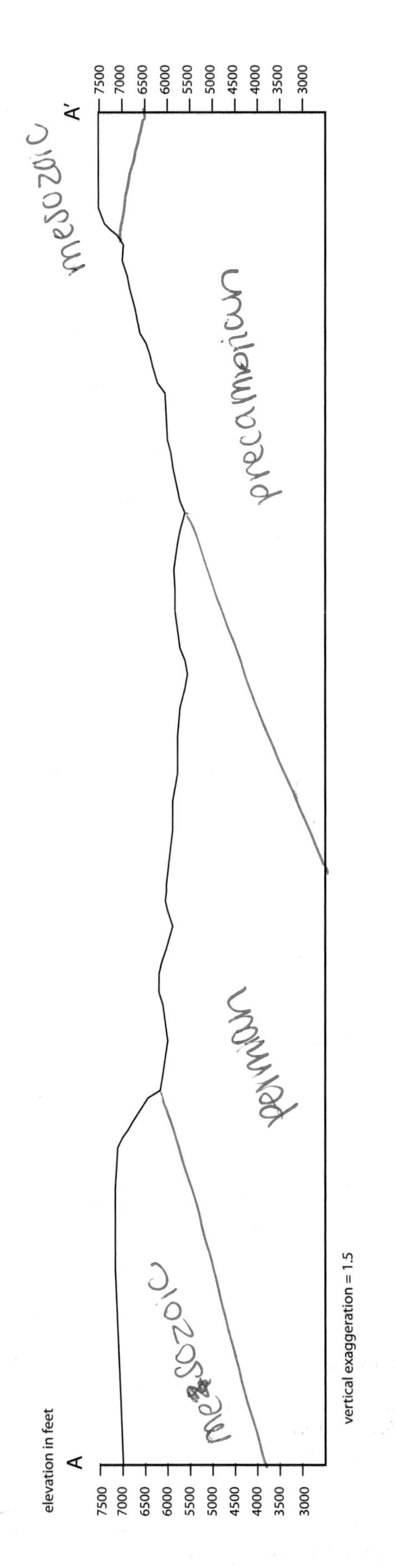

5. What is the nature of the contact between the Cutler and Moenkopi Formations in the southern half of the map? How did this relationship develop?

6. Briefly outline the geological history (tectonic and depositional events) of western Colorado as represented by the rocks and structures in the map area.

Geology of Gateway, Colorado

Quaternary		**Q**
Qal	Stippled yellow	Alluvium
Qg	Green, yellow circles	Terrace gravels
Qfg	Dark yellow stipple	Angular fragments and boulders
Cretaceous		**K**
Kbc	Dark blue green diagonal	Burrow Canyon Formation Sandstone and conglomerate with interbedded green and purplish shale
Jurassic		**J**
Jmb	Green horizontal	Morrison Formation
Jms	Solid blue green	Upper member of rusty red to gray shale Lower member of varicolored shale with buff sandstone and lenses of conglomerate
Js	Olive green, vertical	Summerville Formation Gray, green, and brown sandy shale and mudstone
Jec	Dark olive green, diagonal	Entrada Sandstone and Carmel Formation undivided Fine-grained massive sandstone and red mudstone
Jn	Dark olive green, solid	Navajo Sandstone Buff, gray, crossbedded sandstone
Jk	Olive green, horizontal	Kayenta Formation Interbedded red, gray siltstone and sandstone
Jw	Dark olive, diagonal	Wingate Sandstone Massive reddish brown crossbedded sandstone
Triassic		**₨**
₨c	Light blue, diagonal	Chinle Formation Red siltstone with lenses of red sandstone, shale, and conglomerate
₨ml	Dark blue, horizontal	Moenkopi Formation Sandy mudstone, brown and red with local gypsum
Pennsylvanian		**IP**
IPc	Medium blue, horizontal	Cutler Formation Red to purple conglomerate and sandstone
Precambrian		**p€**
p€	Brown and white, mottled	Gneiss, schist, granite, and pegmatite

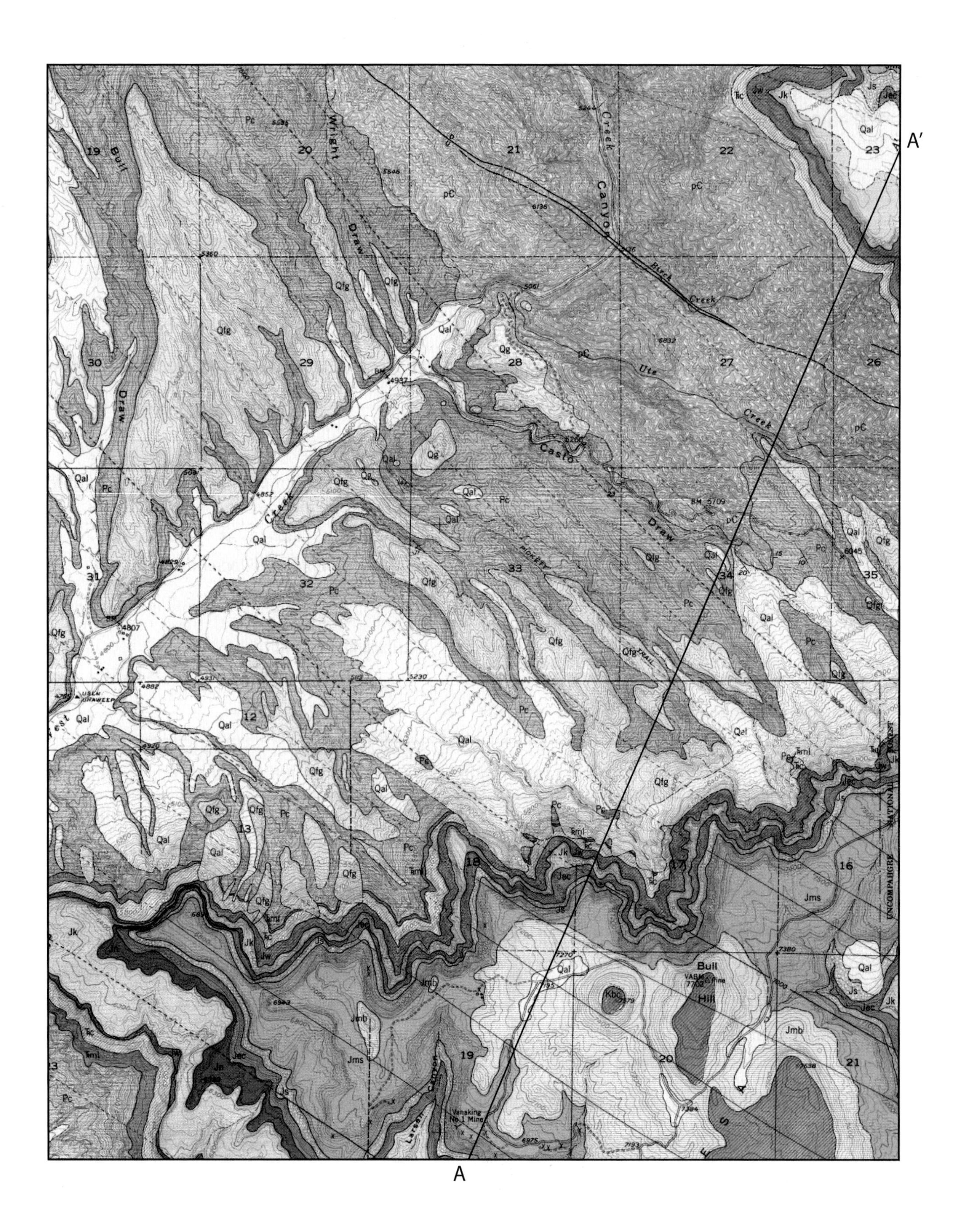

FIGURE 15.5 USGS geological map of the Gateway, Colorado, area.

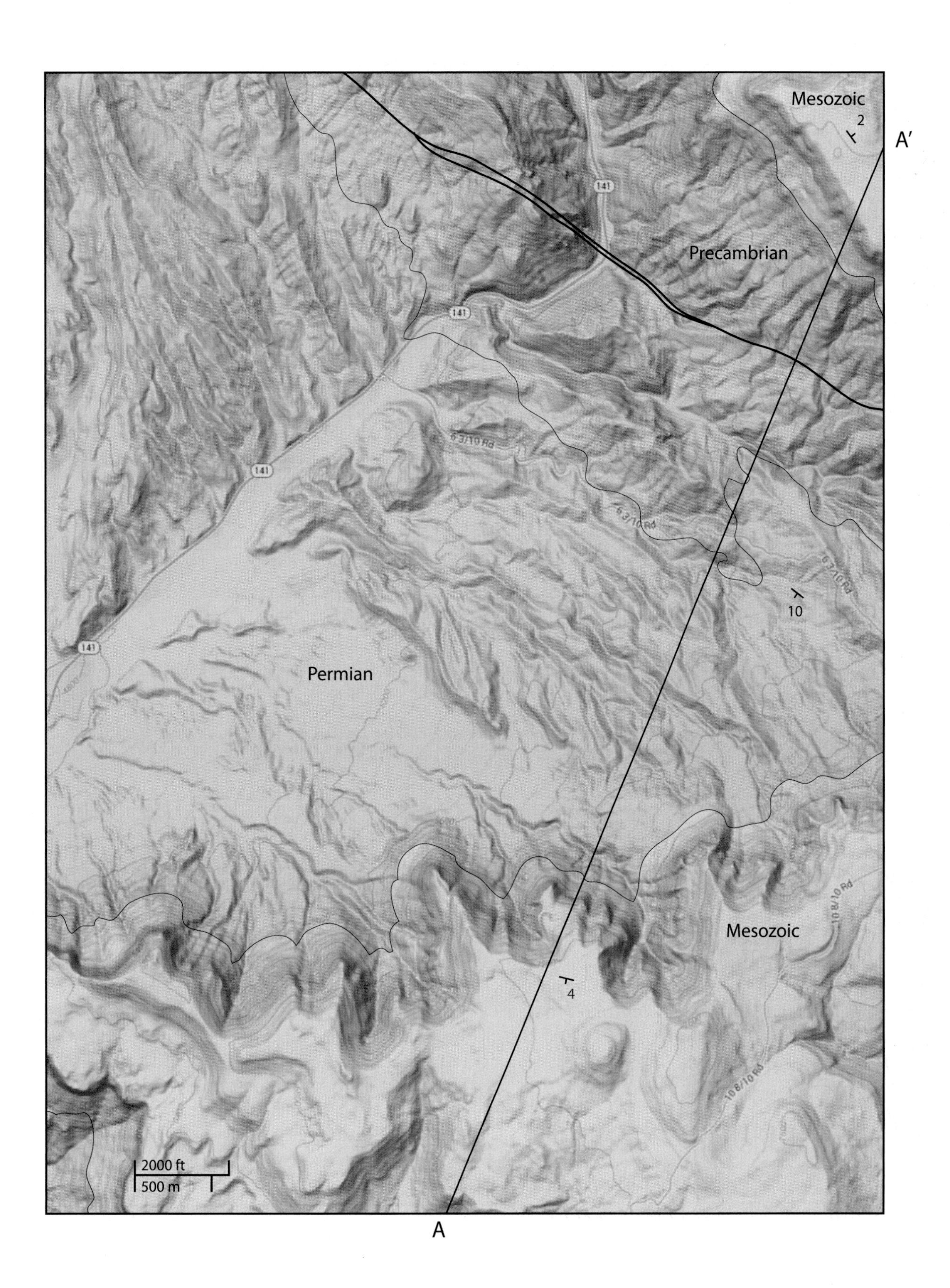

FIGURE 15.6 Regional topography and structural domains of the Gateway, Colorado, area.

Exercise 16 Cordilleran Orogeny

Learning Objectives

After completing this exercise, you will:

1. have gained confidence in your ability to read and interpret geological maps;
2. be better acquainted with rocks and structures that formed during the three phases of the Cordilleran orogeny;
3. recognize rocks and structures that characterize the Nevadan orogeny;
4. recognize structures that characterize the Sevier orogeny;
5. be able to recognize Laramide structures on geological maps; and
6. be better acquainted with the nature of sediments deposited in the Sevier foreland basin (Western Interior Seaway).

This lab will afford you the opportunity to interpret maps and cross sections derived from areas affected by different phases (Nevadan, Sevier, and Laramide) of the Cordilleran orogeny. You will examine geological maps from areas affected by this prolonged episode of tectonic deformation and investigate the nature of sediments that filled the Sevier foreland basin.

The Cordilleran Orogeny

The breakup of Pangea began during the Triassic Period. At that time, tensional stresses dominated the eastern margin of North America, resulting in the formation and filling of rift basins, the emplacement of igneous dikes, and extrusion of basaltic lavas in Connecticut, New York, New Jersey, and other eastern states. As North America moved westward, it overrode the Pacific plate, producing a subduction zone that stretched from Arctic Canada to Mexico. Continued westward motion of North America during the Jurassic and Cretaceous Periods resulted in a wave of eastward-shifting deformation that eventually reached as far inland as modern-day Alberta, Canada, and Colorado (figure 16.1). This episode of subduction-related deformation is known as the **Cordilleran orogeny**. Deformation began first and was most intense in the western Cordillera where the crystalline core complex developed. This phase of deformation, represented by emplacement of the Sierra Nevada, Idaho, and Coast Range (Canada) batholiths and metamorphism of host rocks, is called the **Nevadan orogeny** (figures 16.1 and 16.2).

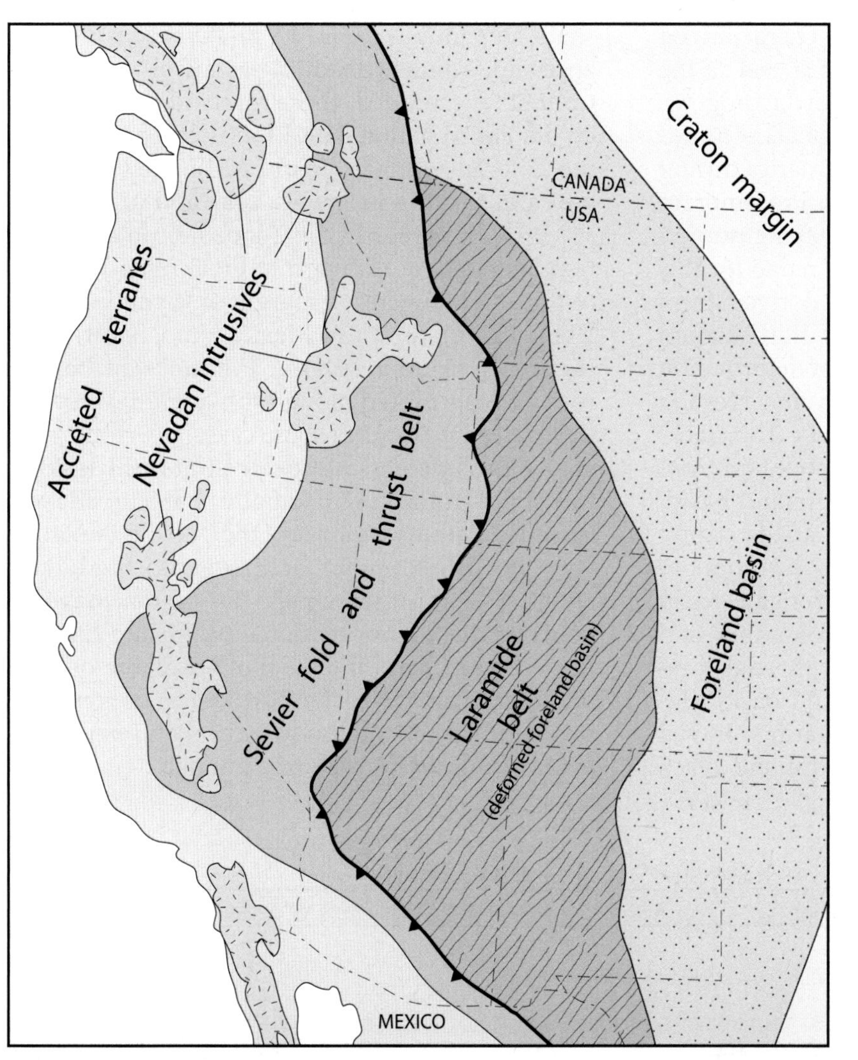

FIGURE 16.1 Tectonic map of the western United States showing areas influenced by the successive phases of the Cordilleran orogeny.

FIGURE 16.2 Intrusive igneous rocks exposed in Yosemite Valley, California. View is eastward from El Portal (see geological map for regional extent of these rock units). Kec = El Capitan Granite, Kic = Illilouette Creek Granodiorite, Ks = Sentinel Granodiorite, Khd = Half Dome Granodiorite.

A second phase of tectonic activity characterized by thrust-related shortening of the area east of the granitic intrusives is known as the Sevier orogeny. During the Sevier orogeny, strata that had accumulated on the passive margin of North America during the Paleozoic Era were detached from Precambrian basement rocks and pushed eastward along numerous, parallel planes of weakness. The name for this phase of the Cordilleran orogeny is derived from central Utah, where "Sevier-type" thin-skinned thrusting is in evidence, but the style of deformation occurs both south and north of Utah and Nevada from northern Mexico to Arctic Canada. The stacking of thrust sheets along the Sevier mountain front resulted in subsidence of the **Sevier foreland basin**, which also stretched from Mexico to Canada (figure 16.1). Subsidence and sedimentation occurred more rapidly along the west side of the elongate basin than on the east side due to crustal loading and proximity to the Sevier uplift. Triassic through Cretaceous rocks deposited in the foreland basin (figures 16.3 and 16.4) range from nonmarine alluvial fan conglomerate to deep-marine chalk and black shale, depending upon the relative rates of subsidence and sediment shedding at any given time. Now uplifted and eroded, these foreland basin strata are on display in national parks in Utah and throughout the Colorado Plateau (figure 16.4). They also contain a wealth of dinosaur bones, coal, and uranium.

By Paleogene (Early Cenozoic) time, deformation had shifted eastward to the foreland basin and to the adjoining craton margin of Wyoming, eastern Utah, Colorado, New Mexico, and South Dakota (figure 16.1). In response, Precambrian basement rocks were uplifted along high-angle reverse faults that deformed, and in some cases, overrode, associated Paleozoic and Mesozoic strata (thick-skinned thrusting) (figure 16.5). Equally characteristic of this so-called **Laramide phase** of the Cordilleran orogeny was the development of basement-cored domes, anticlines, and monoclines. Basins that developed between these compressional uplifts are filled with Early Cenozoic sediments that host large quantities of oil, gas, and coal. The Cordilleran orogeny had a profound impact on the structural and depositional evolution of western North America.

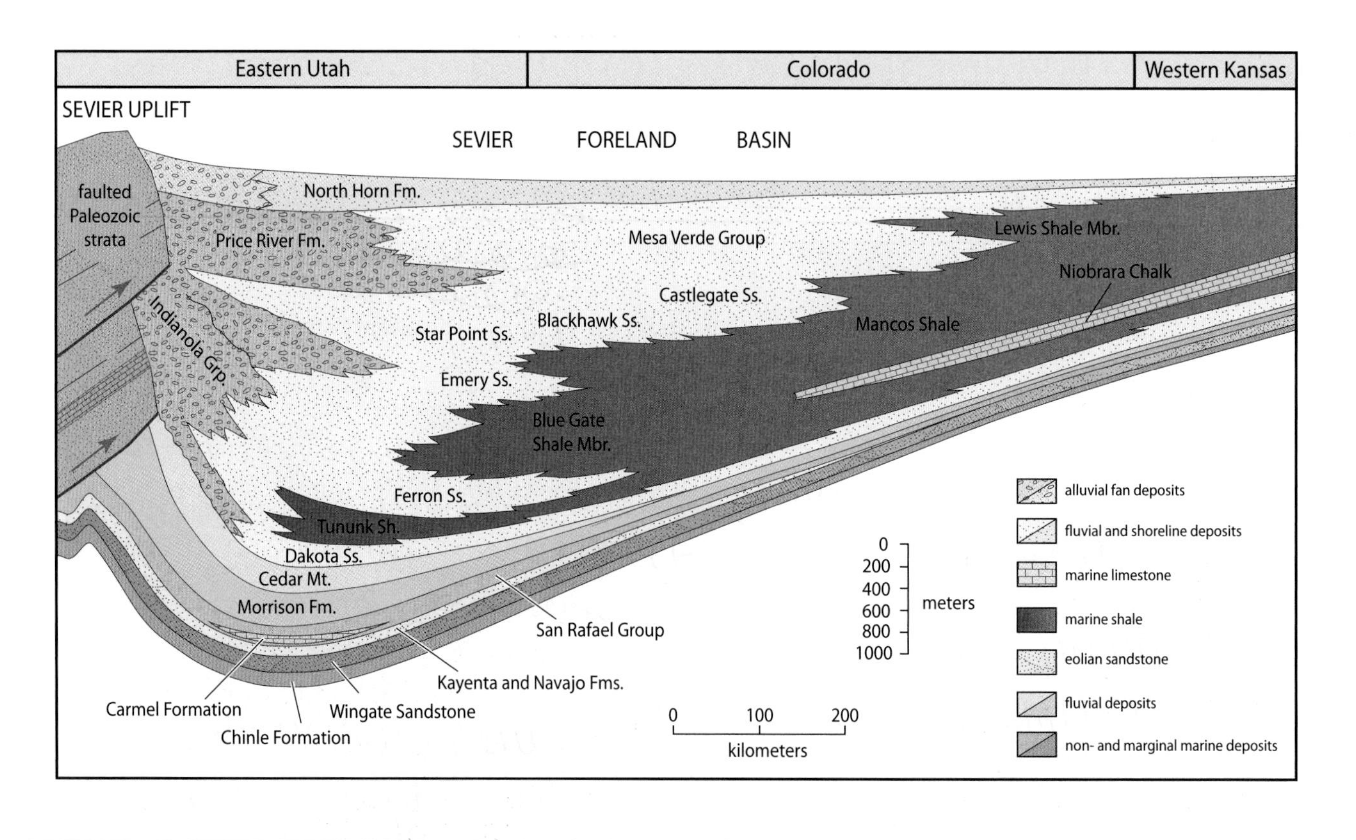

FIGURE 16.3 Stratigraphic cross section showing Triassic through Early Cenozoic rocks deposited in the Sevier foreland basin.

FIGURE 16.4 Strata deposited in the Sevier foreland basin. A. Cretaceous Mancos Shale overlain by sandstone of the Mesa Verde Group, Colorado Plateau, Utah. B. Jurassic Entrada Sandstone, Arches National Park, Utah. *(continued)*

FIGURE 16.4 Strata deposited in the Sevier foreland basin. C. Upper Cretaceous Cedar Mountain Formation overlain by blocks of the Dakota Sandstone, San Rafael Swell, Utah. D. Triassic Moenkopi (grayish tan) and Chinle (red) Formations overlain by the cliff-forming Wingate Sandstone, San Rafael Swell, Utah.

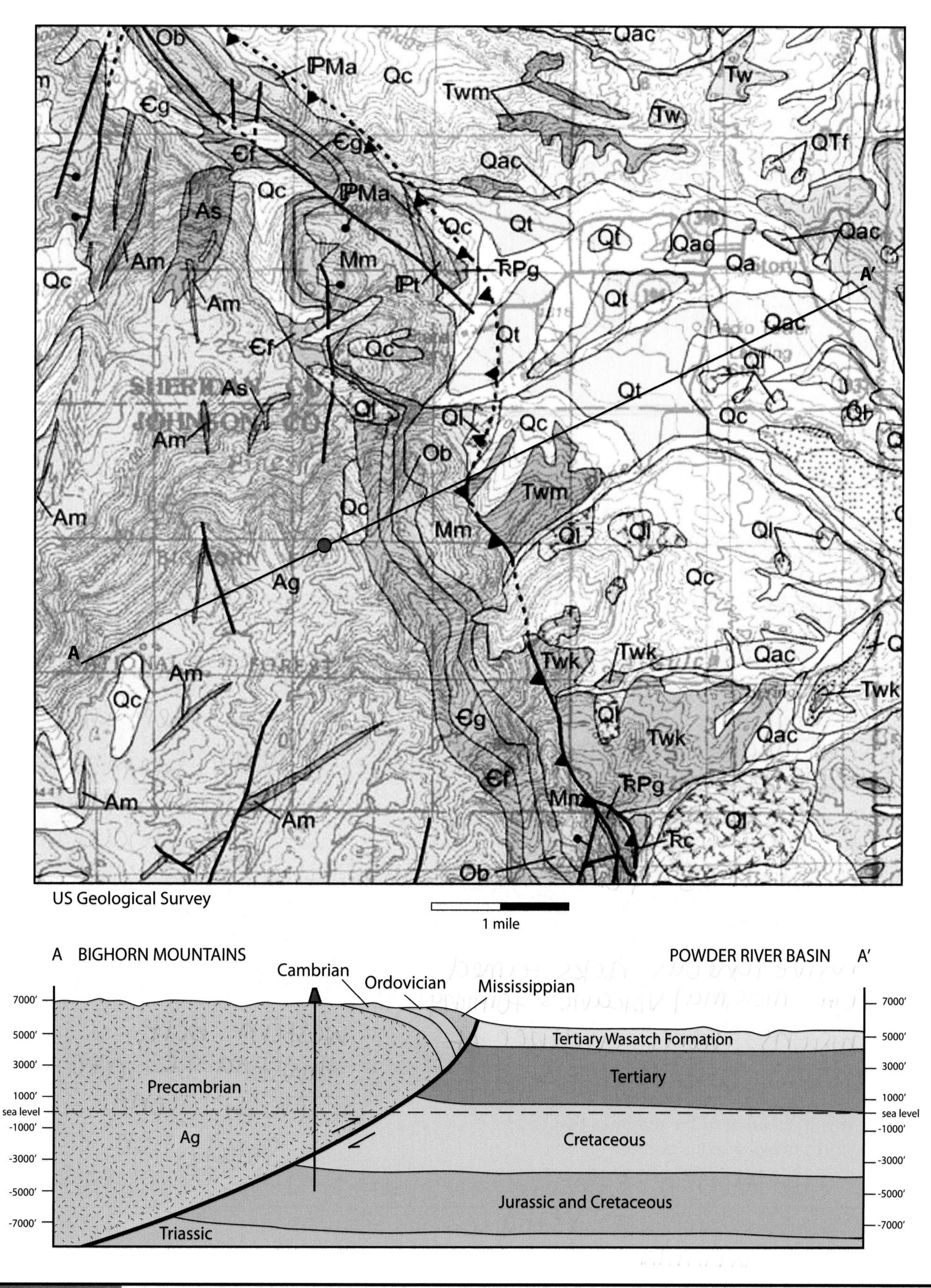

Figure 16.5 Geological map showing structural relationships developed on the northwest side of the Bighorn Mountains, Wyoming. See the geological map of Wyoming (part C of this exercise) for the location of the area shown on this map.

PROCEDURE

Part A

Yosemite National Park, the Sierra Nevada, California

Scale 1 inch = 4 miles

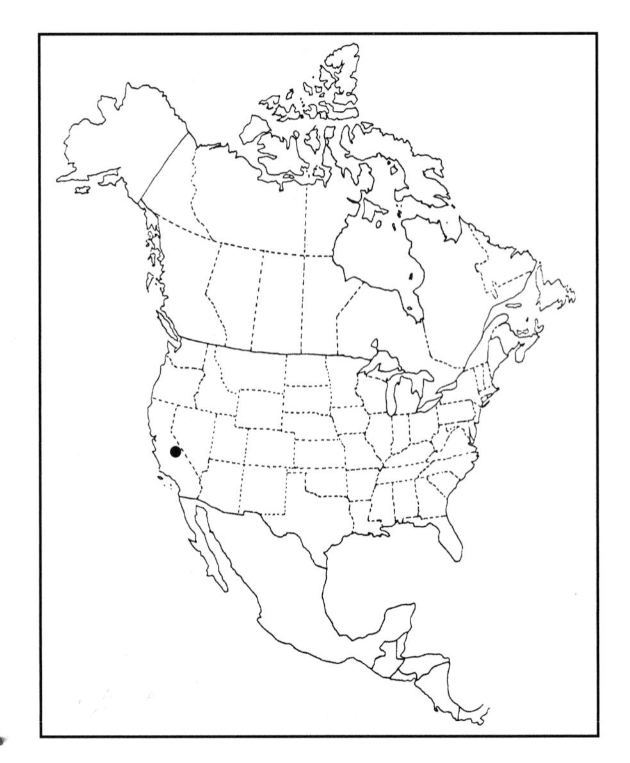

1. See the geological map provided on page 248 (figure 16.6). What rock type predominates in this map area? (circle one)

   intrusive igneous extrusive igneous

   sedimentary metamorphic

2. What is the difference between granite and granodiorite?

   granite - Potassium, feldspar, little dark iron.
   granodiorite - calcium & sodium, more dark minerals

3. During which geological period(s) did the majority of these igneous bodies form?

   igneous rocks formed
   cretaceous period

4. What evidence of multiple episodes of intrusion in the Sierra Nevada is shown on this map and on figure 16.2?

   intrusive igneous rocks formed from magma / volcanos forming intrusions. theres evidence of erosion and weathering.

5. What was the source of the magmas that solidified to form these igneous bodies?

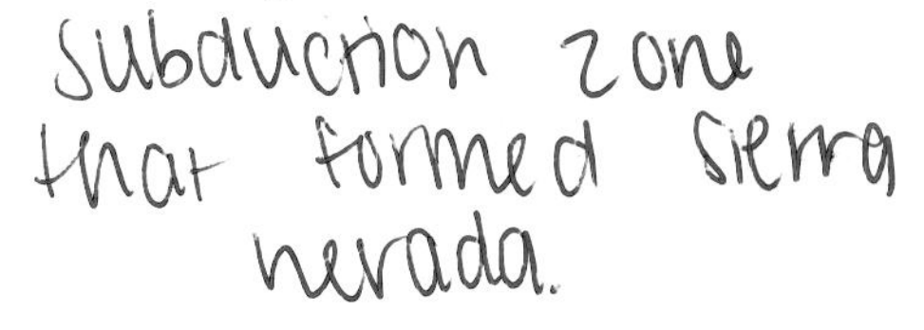

6. What is the nature and significance of the Paleozoic rocks (Pzme?) in the lower left portion of the map?

   Oldest Rock,
   So the metasedimentary Rock were there, igneus rocks intrude them

7. What is the name of the orogenic phase of the Cordilleran orogeny that most greatly influenced this area of California?

   western belt,
   thrusting of oceanic crust along subduction zone.

Key to Geologic Map of Yosemite National Park and Vicinity

Quaternary Units	
Qal	Holocene alluvium
Qtl	Holocene talus
Qti	Pleistocene glacial till
Qta	Pleistocene glacial till
Qpt	Pleistocene glacial till
Tuolumne Intrusive Suite (Late Cretaceous)	
Kap	Aplite dikes and small pegmatite bodies
Khd	Half Dome Granodiorite
Kdc	Granodiorite of Kuna Crest
Sonora Pass Intrusive Suite (Late Cretaceous)	
Ks	Sentinel Granodiorite (K-Ar age about 93 Ma)
Kyc	Yosemite Creek Granodiorite
Buena Vista Crest Intrusive Suite (Middle Cretaceous)	
Kic	Illilouette Creek Granodiorite (U-Pb age about 100 Ma)
Yosemite Valley Intrusive Suite (Early to Middle Cretaceous)	
Kt	Taft Granite (U-Pb age about 96 Ma)
Kec	El Capitan Granite (U-Pb age about 102 Ma)
Fine Gold Intrusive Suite (Jurassic to Early Cretaceous)	
Kar	Arch Rock Granodiorite (U-Pb ages about 116 Ma)
Kga	Gateway Tonalite (U-Pb age about 114 Ma)
Kg	undivided (Cretaceous?) granitic rocks
KJdg	Jurassic to Early Cretaceous diorite and gabbro
Metasedimentary Rocks (Paleozoic to Jurassic)	
Jms	Metamorphosed Jurassic strata
Pzme?	Metamorphosed Paleozoic strata

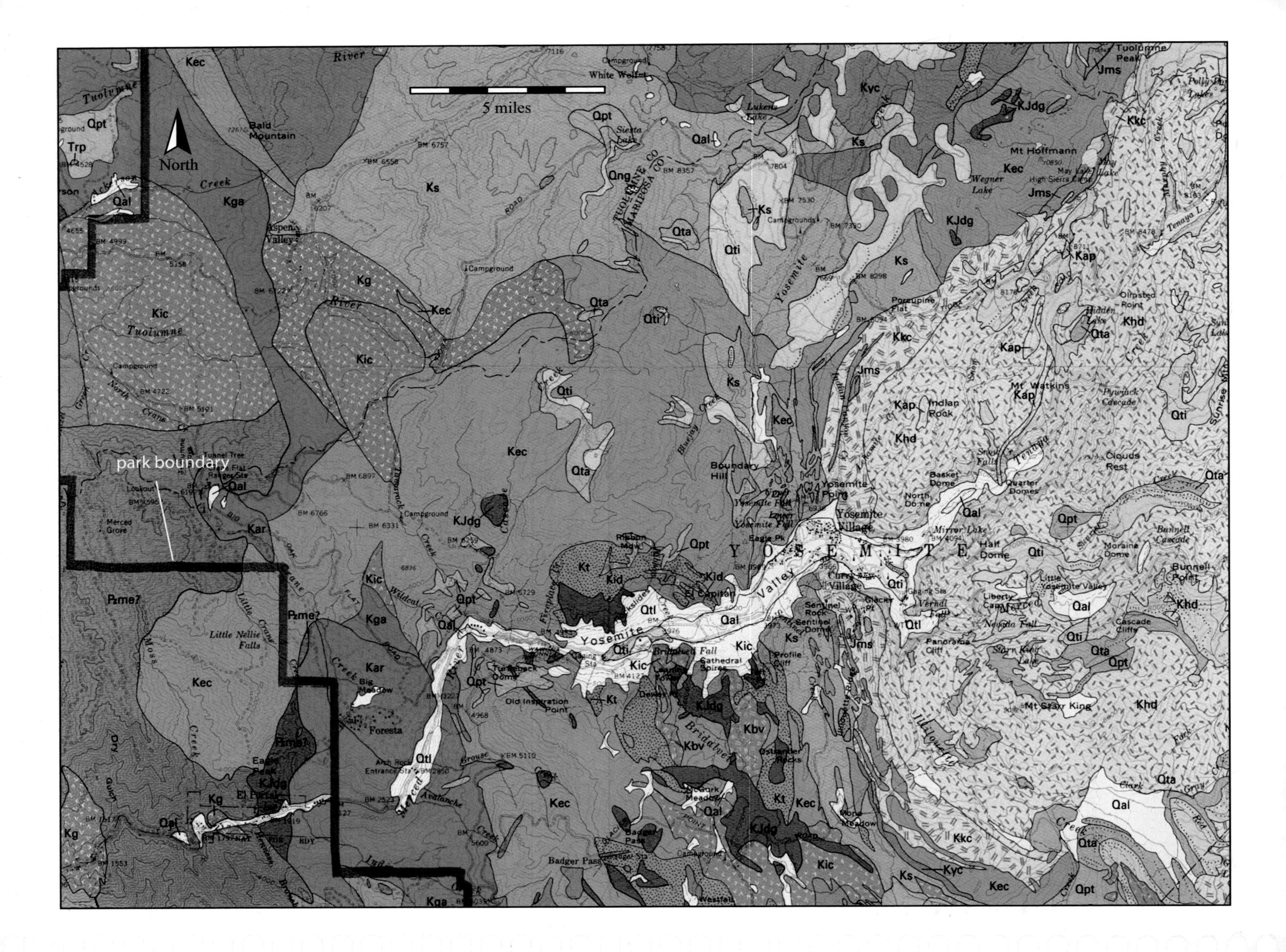
5 miles
North
park boundary
YOSEMITE
Yosemite Valley
Yosemite Village
Yosemite Point
Yosemite Fall
Half Dome
North Dome
Basket Dome
Clouds Rest
Quarter Domes
Mt Watkins
Indian Rock
Mt Hoffmann
Tuolumne Peak
Olmsted Point
El Capitan
Sentinel Rock
Sentinel Dome
Cathedral Spires
Bridalveil Fall
Glacier Pt
Curry Village
Mirror Lake
Liberty Cap
Vernal Fall
Nevada Fall
Panorama Cliff
Mt Starr King
Moraine Dome
Bunnell Point
Cascade Cliffs
Little Yosemite Valley
Old Inspiration Point
Turtleback Dome
Boundary Hill
Big Meadow
Foresta
El Portal
Arch Rock Entrance Sta
Badger Pass
Bald Mountain
Aspen Valley
Little Nellie Falls
Merced Grove
Tuolumne
North Crane
Dry Gulch
Moss Creek
Porcupine Flat
Wegner Lake
Lukens Lake
Siesta Lake
White Wolf
TUOLUMNE CO
MARIPOSA CO
Illilouette
Tenaya
Sunrise Mtn
Kec
Ks
Kic
Kg
Kga
Kar
Kyc
Kkc
KJdg
Jms
Kap
Khd
Kt
Kid
Kbv
Qal
Qpt
Qta
Qti
Qtl
Qng
Trp
Pzme?

PART B

Lake Minnewanka Quadrangle, Alberta, Canada

Scale: 1.25 inch = 1 mile; 2 cm = 1 km

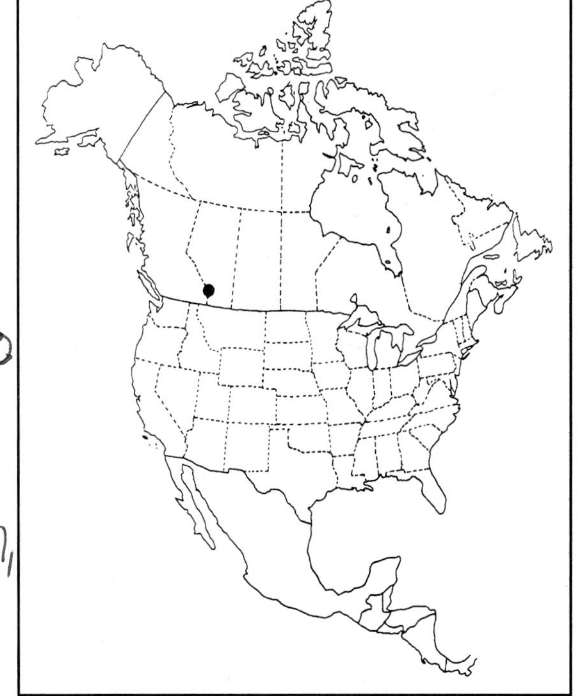

1. See the geological map provided on page 252 (figure 16.7). Generally characterize the rocks shown on this map as to their age and rock type.

   Sedimentary rocks, they are from cambrian, devonian, mississippian, plan, permian, Thasic, Jurrasic, creteceous

2. Characterize the structure of the area as gently, moderately, or complexly deformed. What types of structures are present?

   complexley deformed, a few - many thrusts, faults, & folds.

3. In which direction did the thrust faults move? How does this fit the plate tectonic model?

   Faults move away from each other - convergent boundary / subduction zone.

4. What is the likely age of deformation? What is the evidence for your answer?

   majority of Rocks from cambrian, devonian, mississippian eras.

5. Which phase of the Cordilleran orogeny is represented by deformation in this map area?

   Sevier foreland basin, stretches mex - canada.

Lake Minnewanka Quadrangle, Alberta, Canada

Symbol	Color	Description
Quaternary		**Q**
Qd	Light yellow	Till, alluvium, and colluvium
Cretaceous		**K**
Kbz	Dark green	Brazeau Formation Sandstone, siltstone, mudstone, and minor coal
Jurassic		**J**
Jf	Dark blue green	Fernie Group Shale, siltstone, sandstone, and limestone
Triassic		**TR**
TRwh	Light green	Whitehorse Formation Sandstone, siltstone, and limestone breccia
TRsm	Light yellow green	Sulfur Mountain Formation Siltstone, mudstone, and shale
Permian and Pennsylvanian		**PIP**
PIPrm	Medium blue	Rocky Mountain Group Sandstone, dolomite, and chert
Mississippian		**M**
Met	Light gray	Etherington Formation Limestone and dolomite, cherty
Mmh	Light blue	Mount Head Formation Limestone and dolomite, cherty
Mtv	Dark gray	Turner Valley Formation Limestone and cherty limestone
Msh	Dark pinkish gray	Shunda Formation Limestone and cherty limestone
Mpk	Grayish pink	Pekisko Formation Limestone and cherty limestone
Mlv	Medium pinkish gray	Livingstone Formation Limestone, dolomite, and cherty limestone
Mbfu	Light pinkish gray	Upper part Exshaw and Banff Formations Limestone and dolomite
Mbfm	Medium pink	Middle part Exshaw and Banff Formations Limestone and dolomite
Mbfl	Purple	Lower part Exshaw and Banff Formations Shale, siltstone, and cherty limestone
Mbf	Light pink	Exshaw and Banff Formations undivided Limestone, shale, siltstone, cherty limestone, and dolomite

Lake Minnewanka Quadrangle, Alberta, Canada *(continued)*

Devonian		D
Dpa	Medium blue	Palliser Formation Dolomite and limestone
Dax	Dark blue	Alexo Formation Breccia
Dsx	Brown	Southesk Formation Dolomite
Dcn	Bluish gray	Cairn Formation Dolomite and limestone

Cambrian		Ꞓ
Ꞓlx	Bright blue	Lynx Group Dolomite and shale
Ꞓar	Medium orange	Arctomys Formation Siltstone and dolomite
Ꞓpk	Light orange	Pika Formation Limestone, dolomite, and minor shale
Ꞓel	Medium gray	Eldon Formation Limestone and dolomite
Ꞓst	Dark yellow	Stephen Formation Shale and limestone
Ꞓca	Light bluish gray	Cathedral Formation Limestone
Ꞓmw	Medium grayish pink	Mount Whyte Formation Shale, limestone, and dolomite

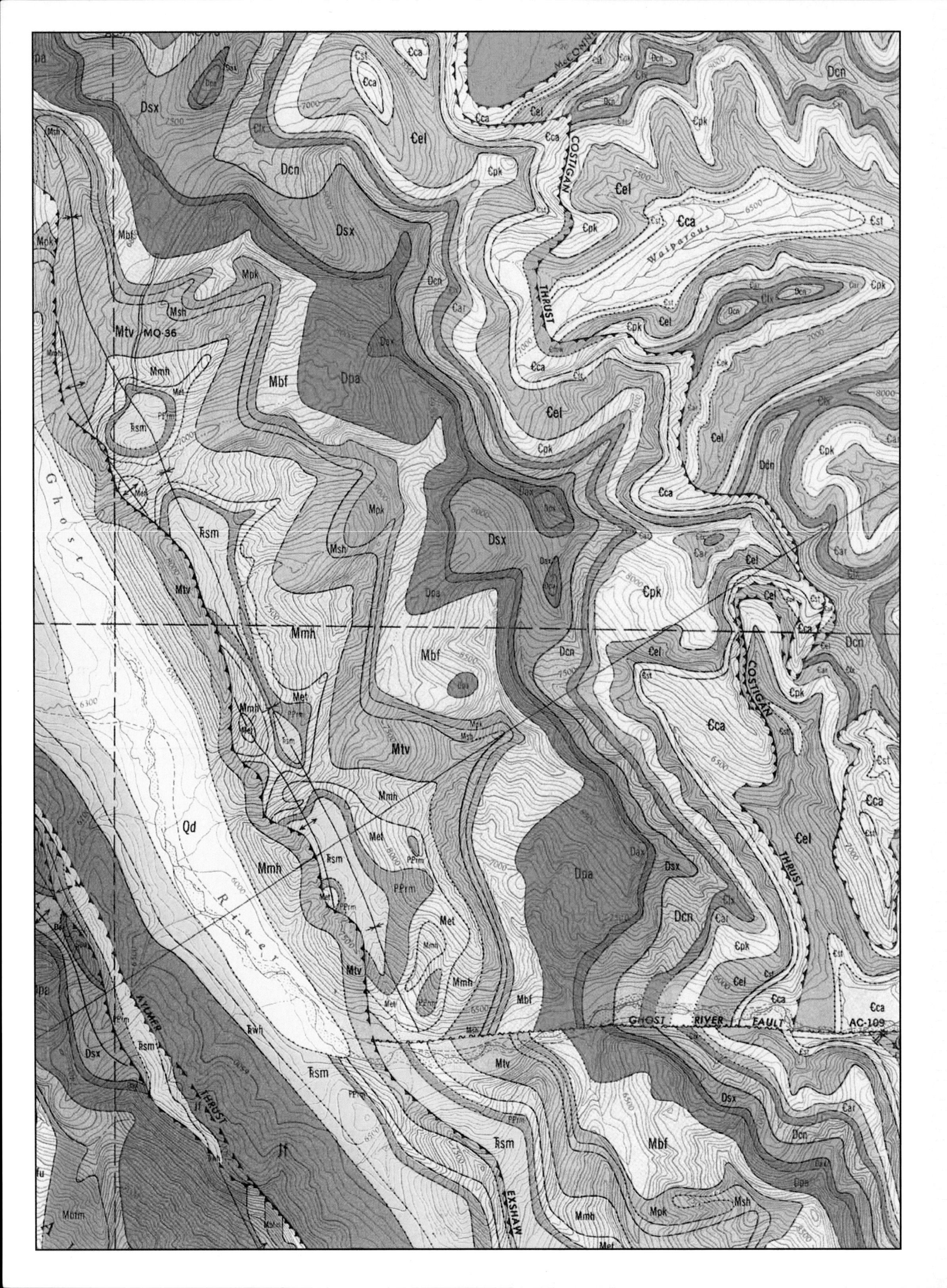

Dsx
Dcn
Єca
Єst
Єel
Єpk
Єar
Єlx
COSTIGAN
THRUST
Waiparous
McCONNELL
Mtv
MQ-36
Mmh
Met
Mbf
Dpa
Mpk
Msh
Ŧsm
PPrm
Qd
Ghost
River
GHOST
RIVER
FAULT
AC-109
AYLMER
EXSHAW
Jf
7000
7500
8000
6500
6300
6000

PART C

Geological Map of Bighorn Mountains, Wyoming

1. See the geological map provided on page 256 (figure 16.8). Draw a generalized structural cross section through the Bighorn Mountains located between Worland and Buffalo, Wyoming. What type of structure is indicated by the patterns of Paleozoic and Mesozoic rocks surrounding the Precambrian core of the Bighorn Mountains?

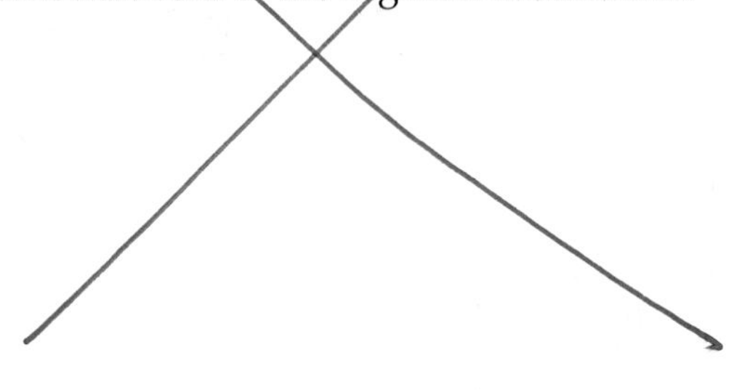

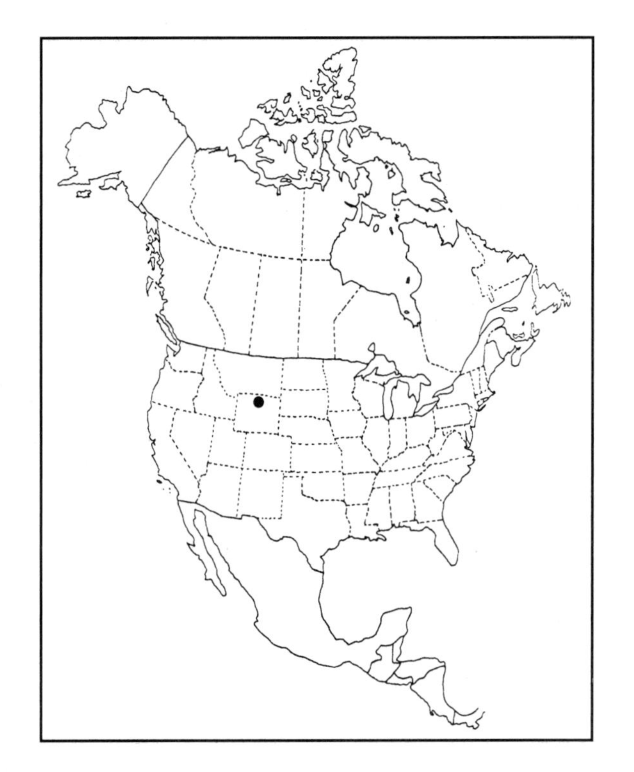

2. Study the outcrops of the Mesozoic (green) and Cenozoic (yellow) rocks in the map area. What structures are indicated by these map patterns? What is the age of the deformation that created these features?

Rocks from triassic to
cretacious periods
younger than
a older Rocks
but older than
pinkish grey rocks.
faults @ high angles.

3. What orogenic phase of the Cordilleran orogeny is responsible for the structures noted in questions 1 and 2 above? Study the outcrop patterns of the White River Group and cite evidence for the timing of this orogenic phase.

Laramide Phase.
took place in wyoming
developed anticlines
domes and monocliner
uplifts filled w/
early cenozoic
sediment.

4. If you were exploring for oil in Jurassic and Cretaceous strata on the northwest side of the Bighorn Mountains and you sited a drill rig at the position shown on both the map and cross section in figure 16.5 (red dot and red polygon, respectively), how many feet of Precambrian crystalline rock would you drill through before encountering Cretaceous sedimentary rock? To which Precambrian province do these crystalline rocks belong and what is their age (see figures 14.2 and 14.3)?

3000 feet below
sea level.
Yavapai provence.
1.74 - 1.68 Ga

Geological Map of Bighorn Mountains, Wyoming

Quaternary		**Q**
Qa	Gray	Alluvium and windblown sand
QPv	Magenta	Younger volcanic rocks
Tertiary (Miocene)		**M**
Mbb	Pink	Browns Peak and Bishop Formations
Moa	Yellow and red stipple	Ogallala and Arikaree Formations
Tertiary (Oligocene)		**Φ**
Φw	Yellowish orange	White River Group
Tertiary (Eocene)		**E**
Eb	Yellow vertical	Bridger Formation
Egr	Yellow diagonal	Green River Formation
Ti	Red	Intrusive rocks
Ews	Yellow	Wasatch Formation
Efu	Yellow diagonal	Fort Union Formation
El	Orange	Lance Formation
Cretaceous		**K**
Km	Yellow and green horizontal	Montana Group
Kmv	Green	Mesaverde Group
Kc	Yellow and green diagonal	Colorado Group
Kdl	Dark green	Dakota Formation and Lower Cretaceous rocks
Cretaceous to Jurassic		**KJ**
KJ	Dark green	Dakota to Morrison Formations
Jurassic		**J**
J	Yellow and green diagonal	Jurassic rocks
Jurassic to Triassic		**JT**
JTr	Green	Jurassic and Triassic rocks
Triassic		**T**
Tr	Dark green	Triassic rocks

Geological Map of Bighorn Mountains, Wyoming *(continued)*

Permian			**C**
Cm	Gray	Lower Permian rocks	
Permian and Pennsylvanian			**C**
Cpp	Yellow and red stipple, green horizontal	Permian and Pennsylvanian rocks	
Pennsylvanian			**C**
Cp	Yellow and red stipple, green diagonal	Pennsylvanian rocks	
Lower Paleozoic			
DC	Purple	Devonian to Cambrian rocks	
€O	Light red	Ordovician to Cambrian rocks	
Precambrian			**A**
Ai	Red	Intrusive rocks	
As	Olive	Metamorphic rocks	
Arg	Red	Granite and other intrusive rocks	
Ar	Brown	Metamorphic rocks	

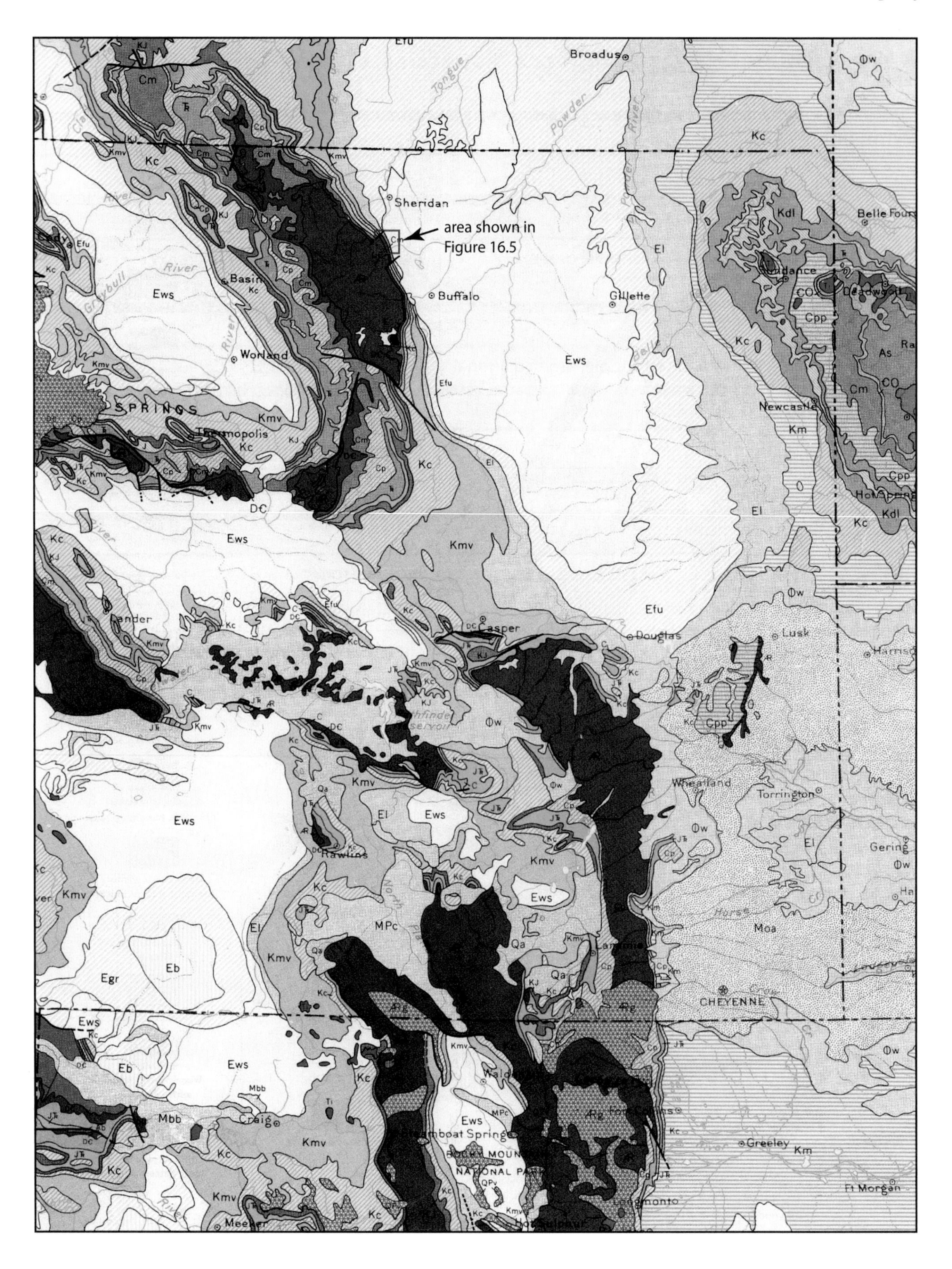
Broadus
Sheridan
area shown in
Figure 16.5
Buffalo
Gillette
Basin
Worland
Thermopolis
SPRINGS
Newcastle
Sundance
Belle Fourche
Lander
Casper
Douglas
Lusk
Wheatland
Torrington
Rawlins
Laramie
CHEYENNE
Craig
Steamboat Springs
ROCKY MOUNTAIN
NATIONAL PARK
Greeley
Ft Morgan
Longmont
Meeker
Tongue
Powder
River
Greybull
North Platte
Horse Cr
Ews
Efu
El
Kc
Kmv
Km
Kdl
Cm
Cp
Cpp
Ow
Moa
Eb
Egr
Mbb
MPc
Qa

Part D

Black Hills and the High Plains, South Dakota State Geologic Map

Scale: 1 inch = 8 miles

1. See the geological map provided on page 260 (figure 16.9). What are the ages of the rocks shown on the map?

2. What type of geologic structure is represented by the belted elliptical outcrops?

3. What kind of rocks are exposed in the core of the Black Hills, along the flank, and in the plains to the east?

4. What is the age of the uplift that produced the Black Hills? Cite your evidence.

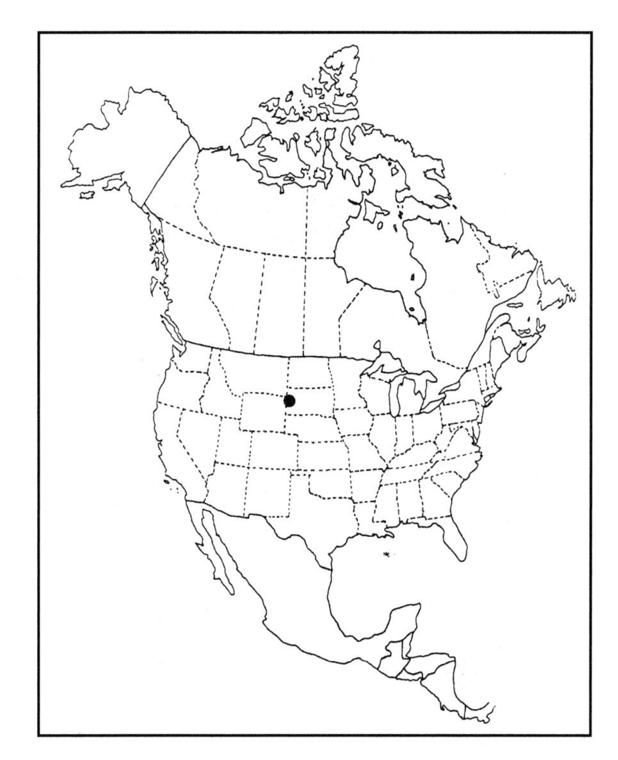

5. Locate and describe a nonconformity, disconformity, and two angular unconformities on this map (see figure 13.2). Indicate the position of the unconformities by listing the strata located above and below them.

6. Where is the most likely area in the Black Hills for commercial mineral production? Why?

Black Hills and the High Plains, South Dakota

Tertiary			**T**
Tw	Light yellow	White River Group	
Cretaceous			**K**
Kp	Light green, horizontal	Pierre Shale	
Kn	Light green, vertical	Niobrara Formation	
Kc	Light green, solid	Carlile Shale	
Kg	Light green, solid	Greenhorn Limestone	
Kbm	Light green, horizontal	Belle Fourche and Mowry Shales	
Ksi	Green, fine vertical	Skull Creek Shale, Inyan Kara Group	
Jurassic			**J**
Jm	Grayish green	Morrison Formation	
Js	Grayish green, horizontal	Sundance Formation	
Triassic			**₮**
₮s	Blue green, diagonal	Spearfish Formation	
Permian			**P**
Pm	Blue, solid	Minnekahta and Opeche Formations	
Pennsylvanian			**C**
Cm	Light blue, fine horizontal	Minnelusa Sandstone	
Mississippian			**C**
Cp	Light blue, diagonal	Pahasapa Limestone and Englewood Limestone	
Ordovician			**O**
Ow	Purple	Whitewood Limestone	
Cambrian			**Є**
Єd	Orange brown	Deadwood Formation	

Black Hills and the High Plains, South Dakota *(continued)*

Precambrian		p€
p€g	Dark brown	Granite and pegmatite
p€a	White, brown horizontal	Basic igneous intrusions
p€sq	Light brown, horizontal, white irregular dashes	Schist and quartz
p€sc	Light brown, diagonal	Sandstone and conglomerate
Igneous Rocks		
Qr	Bright green	Rhyolite and obsidian
Tr	Red and white	Rhyolite
Tp	Red	Light-colored intrusive igneous rocks

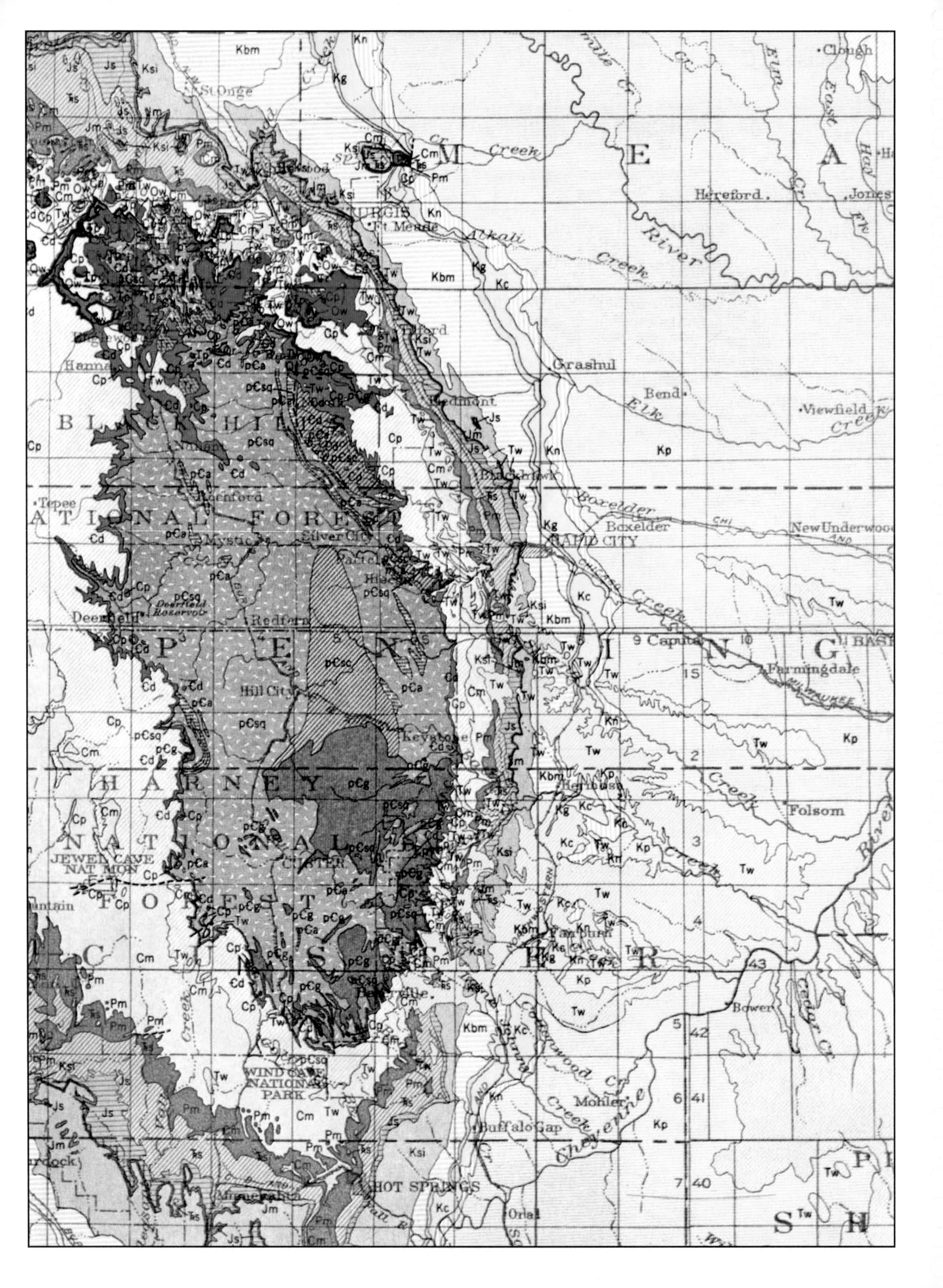

Clough
Hereford
Jones
River
Creek
STURGIS
Ft. Meade
Alkali
Grashul
Bend
Viewfield
Elk
Piedmont
Tilford
Blackhawk
Tepee
Rochford
Mystic
Silver City
Boxelder
RAPID CITY
NewUnderwood
Redfern
Deerfield Reservoir
Hill City
Caputa
Farmingdale
Keystone
Hermosa
Folsom
JEWEL CAVE NAT MON
CUSTER
Fairburn
Bower
WIND CAVE NATIONAL PARK
Mohler
Buffalo Gap
Cheyenne
HOT SPRINGS
Oral
Minnekahta
Cottonwood Cr.
Cedar Cr.
BLACK HILLS
NATIONAL FOREST
HARNEY
NATIONAL FOREST
PENNINGTON
CUSTER
MEADE

Part E
Cordilleran Foreland Basin

The stratigraphic profile shown in figure 16.10 is an east–west series of stratigraphic sections through the Upper Cretaceous rocks of eastern Utah (Mancos Shale and Mesa Verde Group of figure 16.3 and figure 16.4A). The profile shows relationships of the clastic wedge associated with a pulse of the Sevier orogeny in western North America. The marine black shales reflect the presence of the Western Interior Seaway.

1. Construct a restored cross section of these Cretaceous strata.

2. What was the direction of sediment transport?

3. What depositional environment is reflected by the coal deposits?

4. Is the depositional pattern transgressive or regressive? Cite your evidence.

5. When was the most evident pulse of uplift in the source area during this part of the Sevier orogeny? What is your evidence?

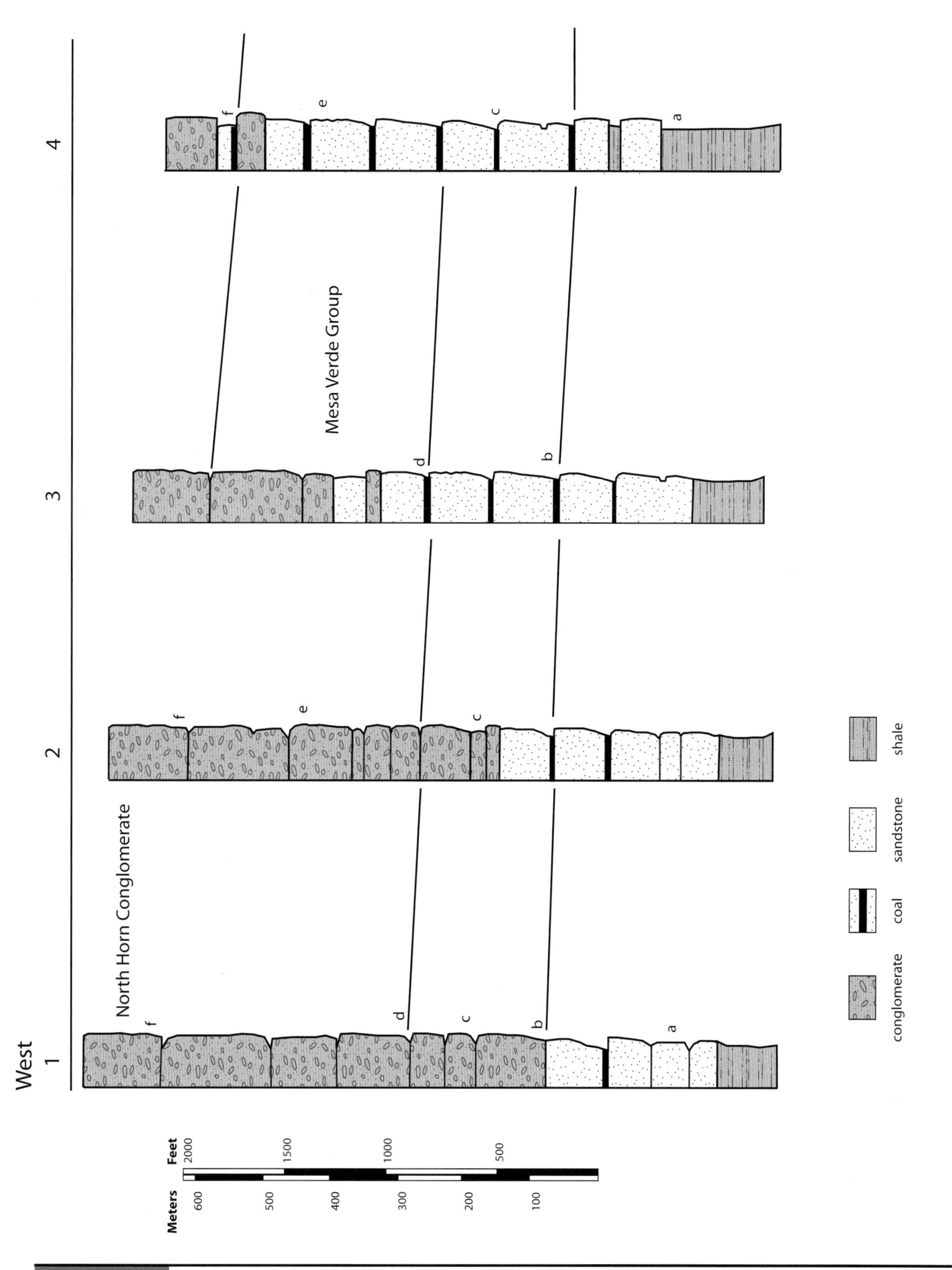

FIGURE 16.10 A series of nine stratigraphic sections through the Upper Cretaceous rocks of eastern Utah. This represents a cross section through a coal-bearing, clastic wedge. Coal beds frequently pinch out over areas of this size.

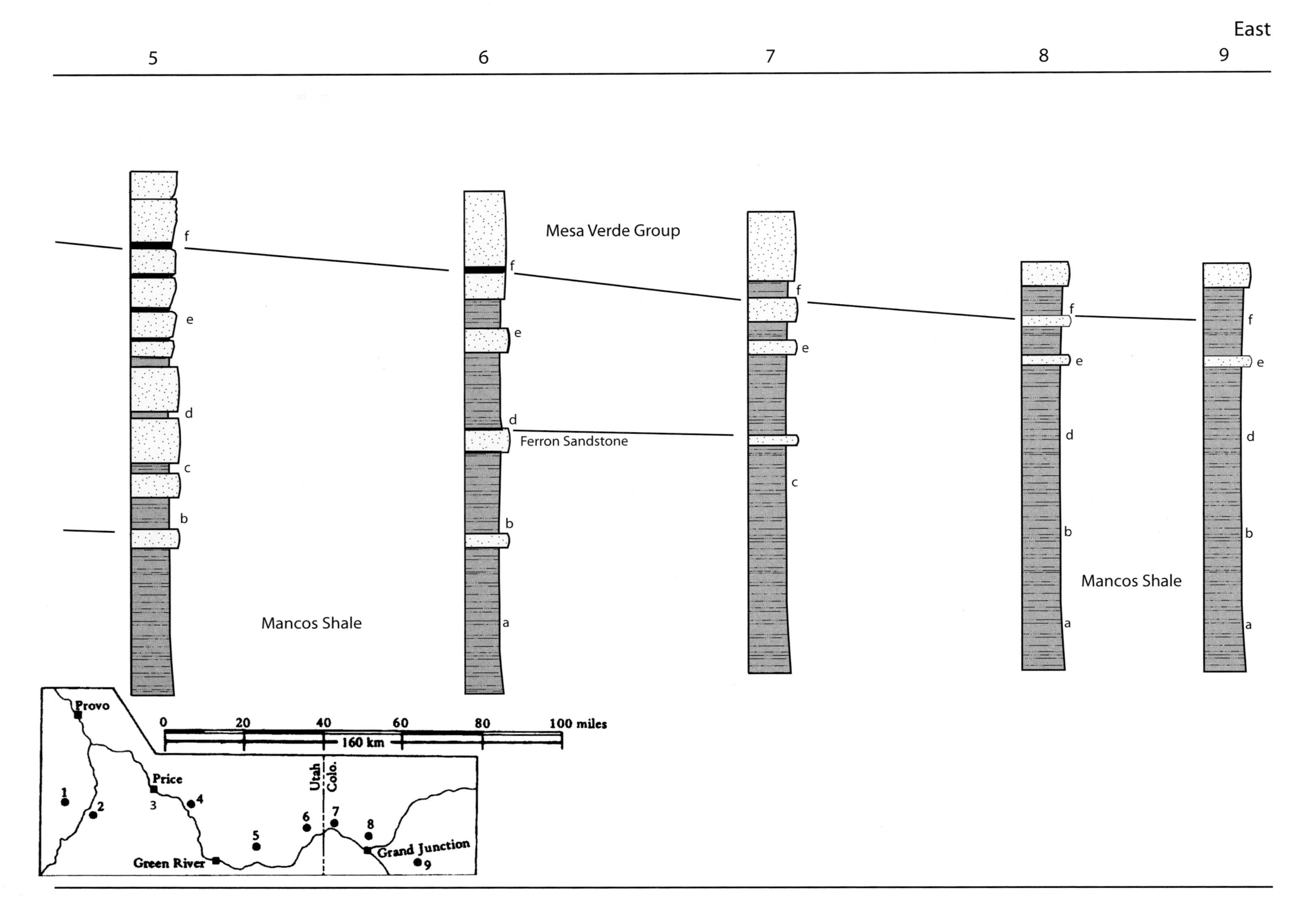
East
5
6
7
8
9
Mesa Verde Group
Ferron Sandstone
Mancos Shale
Mancos Shale
a
b
c
d
e
f
0
20
40
60
80
100 miles
160 km
Provo
Price
Green River
Grand Junction
Utah
Colo.
1
2
3
4
5
6
7
8
9

Part F

If one can document the kind of sediments deposited at any one "brief" interval in time over a wide area, one can construct **lithofacies** maps to show patterns of sedimentation. Figure 16.11 is a map of Upper Cretaceous deposits in the Rocky Mountains. The small circles are points where data are available, and lithology at each is shown by the following series of abbreviations: congl. for conglomerate, ss for sandstone, sh for shale, and m for marl or chalk. The line of zero sediment thickness (dashed line) in the western part of the map represents the position of the Cretaceous shoreline.

1. Construct a lithofacies map by drawing lines separating the various rock types that can be recognized.

p 268

2. Do the facies belts parallel the zero thickness line?

yes, they don't overlap, no perpendicular lines

3. What was the direction of transport of the sediments? What was the probable source area for the coarse sediments along the western border of the map?

deposited from west to East, more course rocks smaller sediments probably traveled further & deposited eastward.

4. Were the Colorado Rocky Mountains present during deposition of these sedimentary rocks? What is the evidence?

gap of sediment in middle of colorado, they were present during this time.

5. Where would you expect the greatest thickness of sediments to have accumulated? Why?

most thick where theres most deposition of shale. eastern side of utah, western side of CO, middle of NM middle of WY

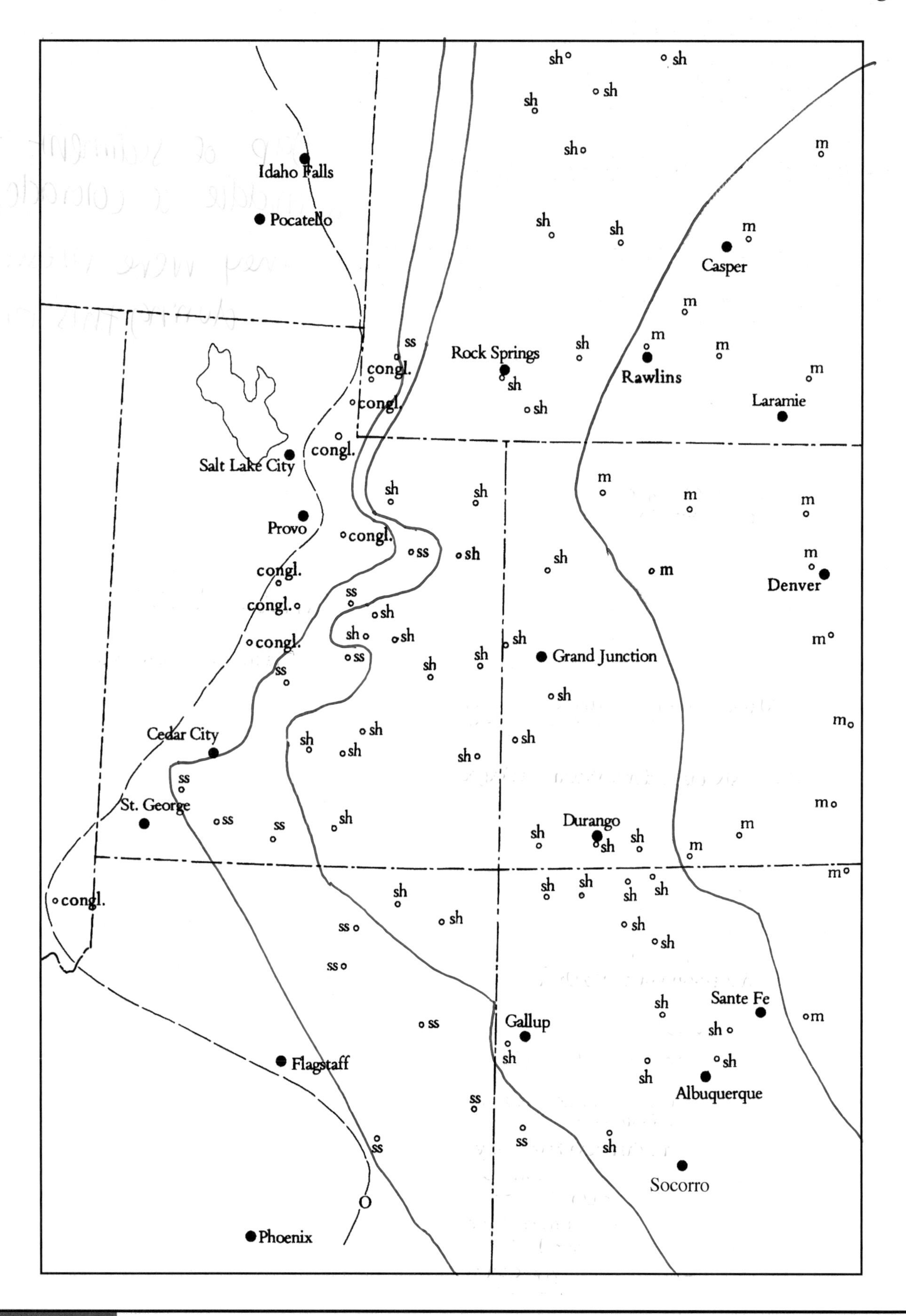

FIGURE 16.11 A map showing distribution of rock types in part of the Upper Cretaceous clastic wedge in the Rocky Mountains. This is the area between the stable region of the continent (east side of map) and the western mobile belt. Abbreviations: congl. = conglomerate, ss = sandstone, sh = shale, m = marl or chalk.

Phanerozoic Geology of North America

A Summary of Major Depositional and Tectonic Events

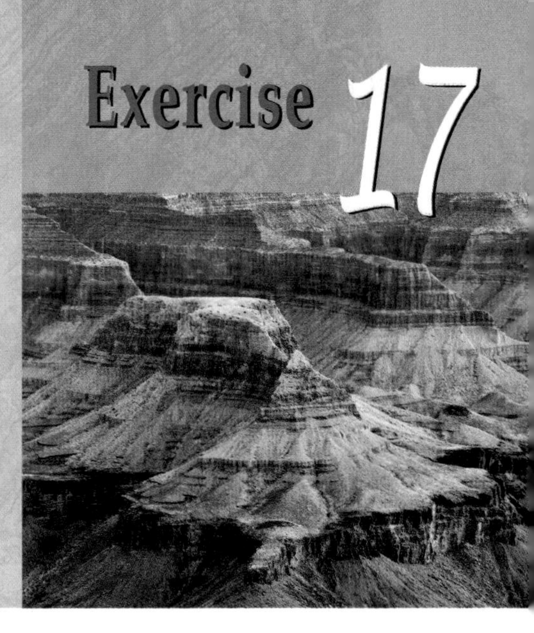

Figure 17.1 summarizes major Phanerozoic depositional and tectonic events that shaped modern North America. The left column is the geological timescale. The next column portrays mountain-building events that occurred on the western or Cordillern margin of North America. The right-hand column shows orogenies that affected deposition and the structural fabric of the East Coast. The center column shows the relative durations of the six **depositional sequences** elucidated by the stratigrapher Larry Sloss. Sloss was the first to discover that the sedimentary cover of the North American craton could be subdivided into six main depositional packages separated by regional or widespread unconformities. He named these transgressive-regressive sequences (Sauk, Tippecanoe, etc.) in honor of Native American groups indigenous to North America. In addition to the Slossian sequences and six orogenic events, figure 17.1 lists 15 other depositional, structural, and paleontological/evolutionary events that played a role in the geologic development of North America.

PROCEDURE

Carefully examine figure 17.1, noting the location of mountain belts, depositional sequences, unconformities, reefs, etc., and then use this figure, your textbook, course notes, or websites that host legitimate science to answer the following questions. Be as concise, yet complete, as possible. Note that the vertical scale represents time, not thickness of sequences or other rock units. We strongly recommend that you view Ron Blakey's excellent paleogeographic maps of North America as you complete this activity at http://cpgeosystems.com/index.html. Examine the suite of maps under the heading entitled North American Paleogeography in the Paleogeography Library. On the map titles, Ma indicates millions of years before present.

1. List the six depositional sequences from oldest to youngest.

2. Why are the sequence-bounding unconformities of greater temporal duration (white areas on figure 17.1) in the center of the craton than on the margins of the craton?

3. How do the Sauk and Tippecanoe sequences differ lithologically? In which of these sequences might you find small archeocyathid reefs?

4. Why were thick accumulations of evaporites deposited in the Michigan Basin during the Silurian Period (feature 2 on figure 17.1)? Examine Blakey's map of Early Silurian (430 Ma) North America. What types of reefs were formed around the margins of the Michigan Basin?

5. During which period of Earth history was black shale (such as the Woodford, Bakken, and Chattanooga Shales) deposited across much of the craton? What was the cause of this widespread "black shale" event?

6. One of the six major mass extinctions occurred at the end of the Devonian Period. How did this change the faunal and floral composition of reefs deposited in the Devonian part of the Kaskaskia Sequence (feature 4 on figure 17.1) versus those deposited in the Mississippian part of the Kaskaskia Sequence (feature 6 on figure 17.1)? Be sure to discuss the nature of Walsourtian mounds as part of your answer.

7. During which geological era was mountain building most active on the east coast of ancestral North America? Circle your answer.

   Paleozoic Mesozoic Cenozoic

8. What is the plate tectonic explanation for the Taconic orogeny? Which plates were interacting and what type of plate interaction (continent–continent, continent–ocean, ocean–ocean) produced this orogeny?

9. What is the plate tectonic explanation for the Acadian orogeny? Which plates were interacting and what type of plate interaction (continent–continent, continent–ocean, ocean–ocean) produced this orogeny?

10. What is the Catskill Delta (feature 3 on figure 17.1) and what was the source of sediments comprising this feature?

11. What is the plate tectonic explanation for the Late Paleozoic Allegheny orogeny? Which plates were interacting and what type of plate interaction (continent–continent, continent–ocean, ocean–ocean) produced this orogeny?

12. **Cyclothems** characterize Middle Pennsylvanian to Early Permian sedimentary deposits worldwide. What is a cyclothem and why are they so characteristic of the Late Paleozoic portion of the Absaroka Sequence? What was the global climate like at this time?

13. What are phylloid algal mounds and during which two periods of Earth history did they form (feature 7 on figure 17.1)? What is the economic significance of phylloid algal bioherms in the Paradox Basin of southeastern Utah and southwestern Colorado?

14. What were the Ancestral Rocky Mountains? When, where, and why did they form? If possible, examine Ron Blakey's map of Late Pennsylvanian (300 Ma) North America as you consider your answer.

15. During which geological era was mountain building most active on the west coast of North America? Explain this pattern in terms of what was happening to the supercontinent of Pangea during this time (see the sequence of Jurassic and Cretaceous maps of North America on Ron Blakey's website).

16. The western portion of the Zuni Sequence (feature 13 on figure 17.1) was deposited in the foreland basin of the Cordilleran orogeny (see Blakey's map of Late Cretaceous North America [85 Ma]). To what do the terms flysch and molasse apply? What economic resource is associated with the shoreline and delta deposits (e.g., Cretaceous Ferron Sandstone and Blackhawk Formation) on the west margin of the Western Interior Seaway?

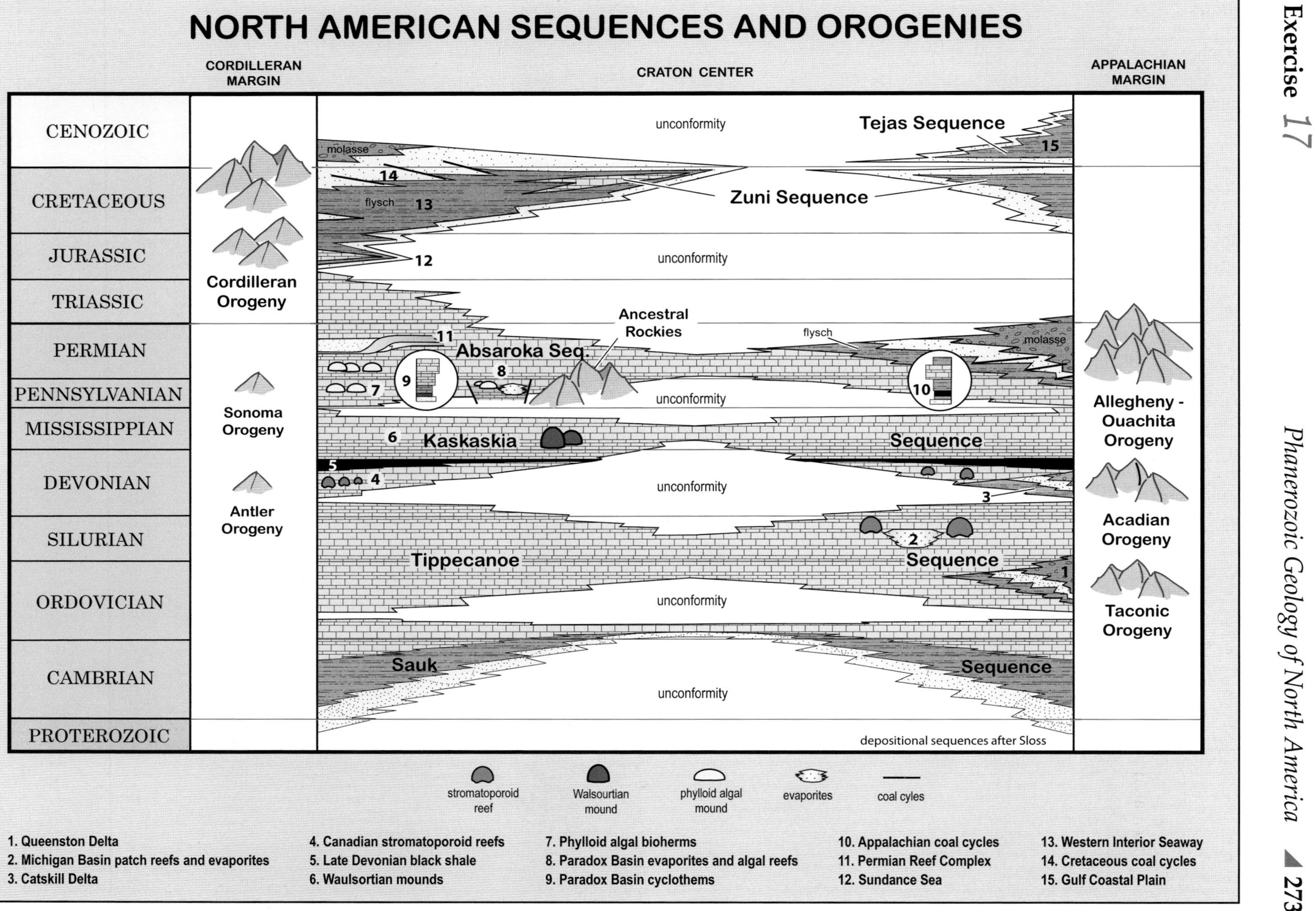

FIGURE 17.1 Summary of key tectonic and depositional events in the Phanerozoic history of North America.

Cenozoic Geology

Exercise 18

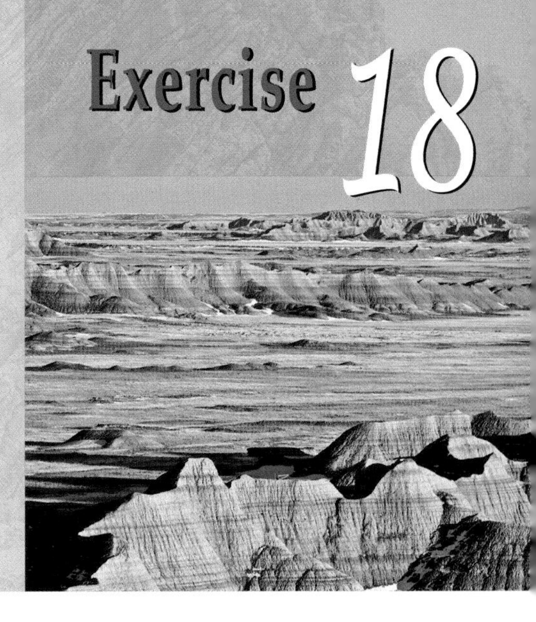

Learning Objectives

After completing this exercise, you will:

1. be able to characterize the landforms and geology of the physiographic provinces of the western United States;
2. be familiar with Laramide structures that characterize Wyoming and adjacent states; and
3. have acquainted yourself with the nature of the extensional tectonics that characterize the eastern margin of the Basin and Range Provinces.

Much of the physiographic character of North America came into being during the Cenozoic Era. The mountain ranges, valleys, river systems, volcanoes, and deserts that are familiar parts of the human experience came into their current configurations during the past 65 million years. Many of these features owe their existence to even more recent events. The purpose of this exercise is to acquaint you with the patterns and origins of the physiographic provinces of western North America. The geological map activities presented here permit you to further see the nature of rocks and structures that developed during Cenozoic time.

PROCEDURE

Part A

Geologists and geographers have subdivided North America into formal physiographic provinces on the basis of bedrock geology, surficial deposits, and landforms. Familiarize yourself with the geographic extent and characteristic landforms of the major western US provinces by filling in the blank boxes on table 18.1. The requested information can be obtained from Wikipedia (http://en.wikipedia.org/wiki/United_States-physiographic_region). You may also wish to visit these areas via Google Earth and explore the landforms and features that occur in each province.

Province	Location	Characteristics
High Plains (Interior Plains)		
Rocky Mountains		
Colorado Plateau		
Basin and Range		
Columbia Plateau		
Sierra Nevada		
Pacific		

TABLE 18.1 Major western provinces of the United States.

PART B

Geological Map of the Central Wasatch Front

The Wasatch Fault separates the Central Rocky Mountain province in the east from the Basin and Range physiographic province to the west. The area represented on the map on page 279 (figure 18.1) straddles this boundary. The developed area on the west half of the map is the eastern portion of Provo, Utah. The eastern half of the map depicts strata and structures characteristic of the central Wasatch Range. Carefully examine the map and answer the following questions.

1. Note the fan-shaped accumulations of sediment (Qal) at the mouths of Rock Canyon and Slate Canyon. What process formed these alluvial fans?

2. If this process continues to be active in the future, what geological hazard might residents of the area anticipate?

3. What kind of fault is the Wasatch Fault? What type of tectonic stresses (compression, tension, shear) produced this faulting as well as faults located elsewhere in the Basin and Range?

4. What type of geological hazard might residents of central Utah anticipate along this active fault?

Central Wasatch Front

Quaternary		Q
Qal	Yellow	Stream valley alluvium Rounded, stream-transported debris
Qrf	Yellow	Recent alluvial fan deposits
Qco	Yellow	Colluvium Angular debris on hill slopes, includes talus
Qls	Yellow	Landslide, avalanche, slump, and mudflow deposits
Qlb	Orange	Lake Bonneville sediments, undifferentiated
Tertiary?		**Tq**
Tqof	Orange	Pre-Bonneville fan and landslide deposits
Mississippian		**M**
Mmc	Green	Manning Canyon Shale
Mgb	Blue	Great Blue Limestone
Mh	Purple	Humbug Formation
Md	Green	Deseret Limestone
Mg	Light blue	Gardison Limestone
Devonian and Mississippian		**DM**
DMf	Brown	Fitchville Dolomite
Cambrian		**Є**
Єm	Red	Maxfield Limestone
Єo	Green	Ophir Formation
Єl	Red	Tintic Quartzite
Precambrian		**pЄ**
pЄmf	Light brown	Mineral Fork Tillite
pЄbc	Dark brown	Big Cottonwood Formation

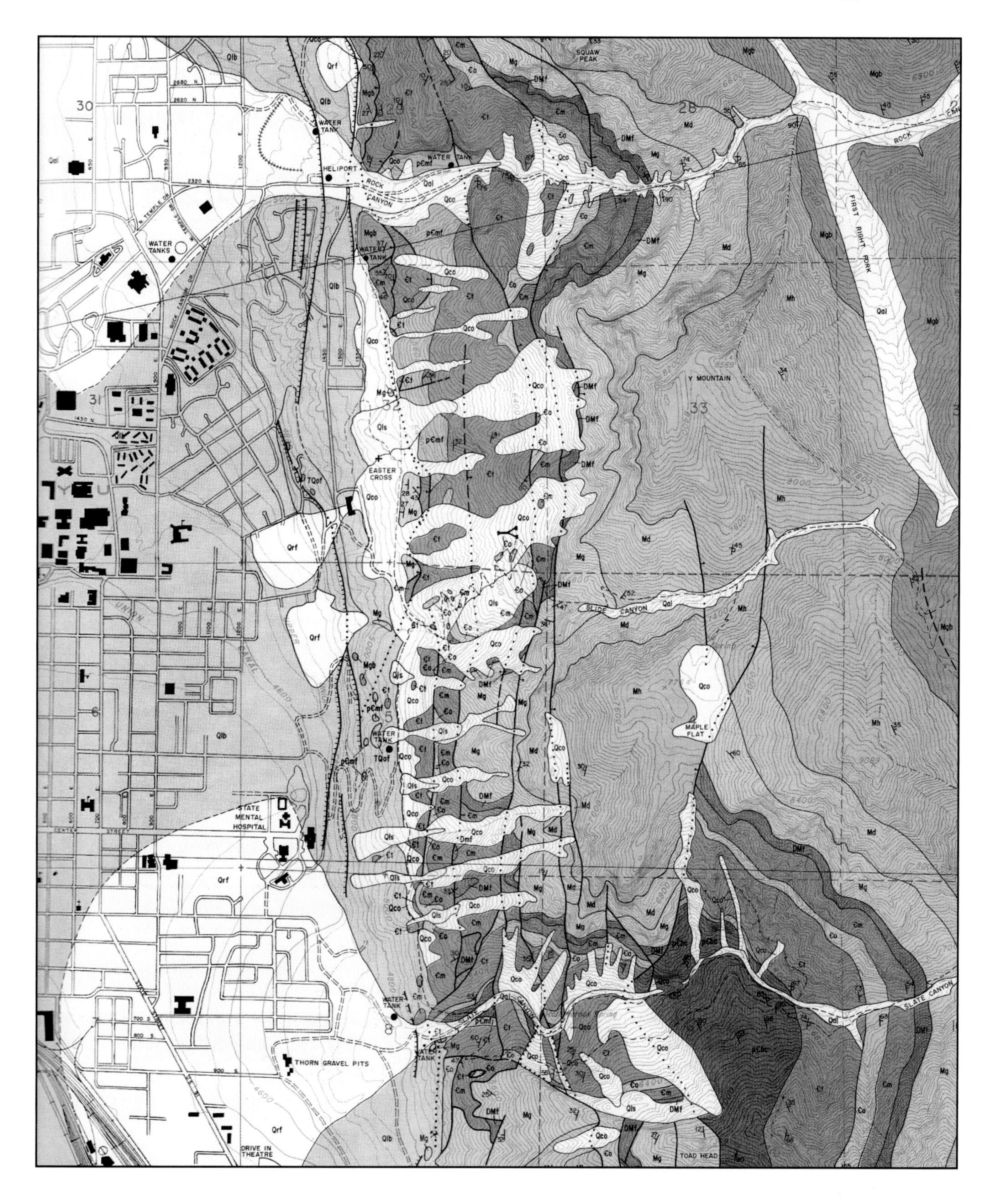

FIGURE 18.1 Geological map of the central Wasatch Range in the vicinity of Provo, Utah.

Exercise 19

Pleistocene Glaciation

Learning Objectives

After completing this exercise, you will:

1. have gained confidence in your ability to read and interpret geological maps;
2. better understand the effects of continental glaciation on the Great Lakes region;
3. know the name and regional extent of each of the four main Pleistocene depositional events; and
4. better understand the effects of valley glaciation on a mountainous region.

PROCEDURE

Part A

Glacial Map of North America

Scale: 1 inch = 72 miles

1. See the geological map provided on page 282 (figure 19.1). On the basis of the principle of superposition, what are the relative ages of the four main Pleistocene till sheets?

2. What was the direction of ice movement in the vicinity of Chicago, Illinois; Fort Wayne, Indiana; and Mankato, Minnesota?

3. What were the likely controlling factors that determined the present position of the Missouri and Ohio Rivers?

4. Explain the lack of glacial deposits in the southwestern corner of Wisconsin.

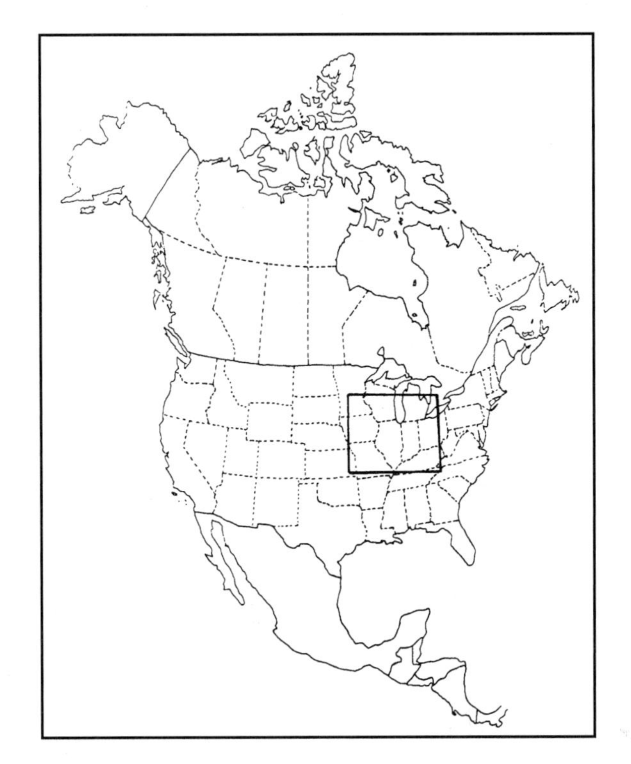

5. Pleistocene lake deposits are illustrated on the map by a horizontal pattern. What is the origin of the Pleistocene lake basin approximately 50 miles south of Chicago?

6. What type of isostatic forces are likely to be affecting this area at the present time? Explain your answer.

Pleistocene		
Pleistocene Lakes	Blue horizontal	
Wisconsin Glaciation	Dark pink	Moraine
	Light pink	Intermoraine
Illinoian Glaciation	Dark green	Moraine
	Light green	Intermoraine
Kansan Glaciation	Light yellow	Glacial deposits
Nebraskan Glaciation	Yellow, vertical	Glacial deposits

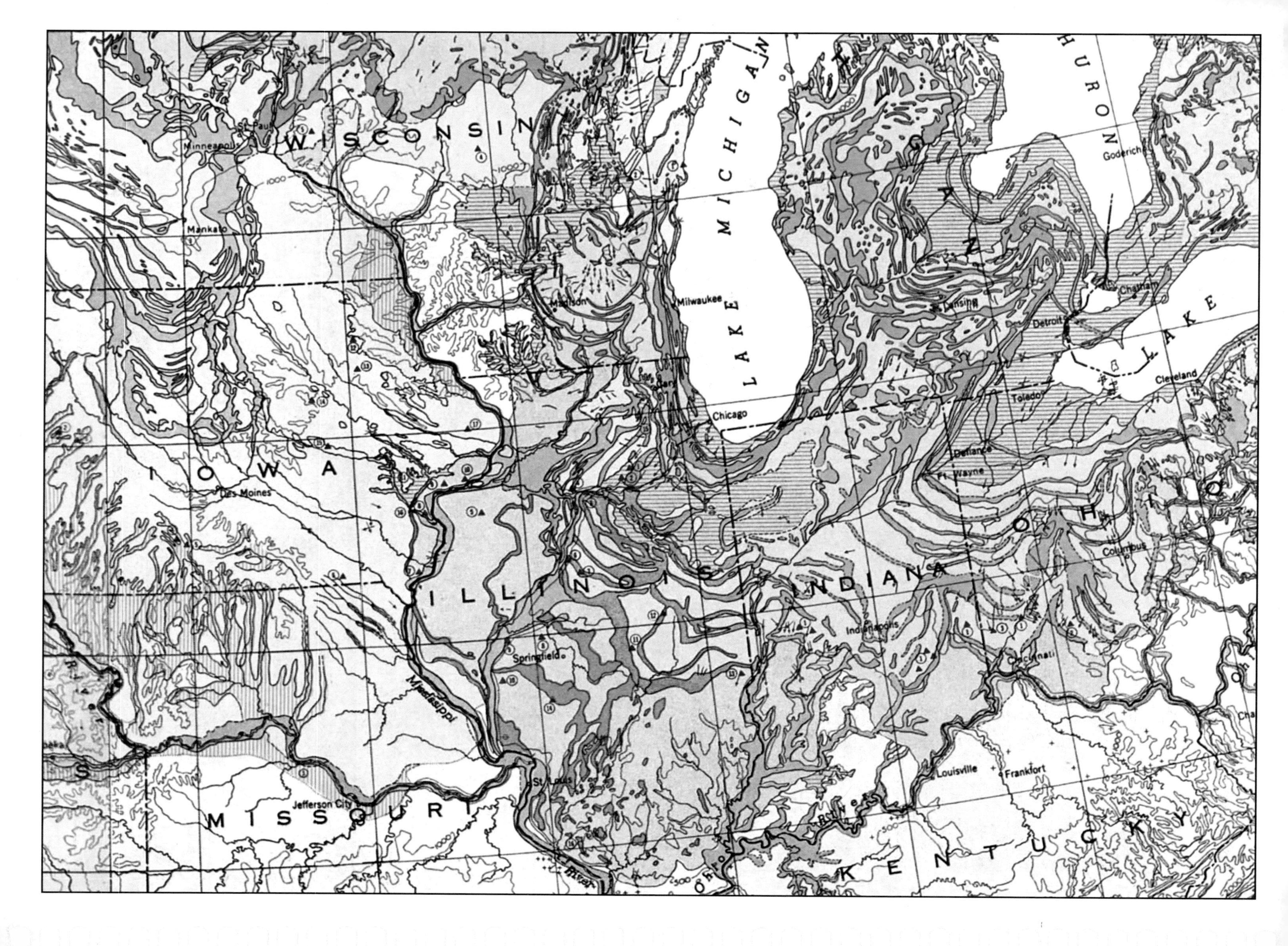

WISCONSIN
IOWA
ILLINOIS
INDIANA
OHIO
KENTUCKY
MISSOURI
LAKE MICHIGAN
HURON
LAKE ERIE
Minneapolis
St. Paul
Mankato
Des Moines
Madison
Milwaukee
Chicago
Springfield
St. Louis
Jefferson City
Mississippi
Indianapolis
Louisville
Frankfort
Cincinnati
Columbus
Ft. Wayne
Defiance
Toledo
Detroit
Lansing
Chatham
Goderich
Cleveland

PART B

Physiographic Effects of Valley Glaciers, Yosemite Valley, California

The stunning beauty of Yosemite Valley in California reflects the power of valley glaciers to shape the landscape. In this part of the exercise you will compare and contrast physiographic features in glaciated (figures 19.2 and 19.3) and nonglaciated (figure 19.4) portions of the Merced River valley. Both map areas are underlain by Jurassic granite, so differences in landforms are a function of erosional processes, not differences in bedrock geology.

1. Construct topographic profiles along lines A to A′ and B to B′ in the spaces provided on figures 19.3 and 19.4. The technique for constructing topographic profiles is explained in exercise 13 (figure 13.7).

2. Compare and contrast the shape of the Merced River valley in the two topographic profiles.

3. What is the gradient (elevation loss per mile) of the Merced River in figure 19.3?

_________________ feet/mile

4. What is the gradient of the Merced River in the area depicted in figure 19.4?

_________________ feet/mile

5. Name a map feature in figure 19.3 that represents a hanging valley.

6. Based on the comparison of the glaciated and nonglaciated stretches of the Merced River valley, discuss how valley glaciers modify the valleys through which they flow.

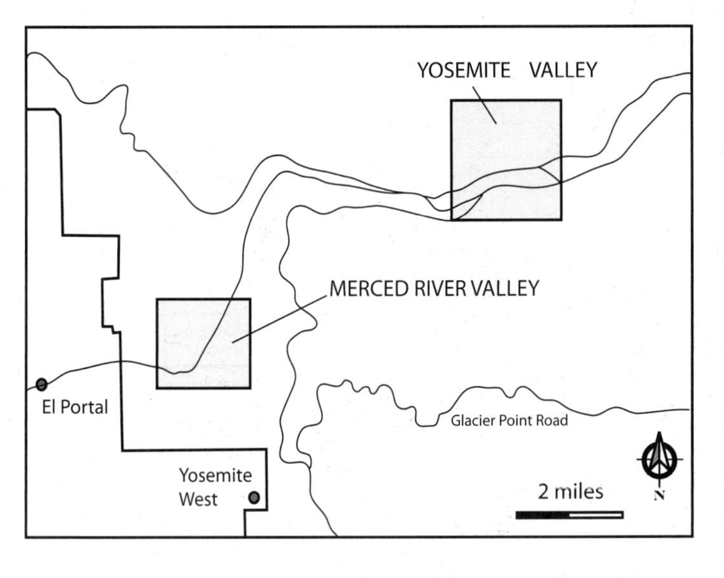

FIGURE 19.2

Index map showing location of figures 19.3 and 19.4 in Yosemite National Park, California.

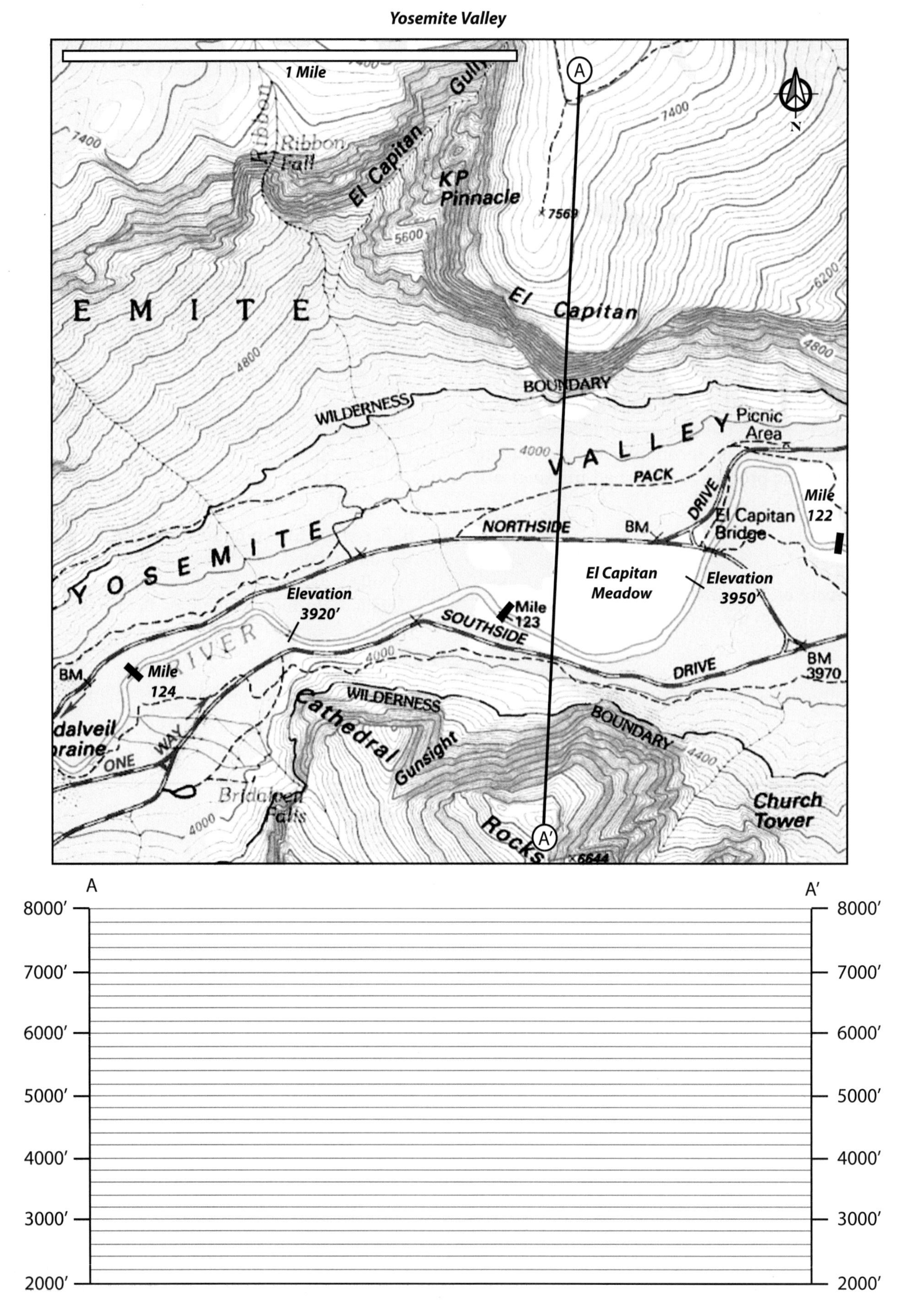

FIGURE 19.3 Topographic map of the western portion of Yosemite Valley, California. See figure 16.2.

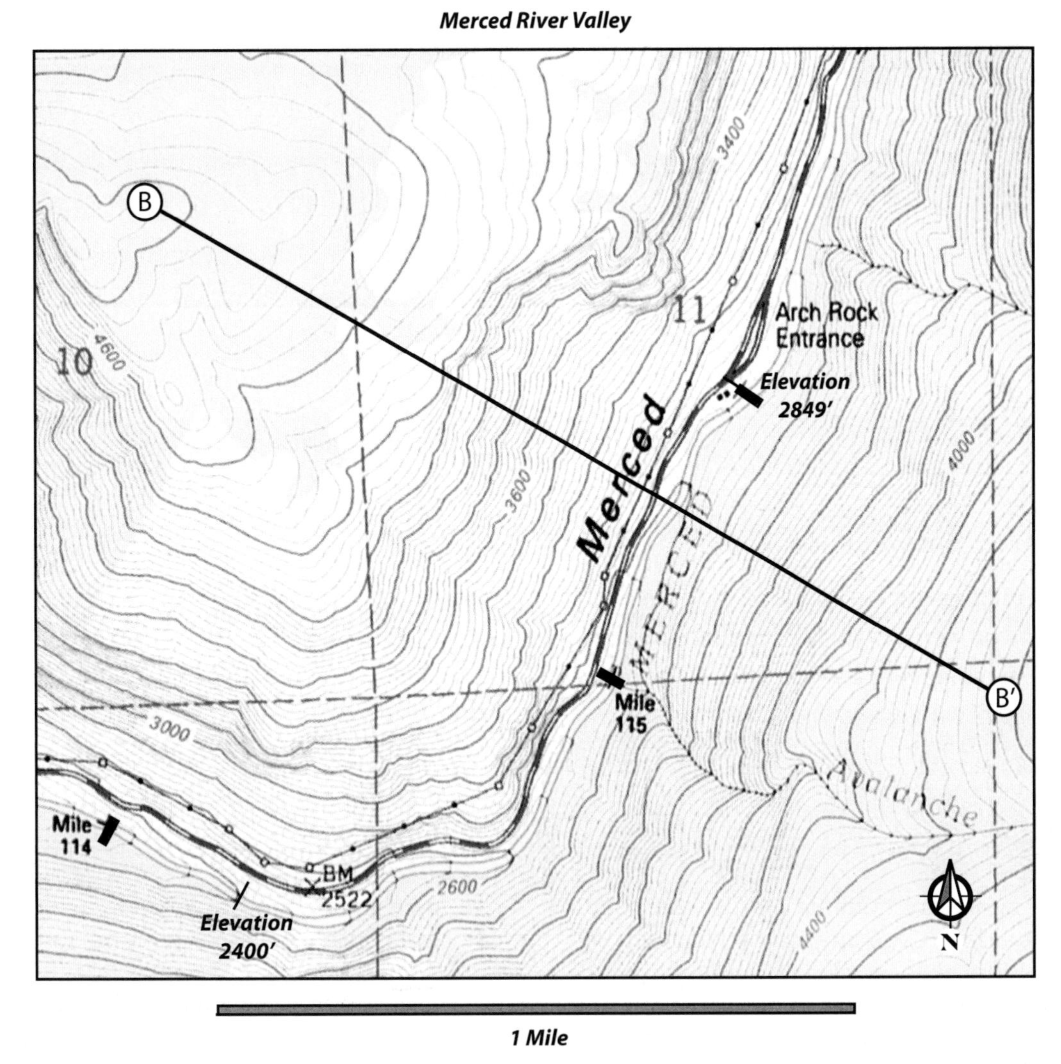

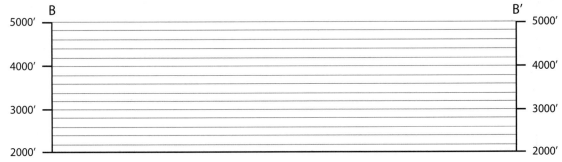

FIGURE 19.4 Topographic map of the Merced River valley in the westernmost portion of Yosemite National Park, California.

Exercise 20

Hominin Fossils

Learning Objectives

After completing this exercise, you will:

1. better understand the nature of the hominin fossil record;
2. be familiar with certain key hominin fossil discoveries;
3. know the difference between the genera *Homo* and *Australopithecus*; and
4. be familiar with diagnostic cranial features.

The first significant hominin fossils were found north of Düsseldorf, Germany, in the Neander Valley in 1856. From then until now numerous finds have expanded the fossil collections of the subfamily Homininae. These fossil remains are among the most valuable objects of antiquity in the possession of humankind. They are literally "national treasures" housed in approximately 25 museums around the world where they are given the utmost protection in safety and environmental controls. There are so few existing specimens that each one can have a significant effect on the theory of hominin development. Examples of single finds that have revolutionized our evolutionary theories include the discovery of the "Taung baby" in Botswana in 1924, *Homo habilis* in Tanzania in 1962, "Lucy" in Ethiopia in 1974, 3.6- to 3.8-million-year-old footprints ("first family") in Laetoli in Tanzania in 1976, the finding of the "black skull" in Kenya in 1985, and "Turkana boy" in Kenya in 1984. Each discovery represents either a profound bit of luck or, in most cases, the result of very expensive, painstaking, and time-consuming excavations whose purpose was to seek hominin fossils in known productive locations. As a collection, they represent the evolutionary record of humankind for approximately the past 5 million years.

The classification of the hominins is in a state of flux. Although there is almost no disagreement among scientists about the fact of hominin evolution, uncertainty exists over the specifics of evolutionary relationships. There is widespread belief that science teaches that humans were derived from the apes. This is not so. Modern hominins did not evolve from modern pongins; both are contemporaneous. They might, or might not, have shared a common ancestral stock. Whether they both evolved from a single common family, or each followed independent lines of evolution, is not clear. The hominin fossils presently available to scholars provide us with a reasonably good look at the evolutionary history of the subfamily. The evolutionary sequence we see from these specimens elucidates hominin origins at a time that was perhaps close to the probable time of divergence, if indeed hominins and pongins do share a common ancestor. The missing information at present is the dearth of fossil pongins over the past 5 million years. If we had access to comparable pongin fossils covering this interval, the evolutionary story of these two families would be much clearer, and the identity of a possible common ancestor would be more predictable.

Most taxonomic schemes recognize two genera of hominins: *Australopithecus* and *Homo*. Within *Australopithecus*, six species have been differentiated: *Australopithecus afarensis, Australopithecus africanus, Australopithecus anamensis, Australopithecus aethiopicus, Australopithecus boisei*, and *Australopithecus robustus*. Seven species are recognized in the genus *Homo*: *Homo rudolfensis, Homo habilis, Homo ergaster, Homo erectus, Homo heidelbergensis, Homo neanderthalensis*, and *Homo sapiens*. A third genus of the Homininae has been described and named: *Ardipithecus*, with a single species: *ramidus*. The material for this taxon was found in Ethiopia in 1994–1995. Table 20.1 presents a tabular summary of the above listed taxa. Figure 20.1 illustrates the general evolutionary relationships thought to exist among hominin species.

Name	Age (m.y.)	Sites	Special Features
Homo sapiens	0.1–0.0	Worldwide	Oldest cave paintings, c. 31,000 years, includes Cro-Magnon.
Homo neanderthalensis	0.3–0.03	Europe, Asia, Africa	Discovered 1856 in Neander Valley in northern Germany. Most productive site in Croatia, where 850 fossils have been found.
Homo heidelbergensis	0.6–0.2	Europe	Discovered in 1921, first fossil hominin found in Africa. First to build shelters.
Homo erectus	1.2–0.4	North and East Africa, Indonesia, China	First to use controlled fire. Includes "Peking Man" and "Java Man."
Homo ergaster	1.8–1.5	East Africa	"Turkana boy," oldest complete skeleton found, stood nearly 6 ft tall.
Homo habilis	1.9–1.6	East Africa	Smallest adult hominin ever found, height estimated to be approximately 1 m.
Homo rudolfensis	2.4–1.9	East Africa	Similar to *H. habilis*. The two lived contemporaneously in East Africa.
Homo sp.	2.5–?	East Africa	A single upper jaw (maxilla) collected from a site otherwise restricted to specimens of *Australopithecus*. Stone tools present.
Australopithecus robustus	1.9–1.0	South Africa	Originally called *Paranthropus robustus*, and many workers prefer the older term.
Australopithecus boisei	2.3–1.4	East Africa	Became famous as *Zinjanthropus*, the nutcracker man, a hyper-robust species.
Australopithecus aethiopicus	2.7–1.9	East Africa	"Black skull," robust australopithecine with a prominent sagittal crest and huge teeth.
Australopithecus africanus	2.8–2.4	South Africa	"Taung child" found in 1924, highly controversial for many years, a fossil gem.
Australopithecus afarensis	3.9–3.0	East Africa, Central Africa	"Lucy" and the Laetoli footprints. Includes *A. bahrelghazali* found in central Africa.
Australopithecus anamensis	4.2–3.9	East Africa	Material exhibits an unusual combination of *Homo* and *Australopithecus* characters.
Ardipithecus ramidus	4.4	East Africa	A third genus within the Homininae, thought to be ancestral to *Australopithecus*.

TABLE 20.1 Taxonomic scheme of hominins.

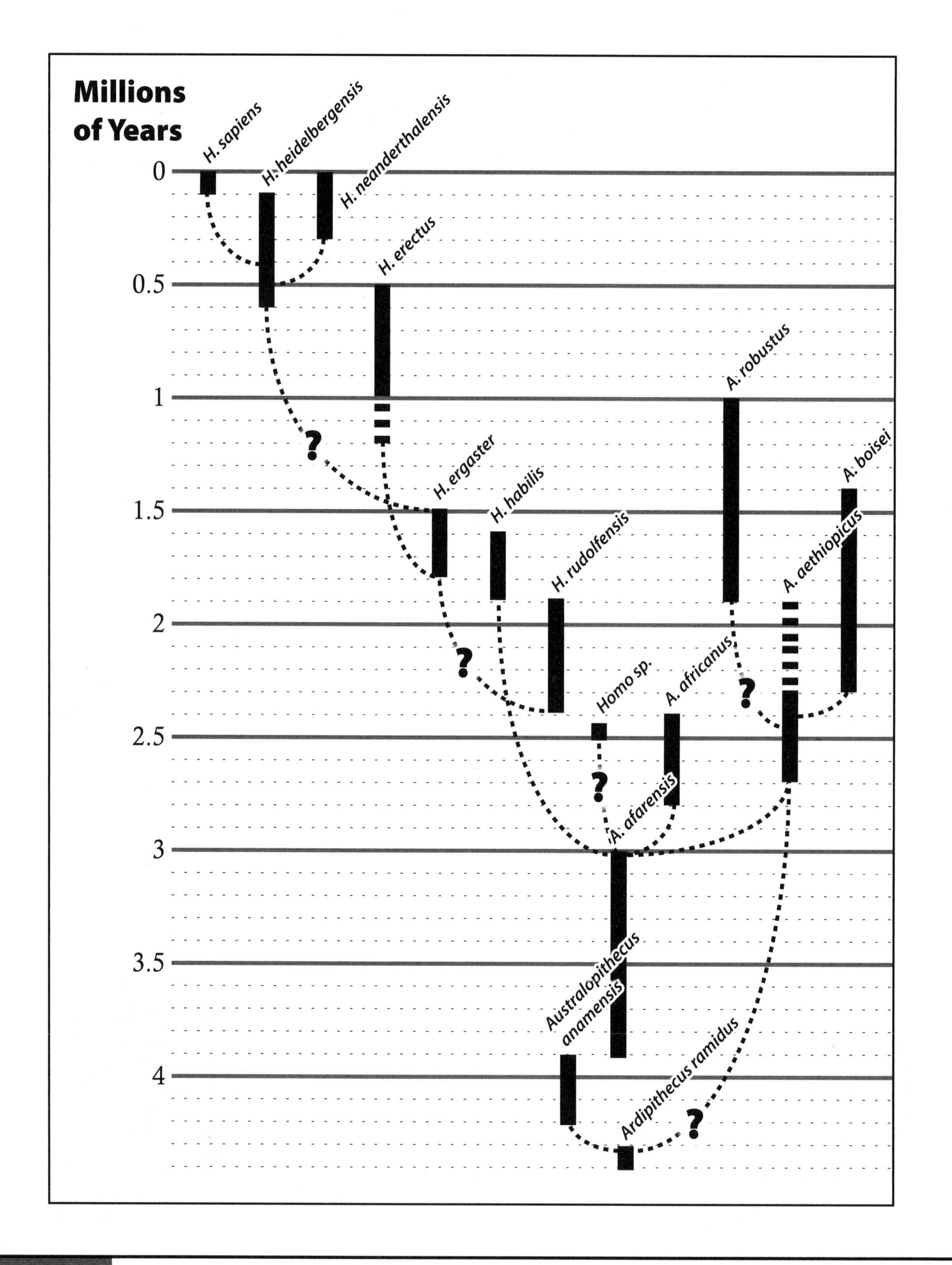

FIGURE 20.1 Diagram showing temporal and evolutionary relationships among fossil hominins. (Reprinted with permission from *Nature*, Bernard A. Wood, "The Oldest Hominid Yet," 371:280–281, 1994, Macmillan Publishers Ltd.)

PROCEDURE

1. Figure 20.2 illustrates a comparison of the main characteristics of skull morphology of a chimpanzee, representing the apes (subfamily Ponginae), and a modern human skull (subfamily Homininae). Study the important differences between these two skulls.

2. Examine the drawings of the six fossil skulls in figure 20.3, noting the similarities with the chimp and modern human skulls found in figure 20.2. The natural stratigraphic order of the fossil skulls has intentionally been altered for your later reconstruction.

3. Reconstruct what you think to be the natural stratigraphic sequence of the fossil skulls based on their evolutionary development. Label the individual skulls from 1 to 6 in order of their similarity to the chimp and human skulls (1 being most apelike and 6 being most humanlike).

4. Compare your evolutionary sequence with that of your instructor's.

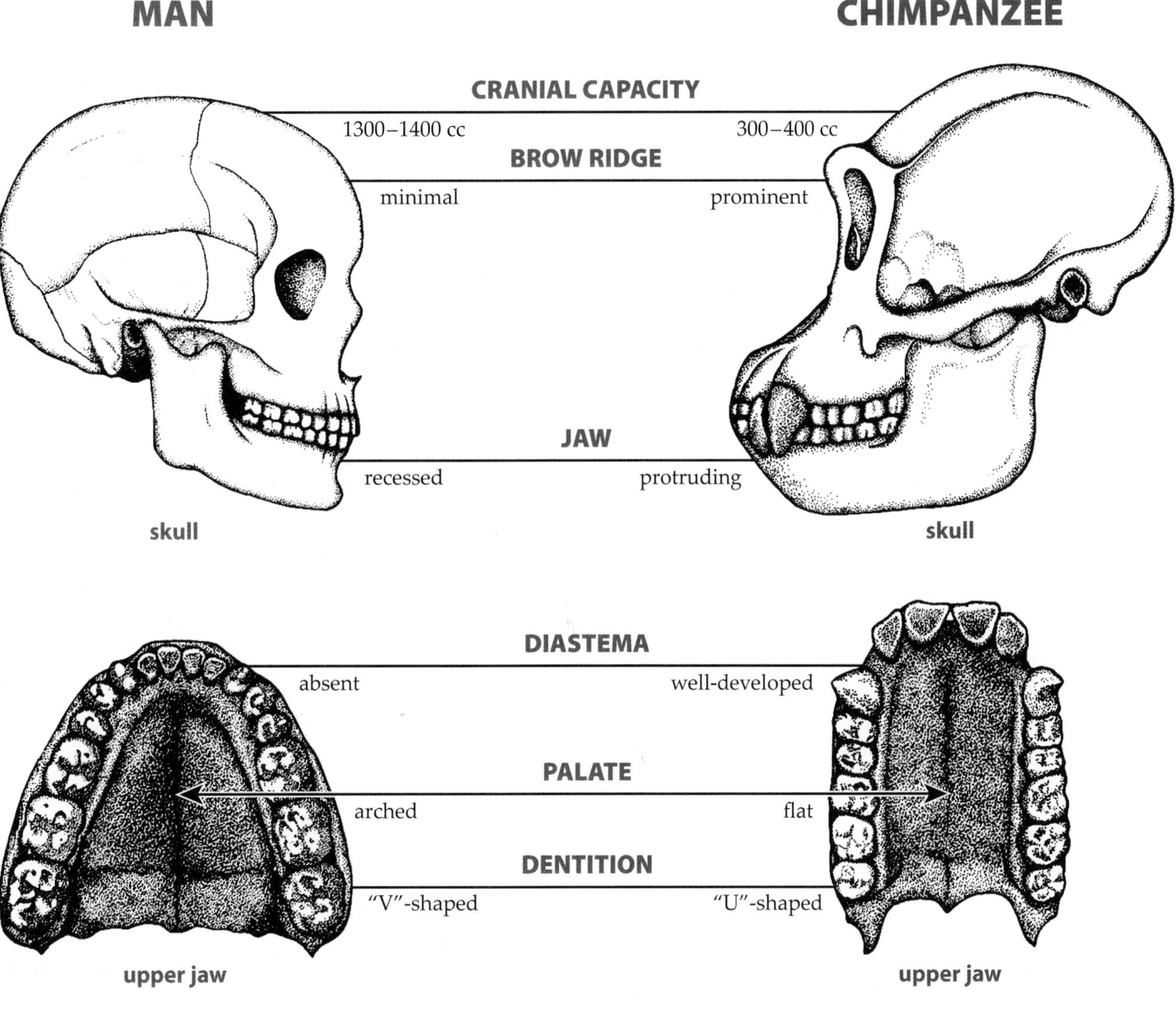

FIGURE 20.2 Characteristics of skull morphology for a human and a chimp skull and upper jaw.

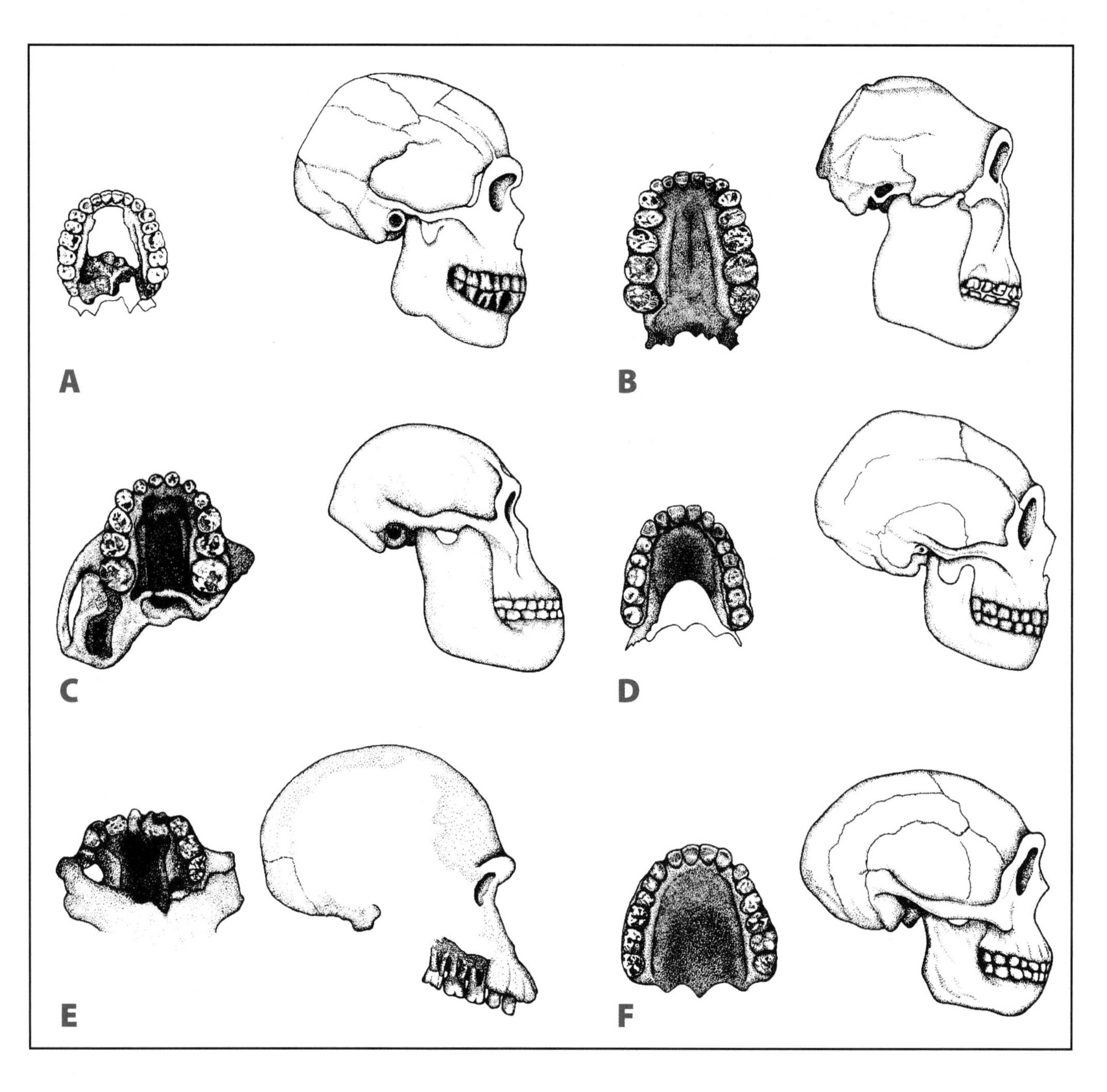

FIGURE 20.3 Selection of fossil hominin skulls and their dentition. The stratigraphic order of their occurrence has intentionally been altered awaiting the student's evaluation. The jaw associated with each skull is located to its left.

References

Barnhardt, W. A., B. D. Andrews, S. D. Ackerman, W. E. Baldwin, and C. J. Hein (2009). *High-Resolution Geologic Mapping of the Inner Continental Shelf; Cape Ann to Salisbury Beach, Massachusetts* (Open-File Report 2007-1373). Washington, DC: US Geological Survey.

Benz, H. M., R. L. Dart, A. Villaseñor, G. P. Hayes, A. C. Tarr, K. P. Furlong, and S. Rhea (2011). *Seismicity of the Earth 1900–2010: Aleutian Arc and Vicinity* (Open-File Report 2010-1083-B). Washington, DC: US Geological Survey.

Dott, R. H. (1964). Wacke, greywacke and matrix; what approach to immature sandstone classification? *Journal of Sedimentary Research* 34(3): 625–632.

Dunham, R. J. (1962). Classification of Carbonate Rocks According to Depositional Texture. In W. E. Ham (ed.), *Classification of Carbonate Rocks* (pp. 108–121). American Association of Petroleum Geologists, Symposium Memoir.

Enos, P., and R. D. Perkins. 1977. *Quaternary Sedimentation in South Florida* (Memoir 147). Boulder, CO: Geological Society of America.

Haq, B. U., J. Hardenbol, and P. R. Vail (1988). Mesozoic and Cenozoic Chronostratigraphy and Eustatic Cycles. In C. K. Wilgus, B. S. Hastins, H. Posamentier, J. V. Wagoner, C. A. Ross, and C. G. St.C. Kendall (eds.), *Sea-Level Changes: An Integrated Approach* (Special Publication 42, pp. 71–108). Tulsa, OK: Society of Economic Paleontology and Mineralogy.

McKee, E. D., and C. E. Resser (1945, October). Stratigraphy and Ecology of the Grand Canyon Cambrian, Part 1. *Cambrian History of the Grand Canyon Region* (Publication 563). Washington, DC: Carnegie Institute of Washington.

McKenzie, D. P., and F. Richter (1976). Convection currents in the earth's mantle. *Scientific American* 235(5): 72–89.

Moore, R. C., C. G. Lalicker, and A. G. Fischer (1952). *Invertebrate Fossils*. New York: McGraw-Hill.

Ogg, J. G., G. Ogg, and F. M. Gradstein (2008). *The Concise Geologic Time Scale.* New York: Cambridge University Press.

Paleontological Institute (n.d.). *Treatise on Invertebrate Paleontology.* Geological Society of America and the University of Kansas.

Powers, M. C. (1953). A new roundness scale for sedimentary particles. *Journal of Sedimentary Petrology* 23: 118.

Stanley, S. M. (1979). *Macroevolution*. San Francisco: W. H. Freeman.

Stanley, S. M. (2008). *Earth System History*. San Francisco: W. H. Freeman.

Teichert, C. (1958). Some biostratigraphical concepts. *Geological Society of America Bulletin* 69(1): 99–119.

Wegener, A. (1912). Die Entstehung der Kontinente. *Geologische Rundschau* 3(4): 276–292.

Whitmeyer, S. J., and K. E. Karlstrom (2007). Tectonic model for the Proterozoic growth of North America. *Geosphere* 3: 220–259.

Wood, B. A. (1994). The oldest hominid yet. *Nature* 371: 280–281.

Notes

Notes

Notes

Notes